WIND FORECASTING IN RAILWAY ENGINEERING

WIND FORECASTING IN RAILWAY ENGINEERING

HUI LIU

ELSEVIER

Elsevier
Radarweg 29, PO Box 211, 1000 AE Amsterdam, Netherlands
The Boulevard, Langford Lane, Kidlington, Oxford OX5 1GB, United Kingdom
50 Hampshire Street, 5th Floor, Cambridge, MA 02139, United States

Library of Congress Cataloging-in-Publication Data
A catalog record for this book is available from the Library of Congress

British Library Cataloguing-in-Publication Data
A catalogue record for this book is available from the British Library

ISBN: 978-0-12-823706-9

Publisher: Glyn Jones
Editorial Project Manager: Naomi Robertson
Production Project Manager: Kamesh Ramajogi
Cover Designer: Miles Hitchen

Typeset by TNQ Technologies

Contents

List of figures

List of tables

Preface

Strong winds along railways greatly affect the lateral stability of railway trains, and even causes serious accidents such as derailment, overturning, etc. China, the United States, Japan, and other countries have experienced severe wind-induced train overturning accidents, causing serious loss of life and property. To ensure the safe operation of trains, it is urgently needed to enhance the wind-proof performance of railway trains.

In the existing railway wind engineering research, wind forecasting along railways is recognized as able to effectively improve the wind-proof performance of trains. A system based on wind forecasting can prevent trains from being exposed to future strong winds to improve safety, and it can also avoid the unnecessary speed limitation of trains improving efficiency. Researchers have proposed several effective railway strong wind prediction systems.

Due to the nonlinearity and nonstationarity of the wind, it is still a difficult problem to realize high-precision spatiotemporal wind speed prediction. The author refines research contents of the past 10 years and completes this book. This book focuses on three key technologies: anemometer layout, single-point wind prediction, and spatial wind prediction. The characteristics of wind flow field, single-point wind, and spatial wind are analyzed. Advanced physical models and data-driven models are introduced with real data demonstration.

There are eight chapters in this book as follows:

Chapter 1: **Introduction**

In this chapter, the typical scenarios of wind engineering are introduced. The key technologies for wind forecasting, including wind anemometer layout, single-station wind forecasting, and spatial wind forecasting, are overviewed.

Chapter 2: **Analysis of flow field characteristics along railways**

In this chapter, real flow fields in the Hundred Miles Wind Area and the Thirty Miles Wind Area are provided as analysis examples. The Moran's I indexes are applied to analyze the spatial characteristics of the flow field. The planar maximally filtered graph is applied to extract the key spatial correlation structure of the flow field. The fast Fourier transform is applied to analyze the frequency spectrum of the flow field, and the main frequencies are discovered. Bayesian fuzzy clustering is used to extract key flow field seasonal templates.

Chapter 3: **Description of single-point wind time series along railways**

In this chapter, firstly, wind anemometer layout optimization methods for single-station wind speed measurement are introduced. Then, the seasonal and heteroskedastic characteristics of wind are analyzed. Finally, the seasonal autoregressive integrated moving average model, autoregressive conditionally heteroskedastic model, and the generalized autoregressive conditionally heteroskedastic model are utilized for wind description.

Chapter 4: **Single-point wind forecasting methods based on deep learning**

In this chapter, three advanced deep learning methods are introduced for wind forecasting. Decomposition methods are applied to further improve performance. Finally, the deterministic forecasting performance of the deep learning methods is analyzed.

Chapter 5: **Single-point wind forecasting methods based on reinforcement learning**

In this chapter, the reinforcement learning methods are introduced for static ensemble weight optimization, feature selection, etc. The Q-learning, deep Q-network, and deep deterministic policy gradient are investigated. Finally, the advantages and disadvantages of the reinforcement learning methods are summarized.

Chapter 6: **Single-point wind forecasting methods based on ensemble modeling**

In this chapter, three mainstream ensemble methods for single-station wind forecasting are introduced, including the multi-objective ensemble, stacking ensemble, and boosting ensemble. The designed ensemble forecasting methods can combine diverse base forecasting models.

Chapter 7: **Description methods of spatial wind along railways**

In this chapter, the spatial wind correlation characteristics are evaluated by four different correlation coefficients. Then, the weather research and forecasting model is built to describe spatial wind. Finally, the performance evaluating indicators of spatial forecasting are introduced.

Chapter 8: **Data-driven spatial wind forecasting methods along railways**

In this chapter, firstly, two statistical spatial forecasting methods are introduced for spatial prediction, which apply mutual information for spatial feature selection. Then, the intelligent spatial forecasting methods are investigated, which are combined with four binary optimization algorithms. Finally, three deep learning methods are applied for spatial prediction, which use sparse autoencoder for feature extraction.

Prof. Dr.-Ing. habil. Hui Liu
Changsha, China

Acknowledgments

The studies in this book are supported by the National Natural Science Foundation of China, the National Key R&D Program of China, and the Innovation Drive of Central South University, China. In the process of writing this book, Mr. Zhu Duan, Mr. Yinan Xu, Mr. Chao Chen, Ms. Shi Yin, Mr. Ye Li, Mr. Yu Xia, Mr. Guangxi Yan, Ms. Jing Tan, Mr. Guangji Zheng, and other team members have done a lot of model verifications and other works. These team members, as mentioned, have made the same contribution to this book. The author expresses his heartfelt thanks to them.

Nomenclature list

A

A3C Asynchronous Advantage Actor Critic
ACF Autocorrelation Function
AdaBoost Adaptive Boosting
ADF Augmented Dickey Fuller
AIC Akaike Information Criterion
ANN Artificial Neural Network
AR Autoregressive Model
ARCH Autoregressive Conditionally Heteroskedastic
ARIMA Autoregressive Integrated Moving Average
ARIMAX Autoregressive Integrated Moving Average With Extra Input
ARMA Autoregressive Moving Average
ARW Advanced Research WRF
ASM Algebraic Stress Model

B

BA Bat Algorithm
BABO Binary Artificial Butterfly Optimization
BCOA Binary Coyote Optimization Algorithm
BDE Binary Differential Evolution
BFC Bayesian Fuzzy Clustering
BFGS Broyden, Fletcher, Goldfarb, and Shanno Quasi-Newton
BGWO Binary Grey Wolf Optimization
BHHO Binary Harris Hawk Optimization
BIC Bayesian Information Criterion
BILSTM Bidirectional Long-Short Term Memory
BOA Butterfly Optimization Algorithm
BP Back Propagation
BPSO Binary Particle Swarm Optimization
BRCGA Binary Real-Coded Genetic Algorithm

C

CEEMD Complete Ensemble Empirical Mode Decomposition
CEEMDAN Complete Ensemble Empirical Mode Decomposition With Adaptive Noise
CFD Computational Fluid Dynamics
CNN Convolutional Neural Network
CTC Centralized Traffic Control
CTCS Chinese Train Control System
CWC Combinational Coverage Width-Based Criterion

D

DBN Deep Belief Network
DDES Delayed Detached-Eddy Simulation
DDPG Deep Deterministic Policy Gradient
DeepESN Deep Echo State Network
DES Detached-Eddy Simulation
DFT Discrete Fourier Transform
DGA Distributed Genetic Algorithm
DGF Double Gaussian Function
DL Deep Learning
DNN Deep Neural Network
DNS Direct Numerical Simulation
DPG Deterministic Policy Gradient
DPMM Dirichlet Process Mixture Model
DQN Deep Q-Network
DRL Deep Reinforcement Learning
DTW Dynamic Time Warping
DWT Discrete Wavelet Transform

E

EEMD Ensemble Empirical Mode Decomposition
EGARCH Exponential GARCH Model
ELBO Evidence Lower Bound
ELM Extreme Learning Machine
EM Expectation Maximum
EMD Empirical Mode Decomposition
ENN Elman Neural Network
ESN Echo State Network
EVM Eddy-Viscosity Model
EWT Empirical Wavelet Transform

F

FCP Fuzzy Cluster Prior
FDL Fuzzy Data Likelihood
FDM Finite Difference Method
FEEMD Fast Empirical Mode Decomposition
FEM Finite Element Method
FFT Fast Fourier Transform
FIGARCH Fractionally Integrated GARCH Model
FVM Finite Volume Method
FVP Finite Volume Particle

G

GA Genetic Algorithm
GAN Generative Adversarial Net
GARCH Generalized Autoregressive Conditionally Heteroskedastic

GBoost Gradient Boosting
GCDLA Graph Convolutional Deep Learning Architecture
GDAS Global Data Assimilation System
GIS Geographic Information System
GLR Generalized Linear Regression
GMCM Gaussian Mixture Copula Model
GMM Gaussian Mixture Model
GOA Grasshopper Optimization Algorithm
GRU Gated Recurrent Unit
GSM Global System for Mobile Communications
GSM-R Global System for Mobile Communications for Railway
GWO Grey Wolf Optimizer

H
HC Hill Climbing
HMM Hidden Markov Model

I
ICEEMDAN Improved Complete Ensemble Empirical Mode Decomposition With Adaptive Noise
IDDES Improved Delayed Detached-Eddy Simulation
IDW Inverse Distance Weighting
IEWT Inverse Empirical Wavelet Transform
IGARCH Integrated GARCH Model
IMF Intrinsic Mode Function
IOM Intrinsic Oscillatory Mode

K
KDE Kernel Density Estimation
KL Kullback-Leibler
KNN K-Nearest Neighbor

L
LBM Lattice Boltzmann Method
LBQ-test Ljung-Box Q-test
LES Large Eddy Simulation
LiDAR Light Detection and Ranging
LM Lagrange Multiplier
LPBoost Linear Programming Boosting
LS Local Search
LSSVM Least Squares Support Vector Machine
LSTM Long Short-Term Memory

M
MA Moving Average
MAE Mean Absolute Error

MAP Maximum A Posteriori
MAPE Mean Absolute Percentage Error
MAR Missing At Random
MCMC Markov Chain Monte Carlo
MGA Micro-Genetic Algorithm
MI Mutual information
MIMO Multiple-Input Multiple-Output
MKDE Multivariate Kernel Density Estimation
MLP Multi-Layer Perceptron
MM5 Fifth-Generation Mesoscale
MODWPT Maximal Overlap Discrete Wavelet Packet Transform
MODWT Maximal Overlap Discrete Wavelet Transform
MOGOA Multi-Objective Grasshopper Optimization Algorithm
MOGWO Multi-Objective Grey Wolf Optimization
MOMVO Multi-Objective Multi-Verse Optimization
MOPSO Multi-Objective Particle Swarm Optimization
MPGA Multi-Population Genetic Algorithm
MPS Moving Particle Semiimplicit
mRMR Minimum Redundancy Maximum Relevance
MSE Mean Square Error
MST Minimum Spanning Tree
MTGP Multi-Task Gaussian Process

N

NCAR National Center for Atmospheric Research
NCEP National Center for Environmental Prediction
NMM Nonhydrostatic Mesoscale Model
NNCT No-Negative Constraint theory
NOAA National Oceanic and Atmospheric Administration
NREL National Renewable Energy Laboratory
NSGA-II Nondominated Sorting Genetic Algorithm II
NWP Numerical Weather Prediction

O

ORELM Outlier Robust Extreme Learning Machine

P

PACF Partial Autocorrelation Function
PANS Partially Averaged Navier-Stokes
PCA Principal Components Analysis
PCC Pearson Correlation Coefficient
PICP Prediction Interval Coverage Probability
PINAW Prediction Interval Normalized Average Width
PITM Partially Integrated Transport Model

PMFG Planar Maximally Filtered Graph
PSO Particle Swarm Optimizer
PSTN Predictive Spatio-Temporal Network

R

RANS Reynolds Average Navier-Stokes
RBC Radio Block Center
RBFNN Radial Basis Function Neural Network
RBM Restricted Boltzmann Machine
RDPG Recurrent Deterministic Policy Gradient
ReLU Rectified Linear Unit
RL Reinforcement Learning
RMSE Root Mean Squared Error
RMSProp Root Mean Square Propagation
RNN Recurrent Neural Network
RSM Reynolds Stress Model
RSTD Regime-Switching Space-Time Diurnal
RTFDDA Real-Time Four-Dimensional Data Assimilation

S

SAE Sparse AutoEncoder
SARIMA Seasonal Autoregressive Integrated Moving Average Model
SARSA State-Action-Reward-State-Action
SAS Scale-Adaptive Simulation
SDE Standard Deviation of Error
SELU Scaled Exponential Linear Unit
SGS Sub-grid-scale Stress
SLFN Single-Hidden-Layer Foreword Network
SMAE Spatial Mean Absolute Error
SMAPE Spatial Mean Absolute Percentage Error
SN Seriesnet
SoDAR Sonic Detection and Ranging
SOM Self-Organizing Map
SPH Smoothed Particle Hydrodynamics
SRMSE Spatial Root Mean Square Error
SRN Simple Loop Network
SVM Support Vector Machine
SVR Support Vector Regression

T

TCC Train Control Center
TDCS Train Dispatching Command System
TDD Trigonometric Direction Diurnal
TLFN Two-Hidden-Layer Foreword Network
TMD Tuned Mass Damper
TSRS Temporary Speed Restriction Server

V

v-SVM v-Support Vector Machine
VMD Variational Mode Decomposition

W

WD Wavelet Decomposition
WINDAS WInd Profiler Network and Data Acquisition System
WMLES Wall-Modeled Large Eddy Simulation
WPD Wavelet Packet Decomposition
WPF Wavelet Packet Filter
WRF Weather Research and Forecasting
WSTD Wavelet Soft Threshold Denoising

CHAPTER 1

Introduction

Contents

Wind Forecasting in Railway Engineering
ISBN 978-0-12-823706-9
https://doi.org/10.1016/B978-0-12-823706-9.00001-6

1.1 Overview of wind forecasting in train wind engineering

Wind is the main natural disaster hazard for rail operation [1]. There have been many rail accidents caused by strong wind [2,3]. In 2005, a passenger train on the Uetsu Main Line in Japan was overturned by strong wind, resulting in 5 deaths and 33 injuries [4]. In 2007, a passenger train on China's Southern Xinjiang Railway was overturned by strong wind, which caused 3 deaths and 34 injuries [5]. In 2019, 25 freight trains in New Mexico, USA, were blown off the bridge by strong wind, causing serious damage to the bridge [6]. The railway strategies of many countries, including China's "The Belt and Road" have emphasized the safety of rail operations in future railway development. So, it is urgent to improve the safety of rail operation under strong wind.

Wind forecasting is one of the effective methods to improve railway operation performance against the strong wind. The system based on wind forecasting can be aware of future wind information. If the future wind exceeds the threshold, the speed of the train will be limited, so as to prevent exposure to future strong wind. The safety can therefore be improved. If the future wind is below the threshold, the train will return to normal operation status, so as to avoid the redundant speed limitation of the trains. Hence, the efficiency can be improved.

Germany, Japan, Italy, and other countries have carried out researches on railway wind prediction systems. The German Federal Railways (Deutsche Bundesbahn in German) has established the Nowcasting wind speed forecast system [7]. The system is based on linear regression and can achieve wind speed prediction 2 min ahead. East Japan Railway Company constructed Wind profiler Network and Data Acquisition System wind speed prediction system, which uses the Kalman filter method to predict wind speed [8]. The system can achieve 36 min ahead prediction. Both of the above systems are data-driven. In contrast, Italy has constructed a physical-based wind speed warning system [9]. This method can achieve an effective prediction of space wind speed, but the computational burden is heavier. Besides, France, Spain, and other countries have also established wind speed warning systems [10].

1.2 Typical scenarios of railway wind engineering

1.2.1 Train overturning caused by wind

Crosswind, as one of the main sources of train lateral force, threatens the overturning stability of trains [11]. There are three important indexes to evaluate the overturning stability, including overturning coefficient, derailment coefficient, and rate of wheel load reduction. These indexes are explained as follows:

(a) The overturning coefficient evaluates the ratio of dynamic wheel load to static load. The equation is presented as follows [12]:

$$D = \frac{P_W}{P'_W} \tag{1.1}$$

where P_W and P'_W are the dynamic and static vertical forces of the wheel. When the overturning coefficient is equal to 1, the dynamic vertical force of the windward wheel is equal to the static force, i.e., the train is about to overturn. To ensure safety margin, the threshold value of the overturning coefficient is set as 0.8 according to Chinese standard "Railway vehicles—Specification for dynamic performance evaluation and accreditation test" (GB 5599-85) and "Specification for strength and dynamic performance of high-speed test train" (95J01-L).

(b) The derailment coefficient Q/P can calculate the ratio between the lateral force Q and the vertical force P. According to the Chinese standard (GB 5599-85), the derailment coefficient is evaluated with two threshold values as follows [13]:

$$\begin{cases} Q/P \leq 1.2 & \text{first threshold} \\ Q/P \leq 1.0 & \text{second threshold} \end{cases} \tag{1.2}$$

where the first threshold is the standard value; the second threshold has more safety margin. To improve the reference value of the derailment coefficient, a new derailment coefficient for crosswind is proposed [14]. This derailment coefficient comprehensively considers crosswind, wheel–rail interaction, and vehicle dynamic.

(c) The rate of wheel load reduction $\Delta P/\overline{P}$ evaluates the ratio of wheel loads' difference to static load.

$$\frac{\Delta P}{\overline{P}} = \frac{P_L - P_W}{P_{st}} = \frac{P_L - P_W}{P_L + P_W} \tag{1.3}$$

where P_L is the vertical force of leeward wheel; P_W is the vertical force of windward wheel; P_{st} is the static vertical force, which is equal to $P_L + P_W$. According to the Chinese standard (GB 5599-85), the rate of wheel load reduction is evaluated with two threshold values as follows [15]:

$$\begin{cases} \Delta P/\overline{P} \leq 0.65 & \text{first threshold} \\ \Delta P/\overline{P} \leq 0.60 & \text{second threshold} \end{cases} \tag{1.4}$$

where the first threshold is the standard value; the second threshold has more safety margin.

The crosswind can cause overturning moment of the train, decrease the vertical force of the windward wheel, and worsen the above indexes [16]. Compared with the derailment coefficient, crosswind has a greater influence on the rate of wheel load reduction and overturning coefficient [17]. In practice, the wheel force cannot be obtained during operation, because the wheel with a vertical force sensor is unable to brake. To infer the overturning coefficient, a new approach is reported in Ref. [18]. This approach can estimate the overturning coefficient based on variable parameters, which was validated by simulation and practical tests.

1.2.2 Pantograph—catenary vibration caused by wind

Due to its strong railway transportation capacity, high energy conversion efficiency, and environmental friendliness, electrification is currently the development trend of railways [19]. At present, the power supply of electric trains is based on the catenary. Since the catenary is directly exposed along the railway, it will be severely affected by natural conditions. Once the pantograph vibrates and falls off, the current collection quality of the train will drop drastically.

According to the wind variation, the wind can be classified as steady wind and fluctuating wind. The first type of wind will cause the catenary to shift, and the second type of wind will cause the catenary to vibrate [20]. Experimental results show that steady wind and fluctuating wind have similar effects on the quality of pantograph—catenary current collection [21]. But when the fluctuating wind has large amplitude or the frequency is similar to the natural frequency of the pantograph—catenary system, the current collection quality will drop sharply. Moreover, under severe weather such as rain and snow, the pantograph—catenary interaction will deteriorate sharply [22].

To reduce the degradation of current collection quality caused by strong winds, a series of improvement approaches have been proposed to optimize the aerodynamic quality of the pantograph. Luo et al. [23] optimized the fixing position of the pantograph to maximize the current collection quality. Song et al. [24] proposed a proportional-derivative controller for pantograph against stochastic wind. This controller is designed based on dynamic model of the pantograph and can decrease variations of the pantograph and improve current collection quality.

1.2.3 Bridge vibration caused by wind

Wind-to-bridge vibration is one of the key issues when designing large bridges. Bridges such as the Tacoma bridge and the Humen bridge have experienced severe wind-to-bridge vibration problems. As the span of the bridge increases, the vibration problem becomes more severe. The vibration of the bridge can be divided into four types, including flutter, galloping, vortex-induced vibration, and buffeting [25]. The details are presented as follows:

(a) The flutter and galloping are both self-excited vibration of wind, which is caused by negative aerodynamic damping [26]. The flutter mainly occurs in the section of the main girder, while the galloping mainly occurs in elongated structures such as cables of cable-stayed bridges. The flutter has a significant impact on structure health. Once the flutter occurs, the bridge will be irreversibly damaged. The galloping usually does not cause serious damage in a short time, but can shorten the fatigue life of the cable.

(b) The vortex-induced vibration has the characteristics of self-excited vibration and forced vibration. The vortex-induced vibration is caused by alternating whirlpools on both sides of the bridge in stable wind fields [27]. In a certain wind speed range, the vibration frequency keeps lock-on. The vortex-induced vibration mainly occurs in the section of the main girder, and is less harmful than the flutter.

(c) The buffeting is a forced vibration, which is caused by the unstable wind field [28]. The amplitude of the buffeting will increase when the wind speed increases. Under the long-term buffet, the significant displacement response will be caused in a long-span flexible bridge.

In the wind-resistant design of bridges, there are different strategies for self-excited vibrations (flutter and gallop) and limited amplitude vibrations (vortex-induced vibration and buffeting). For the self-excited vibrations, the design aim is to keep the critical wind speed larger than the rated wind speed. For the limited amplitude vibrations, the design aim is to keep the amplitude smaller than the rated amplitude [29]. Tuned Mass Damper (TMD) is an important method to dampen vibration. Wang et al. [30] utilized single-side pounding TMD to alleviate the vortex-induced vibration of the bridge. Casalotti et al. [31] applied TMD to mitigate the flutter.

1.2.4 Wind-resistant railway yard design

Wind-resistant performance of station yard is an important part of station yard design. When humping the freight train, the wind resistance counts for more than 30% of total resistance [32]. Strong winds will reduce the stability of the train on the slope of the station yard. If the train in the classification yard cannot get off from the hump, the transportation will be affected. Kun [33] proposed a formula for computing wind resistance when humping. This formula can be used for speed control of the train in the yard. Yu et al. [34] optimized the gradient and length of the slopes for wind-resistant yard.

Besides, if the cargo is coal or other mineral resources, the dust caused by winds can seriously reduce air quality. To prevent the wind effect on the cargo, wind-break wall is widely developed. Tian [35] designed a meshed wind-break wall, which is validated in the aspect of the structural strength. Wang et al. [36] proposed a comprehensive strategy to control dust. The dust control percentage can achieve 85%.

1.2.5 Wind-break wall design

The wind-break wall can resist the effect of strong wind and improve the overturning stability of the train. The windshield facilities on China's Lanzhou-Xinjiang line effectively increased the safe wind speed limitation. In areas with open wind-break walls, the safe wind speed is increased by 10 m/s, and in areas with closed wind-break walls, the safe wind speed is increased by 20 m/s [37]. The current research hotspots of wind-break wall are presented as follows:

(a) Optimization of wind-proof performance. The types of the wind-break wall contain reinforcement type, concrete type, earth embankment type, etc. [38]. Zhang et al. [38] compared the performance of these types with numerical simulation, and found the earth embankment type has the worst performance. When selecting the parameters of the wind-break wall, selecting solution from the simulation results directly may lead to local optimality. To achieve global optimality, a surrogate model is applied. Xiang et al. [39] utilized support vector machine as the surrogate model to fit the simulation results, and found the best heights and porosities.

(b) Avoid sand particle accumulation. In desert railway, sand particles can accumulate around windshields and tracks, severely hindering the normal operation of trains. The accumulation is caused by the backflow area behind the wall [40]. To prevent sand from accumulating, Xin et al. [41] added an additional wind-break wall to alleviate sane particle accumulation. This two-wall structure can effectively reduce the density and deposition rate of the sand, and can be applied on existing wind-break walls.

(c) Avoid fatigue damage. As the train is running, the front and rear of the train have opposite wind pressures. Therefore, the windshield will vibrate violently when the train passes the windshield. The impact of the wind dynamic load on wind-break wall has been widely investigated [42]. Tokunaga et al. [43] proposed two methods to estimate the dynamic response of the wind-break wall.

1.2.6 Other scenarios

Under high speed, aerodynamic resistance and noise take up the majority of train resistance and noise. The aerodynamic design of the high-speed train affects the resistance, noise, and overturning-proof performance. The train shape is controlled by dozens of parameters and is constrained to ensure feasibility for application. The optimization objective functions are various. The optimization of the high-speed train shape is a difficult task. The multi-objective optimization methods are widely used for aerodynamic design. The multi-objective optimization algorithms can compromise the contradictions between different objective functions and achieve synchronous global optimization. The Non-dominated Sorting Genetic Algorithm II (NSGA-II) is used to optimize the aerodynamic resistance and noise [44]. Yao et al. [45] combined the Kriging model with the NSGA-II method to optimize the volume and lift force. The Kriging surrogate model was applied to construct the response surface and reduce computation time.

In winter of high latitude areas, high-speed trains will face serious problems with bogie snow. Since the bogie is directly exposed to the air, the snow is blown onto the bogie by the wind. The snow in traction motors and other parts will be transformed into ice after melting and freezing. The bogie is important for ensuring safe train operation. Snow-covered brake pads, traction motors, and gearboxes affect the performance of the bogie. To improve the adaptability of the train in severe cold areas, it is necessary to promote anti-icing performance of the bogies. Li et al. [46] designed an anti-ice dome. Wang et al. [47] added two deflectors to avoid the snow accumulated in the bogie. To improve the anti-snow performance, the angles for these two deflectors were optimized.

1.3 Key technical problems in wind signal processing

From the analysis of railway wind engineering in Section 1.2, it can be seen that strong wind threatens the railway train operation safety greatly, and many researchers have proposed many solutions based on traditional fluid dynamics theory. Different from these traditional methods, this book focuses on security technology based on wind signal processing.

To improve the safety of the train under strong wind, the railway command department will instruct the train to reduce the speed or stop running to avoid the train overturning. According to the Chinese standard "Grade of high impact weather conditions for high-speed railway operation" (QX/T 334-2016), China's wind train situation rules are presented in Table 1.1, where $\overline{V}$ means 2-min averaging wind speed and V means instantaneous wind speed.

In China, the "10-minute rule" is required to cancel the warning and resume normal operation, when a train slows down or stops due to strong winds [48]. In other words, only when the wind speed along the railway no longer exceeds the limit in the next 10 min, the alarm can be canceled. Similar to China, Japan has also adopted a wind speed driving standard based on the "30-minute rule" [49]. In actual operation, this method based on the static operating restriction time has limitations, because it is difficult to specify the appropriate time [50]. If the operating restriction time is too short, the train can quickly resume normal operation. But in the face of continuous strong winds, the train's resistance capacity to overturning is weak. If the operating restriction time is too long, the train can continue to run after the strong wind disappears completely, but the operating efficiency will be reduced.

Table 1.1 China's high-speed train operation rules under wind

Grade	Wind speed	Operation method
Grade 0	$\overline{V} \leq 15m/s$ or $V \leq 20m/s$	Normal operation
Grade 1	$15m/s < \overline{V} \leq 20m/s$ or $20m/s < V \leq 25m/s$	Train speed should be smaller than 300km/h
Grade 2	$20m/s < \overline{V} \leq 25m/s$ or $25m/s < V \leq 30m/s$	Train speed should be smaller than 200km/h
Grade 3	$25m/s < \overline{V} \leq 30m/s$ or $30m/s < V \leq 35m/s$	Train speed should be smaller than 120km/h
Grade 4	$\overline{V} > 30m/s$ or $V > 35m/s$	Stop operation

The wind speed prediction method can effectively solve the dilemma caused by the fixed operating restriction time. With the predicted wind speed, the train command system can adaptively adjust the operating restriction time based on the trend of future wind speed changes. If the future wind speed will not exceed the limit, the train will be guided in time to resume normal operation. If the future wind speed still exceeds the limit, the train will continue to limit the speed until the predicted wind speed meets the normal operation requirements. The train operation command system with wind speed prediction system can realize active alarm. In the face of sudden strong winds in the future, the train can take advance measures instead of passively waiting for the real occurrence of the strong winds. When the train speed is limited, the command center can actively adjust the train operation plan according to the wind prediction results, instead of passively waiting within the static operating restriction time. In this manner, the operating efficiency of the train is improved. The effectiveness of the wind prediction method has been well proved [50], the averaging operating restriction time can be decreased by about 20%–50% under the premise of equal safety level [8].

To achieve effective wind speed prediction and control, data-driven methods based on signal processing are the international mainstream. The key technical problems of signal processing can be divided into measurement, identification, forecasting, and control, which are explained in the following sections.

1.3.1 Wind measurement technology

Wind speed measurement is the basis of wind speed signal processing. The purpose of wind speed measurement technology is to accurately collect wind speed data along railways in real time. The technologies of anemometers selection and data preprocessing are the focuses of this section.

1.3.1.1 Anemometers selection

The commonly used anemometers contain thermal anemometer, ultrasonic anemometer, cup anemometer, Light Detection and Ranging (LiDAR), Sonic Detection and Ranging (SoDAR) etc. According to Chinese standard (TJ/GW089-2013), the anemometer can measure wind speed within 0–60 m/s and wind direction within 0–360°. These anemometers have diverse characteristics as follows:

(a) The thermal anemometer can measure the temperature change of the hot wire, and then indirectly measure the wind speed. There are two different types of thermal anemometers, including constant temperature type and constant current type, the latter of which is more widely used. The thermal anemometer is characterized by high sensitivity, small size, and fast response.

(b) The ultrasonic anemometer can measure the wind speed by Doppler principle. The ultrasonic sound wave can spread faster in the downwind. According to the spread time difference of the ultrasonic sound wave between the downwind and headwind, the wind speed can be calculated. The ultrasonic anemometer has no mechanical wear, and fast response speed.

(c) The cup anemometer can observe single-point wind speed by measuring the cup's rotation. The measurement performance is affected by cup shape, rotation radius, and so on [51]. Due to the inertia, the cup anemometer has slow response speed.

(d) The LiDAR and SoDAR are two remote sensing devices for wind speed measurement. The LiDAR is based on the particle trajectory, while the SoDAR is based on the sound reflection [52]. These two devices can achieve three-dimension scan of the wind field, and have high spatial and temporal resolutions.

In China's Lanzhou-Xinjiang railway, a redundant wind measurement system is applied, which is composed of cup anemometers and propeller anemometers [53]. In German railway, two ultrasonic anemometers are installed at each single measurement station [7]. This redundant design of the anemometers can improve system robustness. If one of the anemometers fails and the wind speed data cannot be output, the other anemometer can be used to generate the wind speed measurement results. Besides, the difference between the output values of the two anemometers can be used to determine whether the output of the anemometer is reliable. If the difference exceeds a certain limit, there is a problem with an anemometer. At this time, the condition monitoring method can be used to determine which anemometer has failed, and the output value of the other anemometer can be used as the wind speed measurement results.

1.3.1.2 Data preprocessing

Due to the fault and noise of the anemometer, the obtained wind speed data has missing values and outliers. These abnormal values can damage the information of wind speed sequence. To improve the information quality of measured wind speed signal, missing data imputation method and outlier detection method are commonly used in wind speed data preprocessing. The classification of the data preprocessing methods is shown in Fig. 1.1.

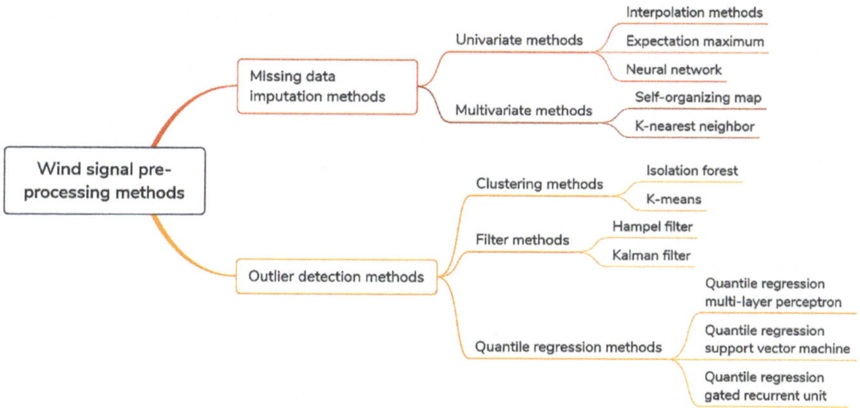

Figure 1.1 The classification of the data preprocessing methods.

The missing data imputation methods can be divided into univariate methods and multivariate methods as follows:

(a) The univariate methods can generate missing wind speed data with only wind speed data. The widely used univariate methods for wind speed contain interpolation methods, Expectation Maximum (EM) methods, neural network methods, etc. The interpolation method is the simplest way to complete missing values. According to interpolation functions, interpolation methods can be divided into linear interpolation, spline interpolation, Hermite interpolation, and so on. Due to the simplicity of this method, when the amount of data is small, good missing value completion performance can be obtained. The EM methods can estimate missing data to maximize the likelihood function. This method assumes the whole dataset obeys certain conditional distribution, and generates the parameters of the distributions to optimize the likelihood function. The EM is suitable for the Missing At Random (MAR) mode with large dataset, to generate unbiased estimation for the parameters. Liu et al. [54] applied the EM algorithm to optimize the Gaussian Mixture Model (GMM) for missing data imputation. Similar to the EM method, the neural network assumes a certain autocorrelation function holds for the whole dataset. The strong learning capacity of the neuron networks makes it possible to fit nonlinear autocorrelation function. Fallah et al. [55] utilized the Multi-Layer Perceptron (MLP) to estimate missing data. Qu et al. [56] proposed Generative Adversarial Net (GAN) imputation method, which was better than the classical neural network.

(b) The multivariate missing data imputation methods need wind speed series in other multiple stations or meteorological parameters. The multivariate methods focus on the spatial correlation or meteorological correlation, while the univariate methods focus on the temporal autocorrelation. The Self-Organizing Map (SOM) is a commonly used multivariate missing data imputation method. The SOM can generate templates by completive learning. In each multivariate vector with missing data, the missing data can be estimated by minimizing the Euclidean distance between the valid data and templates. The advantage of the SOM is that it is independent of the location of missing data. This is to say, a well-trained SOM model can be applied to impute data for many different measurement stations without any modification. In addition, the SOM model is suitable for online imputation, because only valid data at the same time are needed and no further information is involved. The disadvantage of the SOM is that if all data in the multivariate vector are missing, the SOM cannot generate missing data estimation results. In this case, only univariate methods are available. The K-Nearest Neighbor (KNN) methods are also an effective method for multivariate missing data imputation. The KNN methods can estimate the missing values with the nearby multivariate vectors. Oehmcke et al. [57] applied Dynamic Time Warping (DTW) as the distance matrix, and used an ensemble KNN algorithm for multivariate missing data imputation.

The outlier detection methods contain clustering methods, filtering methods, quantile regression methods, and so on. The details are explained as follows:

(a) The clustering methods can detect the outliers according to the fact that the outliers have different behavior from the normal data. The generated clusters after the clustering computation have different characteristics. The clusters that are different from normal clusters can be regarded as outliers. The essence of the clustering methods is how to define the abnormal clusters. The isolation forest method defines the abnormal score according to the isolation degree [58]. The outlier data can be isolated by fewer binary trees than the normal data. This is because the outlier data have a small number and are independent of the normal data. For the K-Means method, if the outlier cluster is not concluded, the clustering performance is better. By optimizing the clustering performance, the outlier clusters can be found [59].

(b) The filter methods can discard the outliers according to the local temporal characteristics. The Hampel filter is a modification of the three-sigma theory, which can detect outliers according to median values and standard deviation of local time series. The values beyond the lower and upper bounds are regarded as the outliers. Marti-Puig et al. [60] applied the Hampel filter to process the Supervisory Control and Data Acquisition (SCADA) data. The Kalman filter is also an effective method. According to the distance between actual data and the estimated data of Kalman filter, the outliers with large distances can be detected [61].

(c) The quantile regression methods can estimate the lower and upper bounds according to certain quantile. Different from the Hampel filter, the bounds are obtained by autocorrelation function. The outliers can be detected as the values beyond the bounds. The neural networks are widely used for quantile regression, which can generate results to minimize the pinball loss. The commonly used quantile regression methods contain the MLP [62], Support Vector Machine (SVM) [63], Gated Recurrent Unit (GRU) [64], etc.

1.3.2 Wind identification technology

The wind speed identification method can recognize the key influencing factors, and construct the wind speed descriptive model. The constructed equation can identify the dynamic performance of wind speed and explain the volatility behavior. The wind speed identification methods include model-driven methods and data-driven methods. For the model-driven methods, the wind speed behavior can be described by simulating physical systems; data-driven methods can fit correlations between parameters. As the most commonly used model-driven method, Numerical Weather Prediction (NWP) model has deep understanding of the wind. However, due to its heavy computational burden, the practical value is limited. The data-driven models have the advantages of low computational complexity and high generalization performance, which are the mainstream wind identification models.

Its key technologies include feature recognition and descriptive model construction. The feature recognition can find the important parameters that affect wind behavior. If redundant features are included, the model may be overfitting; if the features are insufficient, the model has poor description capacity. With the obtained features, the correlation between the features and the wind speed should be described. The descriptive model should have an explicit equation, to explain the causes of wind behavior.

1.3.2.1 Feature recognition

The feature recognition methods can be divided into filter methods, wrapper methods, and embedded methods. The filter methods can select features by the correlations. The greater the correlation between features and wind speed, the more important the features are. Commonly used filter methods contain mutual information, Pearson's test, Minimum Redundancy Maximum Relevance (mRMR), etc. [65]. Although the computation of the filter methods is simple, these methods do not consider the match between the selected features and model. To improve the match performance between the selected features and description model, the wrapper methods are proposed. The wrapper methods can evaluate the selected features by the description accuracy. By optimizing the accuracy, the optimal features can be selected with consideration of the description model. The binary optimization methods include Binary Grey Wolf Optimization (BGWO) [66], Binary Artificial Butterfly Optimization (BABO) [67], Binary Coyote Optimization Algorithm (BCOA) [68], etc. The wrapper methods can select features according to ensemble models. The Gradient Boosted Regression Trees (GBRT) is widely used for feature selection [69].

Wind speed is a spatially meteorological time series, and its related features include spatial parameters, meteorological parameters, and temporal parameters [70]. For these parameters, there are specific feature selection methods as follows:

(a) The wind measuring stations along the railway can measure the spatial wind speed. The wind speed between different stations has a significant correlation; there is a time delay between the wind speed of the upwind station and the downwind station [71]. Because of this characteristic, using spatial wind speed for description is a common method. Distance is the most obvious way to select spatial features. Filik [72] predicted wind speed according to nearby stations. However, due to the influence of terrain and other aspects, it is not reasonable to select spatial parameters based on only distance. The spatial correlation analysis has been widely used for spatial parameter selection. Liu and Chen [73] utilized three-stage spatial feature selection method, which included Pearson's test, mutual information, and binary optimization. After the hybrid feature selection computation, the most correlated stations can be obtained. The clustering method is also effective for spatial feature selection. Yu et al. [74] proposed a sequential spatial feature selection method with clustering. The clustering method can find similar stations of a specific station as the spatial feature.

They found the spatial clustering method outperforms the other three methods. Essentially, the spatial correlation of wind speed is because the stations are located in the same trace. The flow trace is affected by the terrain significantly. As the results, the terrain parameters are important for spatial description of the wind speed. Noorollahi et al. [75] extracted terrain features around the station within a 5 km radius, which contained maximum, minimum, and standard deviation of heights. In addition to the terrain, the wind speed amplitude and wind direction also affect the spatial wind speed correlation. With larger wind speed amplitude, the spatial correlations are higher; the spatial correlation is significantly high in certain orientations [76].

(b) Due to the improvement of sensing technology, the wind measurement station can collect complete meteorological parameters, including temperature, humidity, pressure, etc. In the aspect of the data, the meteorological parameters can be selected by Granger causality test, which can evaluate the contribution of meteorological features to wind speed description. Zhang et al. [77] applied Pearson's test and maximal information coefficient to calculate correlation, and the causality relation between the meteorological features and wind speed was discovered. They divided the Ganger causality typology into center causality, ring causality, chained causality, etc. Ouyang et al. [78] utilized Autoregressive Moving Average (ARMA) to evaluate the causality significance of the meteorological features. The above Granger causality test is static, which does not match time-variant wind speed series. To explore the time-variant causality of the meteorological features, Jiang et al. [79] proposed a causality network to describe the evolution of the causality.

(c) For the temporal parameter of the wind speed, the Autocorrelation Function (ACF) and Partial Autocorrelation Function (PACF) are two commonly used methods. For the classical time series analysis, the ACF and PACF can determine the time lag of the ARMA model. The ACF and PACF can evaluate the temporal dependency of the wind speed. The ACF calculates the correlation of the time series with different time lags. The PACF estimates the correlation without the effect of redundant variables. Typically, the ACF and PACF are proposed for time series analysis models. In recent researches, they are also applied for machine learning methods with good performance, including Extreme Learning Machine (ELM) [80], Kernel Density Estimation (KDE) model [81], etc.

1.3.2.2 Descriptive model construction

The description model should have the capacity to explain the correlation between the parameters and wind speed. The intelligent methods are black boxes, which have poor interpretability. The most commonly used methods contain ARMA, Autoregressive Integrated Moving Average (ARIMA), Autoregressive Conditional Heteroscedasticity (ARCH), Generalized Autoregressive Conditional Heteroscedasticity (GARCH), Autoregressive Integrated Moving Average with Extra Input (ARIMAX), Kalman filter, etc. These methods have explicit equations and can explain how the parameters affect wind speed. The details are presented as follows:

(a) The ARMA can describe the temporal relationship of wind speed. In the aspect of temporal characteristics, the wind speed series is nonstationary. The stationary characteristics of the time series can be defined in two aspects, including strictly stationary process and weakly stationary process. The strictly stationary process requires the time series have the same joint distributions at any time; the weakly stationary process requires the time series to have the same mean and variance at any time. The strictly stationary characteristic is hard to measure, so the weakly stationary process is widely used in current researches. According to the definition of the nonstationary characteristics, the nonstationary time series is time-variant [82]. The Augmented Dickey-Fuller (ADF) test can validate the stationary characteristics. The typical ARMA cannot describe nonstationary series, so the ARIMA is proposed. The ARIMA method can use the differential operation to make the series stationary, and then utilize the ARMA model to describe the differenced time series. The wind speed is also heteroscedasticity, which can be validated by White test [83]. The ARCH and GARCH models can fit the autocorrelation function of the time series innovations, and describe the heteroscedasticity.

(b) With consideration of the spatial and meteorological parameters, the ARIMAX can be applied for prediction. Compared with the ARIMA model, the ARIMAX method is more complete, which can utilize dependent variables to improve description accuracy. To further improve the ARIMAX model, Robles-Rodriguez and Dochain [84] proposed a threshold ARIMAX model. The added threshold feature enables the ARIMAX model to fit nonlinear and asymmetric time series.

(c) The Kalman filter method can describe the wind speed according to the state equation and measurement equation. The time series state equation can be produced by the ARIMA model [85]. The measurement equation assumes the noise obeys Gaussian distribution. The Kalman filter assumes the time series system is linear, which does not match real nonlinear wind speed series. To deal with the nonlinear system, the extended Kalman filter and unscented Kalman filter method are utilized. The extended Kalman filter method uses first-order Taylor expansion to approach the nonlinear system; the unscented Kalman filter method applies an unscented transformation to estimate statistics after nonlinear transformation.

1.3.3 Wind forecasting technology

The wind forecasting is the essence of the wind engineering. There are many different classification methods for wind speed prediction methods, including spatial scale, temporal horizon, uncertainty, mechanism, and so on. The classifications of the forecasting methods are shown in Fig. 1.2. The details are presented as follows:

(a) Spatial scale: According to the spatial scale, the wind forecasting methods can be divided into single-point wind forecasting and spatial wind forecasting. The single-point forecasting focuses on a certain station; the spatial wind forecasting can provide forecasting results

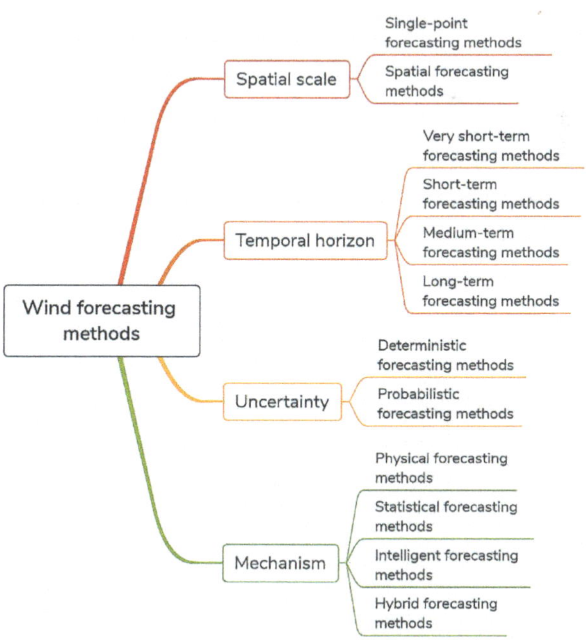

Figure 1.2 The classifications of the forecasting methods.

over a certain area. In the railway wind engineering, these two forecasting types are both important. These single-point results are significant for wind-proof scheduling in the strong wind section. The wind speed prediction results produced by the spatial prediction are of great significance for guiding the large-scale operation scheduling.

(b) Temporal horizon: According to the temporal horizon, the prediction can be divided into very short-term forecasting, short-term forecasting, medium-term forecasting, and long-term forecasting [86]. The very short-term forecasting focuses on several seconds to 30-min forecasting. This kind of forecasting is the most important for the railway wind engineering, because the immediate strong wind is the main reason for train overturning [49]. The short-term forecasting means 30-min to 6-h forecasting, which is important for advance train operation scheduling. The medium-term forecasting means 6-h to 1-day forecasting; the long-term forecasting means 1-day to 1-weak forecasting.

(c) Uncertainty: The wind forecasting contains deterministic and probabilistic forecasting. In the aspect of uncertainty, most of the current wind forecasting studies are deterministic, which can only generate a deterministic value for each prediction. However, the deterministic forecasting results cannot be always accurate. The unreasonable schedule will be made according to the inaccurate deterministic results. To improve the stability of the scheduling, the probabilistic forecasting is necessary, which can generate a prediction interval. The real value should locate in the prediction interval with a large confidence level. In this manner, the manager can estimate the uncertainty of the prediction, and make reasonable schedule with consideration of all scenarios. In the aspect of the time series theory, the wind speed series can be divided into certain components and uncertain components. The certain components have significant autocorrelation, which are predictable. On the contrary, the uncertain components are unpredictable. The deterministic forecasting can only forecast the certain components; the probabilistic forecasting can only fit the uncertain components. If the deterministic methods are applied to fit the uncertain components, the forecasting performance will be reduced, because the uncertain components are unpredictable. If the probabilistic methods are applied to fit the certain components, the uncertainty of the wind speed will be overestimated, because the certain trend is regarded as the uncertainty fluctuation.

(d) Mechanism: According to the forecasting mechanism, the current wind forecasting methods can be divided into physical methods, statistical methods, intelligent methods, and hybrid methods. The commonly used physical methods contain Weather Research and Forecasting (WRF), Fifth Generation Mesoscale (MM5), etc. The physical methods can generate forecasting results by simulating the physical evolution of the atmosphere, which are effective for the large horizon forecasting. However, the computational burden of the physical methods is heavy. For short-term forecasting, the statistical methods are useful because of the fast-computational performance. Different from the physical methods, the statistical methods produce forecasting results by regarding the wind speed as time series. The statistical methods are linear, which have limited accuracy to the nonlinear wind speed. The intelligent methods can fit the nonlinear functions of the wind speed. The neural networks are the most commonly used intelligent methods. Due to the nonstationary characteristics of the wind speed, the performance of the single intelligent methods is limited. The hybrid methods can apply preprocessing methods and postprocessing methods to improve forecasting performance.

1.3.4 Wind control technology

Like an auxiliary system of the Centralized Traffic Control (CTC), railway wind control system takes train schedule as the basis and integrates railway condition information (such as bridges, tunnels, large bending radius, etc.) and predicted wind speed, to obtain safe train speeds of different wind speed levels. In China, the railway wind control system can be divided into railway disaster prevention and safety monitoring system, CTC, real-time information transmission, and Chinese Train Control System (CTCS).

The railway disaster prevention and safety monitoring system can monitor the wind speed and direction along railways; the CTC system can offer real-time location of the train; real-time information transmission system can provide wind warning message to the trains within strong wind areas. The details of these systems are provided as follows:

(a) Railway disaster prevention and safety monitoring system can monitor environmental variables along the railway, and generate warning information. The railway disaster prevention and safety monitoring system can monitor wind, rainfall, snow, earthquake, intruding obstacles, etc. [87]. This system has three levels, including monitoring cell, data processing center, and railway bureau scheduling office [88]. The monitoring cell is composed of a set of DAE (Data Acquisition Equipment) for various environmental variables. The measured data are transformed into the data processing center. The data processing center can collect, store, analyze, and visualize the railway environment data. The prediction computation involved in this book is carried out in the data processing center. Besides the data from the monitoring cell, the data processing center also receives data from China meteorological administration, China earthquake administration, adjacent railway bureaus, etc. According to the data processing results, the railway bureau scheduling office can issue warning message to reschedule trains operation.

(b) The CTC system is developed from Train Dispatching Command System (TDCS). The CTC system can manage train operations, draw train operation diagrams, etc. In the wind control, the CTC system is used to provide the real-time position of the trains. According to the position, the strong wind warning information can be issued to the trains under strong wind.

(c) To ensure the trains can be informed the warning message stably and precisely, the real-time information transmission system is important. Global System for Mobile Communications for Railway (GSM-R) is the commonly used communication method in Europe and China. The GSM-R method is developed from Global System for Mobile Communications (GSM). For the high-speed trains, the redundant GSM-R system should be designed [89]. The double-layer structures are commonly used for the redundant system. The double-layer structures can be divided into cosited double-layer redundant coverage and intercross double-layer redundant coverage [90].

(d) The CTCS system can control the railway trains' operation status. According to the wind warning message, the CTCS system should be able to limit the speed. In the CTCS-2 system, the speed restriction information is issued through Train Control Center (TCC); in the CTCS-3 system, the speed restriction information is issued through Radio Block Center (RBC). To ensure the consistency of the CTCS-2 and CTCS-3 systems, the Temporary Speed Restriction Server (TSRS) is designed [91]. According to the speed restriction information from the TSRS, the on-board equipment of the CTCS system can control the train operation status to ensure safety.

1.4 Wind forecasting technologies in railway wind engineering

Among the key issues introduced in Section 1.3, prediction technology is the focus of this book. The current key forecasting technologies include location selection methods for wind measurement stations, single-point forecasting methods, and spatiotemporal forecasting methods.

1.4.1 Wind anemometer layout along railways

To obtain accurate early warning information, the wind speed monitoring system should use accurate wind speed values across the entire railway line. However, since the anemometer must be installed outside the railway clearances and can only measure wind speed at a single point, it is difficult to obtain accurate wind speed data along railways.

Anemometers installed out the railway will be affected by other facilities and cannot generate accuracy measurement for railway wind speed. The locations of the anemometers are proved to have a significant effect on measurement results [92]. Different countries have proposed different installation plans for wind measurement stations. Germany installed the anemometer on a place that has a horizontal distance of 4 m and a height of 4 m from the track center [7]; China's Urumqi Railway Bureau stipulated that the anemometer should be installed at 4.5 m above track surface [93]; Italy installed the anemometer at approximately 2 m above track surface [94]. To further improve the wind speed measurement accuracy, it is necessary to design special wind measurement stations for different types of railways. Researchers use aerodynamic methods to study the characteristics of the flow field along the railway. By studying the relationship between the measured wind speed of the anemometer and the actual wind speed of the train at different horizontal and vertical positions, the most suitable installation point can be obtained. Gao et al. studied the flow field of the railway with wind-break wall. They found the anemometer should be installed at 5 m from the wind-break wall [95]. Miao [96] studied the anemometer location of the railway bridge, and concluded the anemometer should be installed at mountain pass to ensure accuracy.

Realizing the monitoring of a single point along the railway may result in the inability to accurately and globally monitor strong winds along railways. For the entire railway line, monitoring a single point cannot effectively monitor the overall wind speed. This nonglobal monitoring method may miss local strong winds and affect train safety. If wind measuring stations are installed on the whole line, it will greatly increase the cost of installation and maintenance, which is impractical. To solve this dilemma, Italy proposed a two-step wind speed transformation method. This method utilized the wind measuring stations used for aircraft traffic as the input, and obtained the wind speed along railways [94]. In the transformation, a reference point was designed to connect the two-step method. The wind speed in air traffic stations was transformed into the reference point in the first step by weighted sum; the wind speed in reference point was transformed into the railway line in the second step. Spatial interpolation methods are also very useful. These methods can regress the wind speed on the whole railway according to measurement of stations. Commonly used interpolation methods are Kriging method, Inverse Distance Weighting (IDW), etc. These methods can infer the spatial wind speed from point wind speed. Friedland et al. [97] proposed a wind speed inference method between two diverse regions based on the Kriging method. Ravazzani et al. [98] proposed a spatial interpolation method for complex terrain with the IDW.

Constructing a distributed or on-board wind speed monitoring network is an effective method to cover the whole railway. The distributed measurement network consists of numerous anemometer nodes, and has a center to summarize the measured wind speed. The commonly used distributed anemometers networks are based on fiber optics. With the transmission of different frequency signals, a single fiber can be used to realize large-scale distributed measurement. Liu et al. [99] proposed an optical Venturi tube to achieve distributed measurement. Garcia-Ruiz et al. [100] designed an optical thermal wind speed measurement system for catenary. The measurement system is distributed, which can monitor wind speed with 10 m resolution in a 20 km railway. On-board anemometer of the train can observe the real-time wind speed of the train. Scanning wind speed along the railway can be realized through train operation. Sakuma et al. [101] designed an on-board measurement system based on hot-film anemometer and Pitot anemometer. According to the measured on-board data, the spatial map of the wind speed can be obtained [102].

1.4.2 Single-point wind forecasting along railways

According to the method mechanism, the single-point wind prediction technologies can be divided into physical methods, statistical methods, intelligent methods, and hybrid methods. The structure is shown in Fig. 1.3.

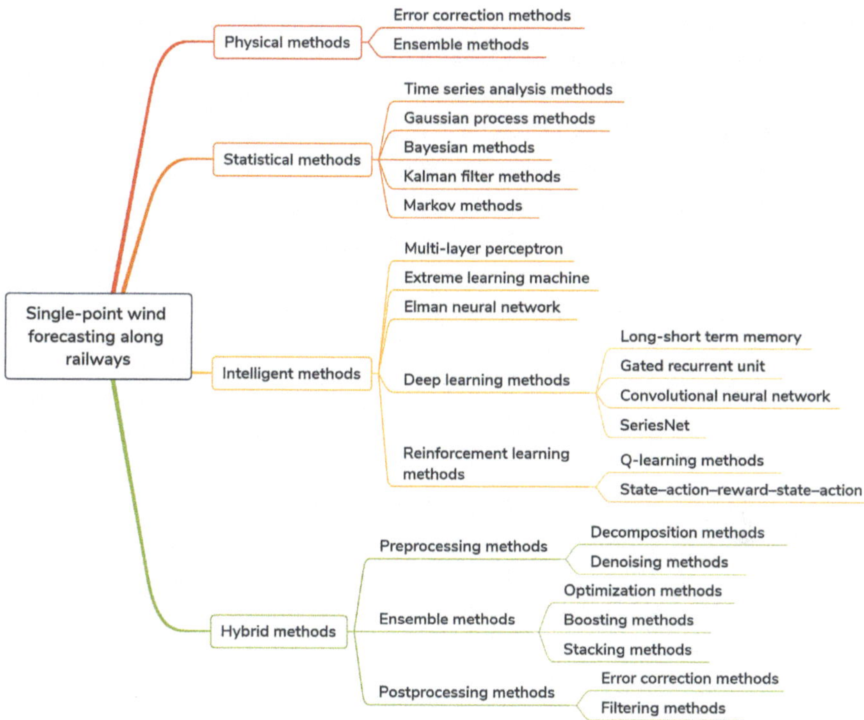

Figure 1.3 Structure of single-point wind forecasting methods.

The physical NWP methods can generate wind speed prediction results according to meteorological parameters and equations. The NWP methods can generate spatial results naturally, the single-point results can be extracted from them. Because of the sensitivity to the initial condition, the results of the physical methods are not stable. To improve the performance, the optimization methods for the NWP parameters are proposed. Because the number of parameters in the NWP model is large and the computation of the NWP model is heavy, the classical optimization methods cannot be applied directly. To make the optimization feasible for application, the surrogate methods were applied [103]. The surrogate methods can construct the response surface of the NWP model to the parameters, which can estimate model performance without NWP computation. In this manner, the computational burden can be reduced. Besides, the NWP forecasting results can be postprocessed to improve performance. In recent researches, the error correction and ensemble methods are widely used.

(a) The error correction methods can apply the statistical methods to predict forecasting residuals of the NWP methods. The single NWP model can hardly achieve accurate results, the generated forecasting residuals still contain predictable components. The error correction methods can further grasp the predictable components and improve performance. Wang et al. [104] proposed an error correction algorithm for the NWP model in sequence. In this study, five different models were applied to correct the NWP results. Zhao et al. [105] segmented the NWP results according to the wave characteristics, and applied a fuzzy clustering method to analyze the characteristics of the forecasting results. For different types of forecasting results, the optimal rules are generated to improve forecasting performance. Cai et al. [106] used the Support Vector Regression (SVR) model to correct the NWP results, and utilized Multi-Task Gaussian Process (MTGP) method to further generate the probabilistic results. Xu et al. [107] applied a hybrid model for error correction, which outperformed classical statistical methods. The hybrid model was composed of Variational Mode Decomposition (VMD), Principal Components Analysis (PCA), and Long Short-Term Memory (LSTM).

(b) The NWP model has inherent diversity. Different parameter configurations lead to diverse forecasting results. According to the ensemble model theory, the ensemble model can combine diverse results from weak learners into a more accurate strong learner. For the NWP results, the ensemble methods can combine several sets of the NWP results and generate better results. Yang et al. [108] developed two ensemble physical models, which applied linear regressor and Micro-Genetic Algorithm (MGA) to combine NWP prediction results. Singh et al. [109] constructed an optimal averaging ensemble model to combine four base models. Zhao et al. [110] constructed several NWP models with diverse initial conditions, applied membership degrees to select the best base NWP models, and used a fuzzy system to combine them.

The statistical methods can fit statistical rules of the wind speed data, and generate forecasting results. The commonly used statistical methods contain time series analysis methods, Gaussian process methods, Bayesian methods, Kalman methods, Markov methods, etc. The details are presented as follows:

(a) The Autoregressive model (AR), Moving Average (MA), ARMA, ARIMA, and ARIMAX methods are the most commonly used statistical models. These methods can fit the autocorrelation functions of the observations and innovations. The ARMA is composed of the AR part and MA part, which can fit the autocorrelation function of the observations and innovations respectively. The ARIMA is the ARMA model with an additional differential operation. Aasim et al. [111] proposed the very short-term ARIMA model, and proved the superiority. Due to the accumulative error, the large-step forecasting performance of the ARIMA is limited. To overcome this drawback, Liu et al. [112] proposed a recursive ARIMA model, which can be updated in real time to ensure accuracy. To improve performance to long-range dependency, fractional-ARIMA method was proposed [113]. In the fractional-ARIMA model, the differential degree can be set as a fractional value within −0.5 to 0.5.

(b) The Gaussian process models are nonparametric. The nonparametric methods have an undefined number of the parameters. In the Gaussian process models, the number of the parameters increases with the number of data, which is more flexible than the classical nonparametric models. Hu and Wang [114] used the Gaussian process model to fit the autocorrelation function of the observations, and proved the superiority of the Gaussian process models. Yu et al. [115] proposed an ensemble Gaussian process model based on the Gaussian Mixture Copula Model (GMCM). Hu et al. [116] applied t-observation to improve the Gaussian process model. Fang and Chiang [117] used the Gaussian process with NWP parameters for prediction. The Gaussian process models are probabilistic, which can generate prediction interval results. With this advantage, the Gaussian process models are widely used for postprocessing which can turn deterministic results into probabilistic results. Wang and Hu [118] applied the Gaussian process method to generate probabilistic results by combining several deterministic models. Zhang et al. [119] applied the Gaussian process model to correct the AR model, and verified both the deterministic and probabilistic performance can be improved by the Gaussian process. Zhang et al. [120] applied the Gaussian process model for secondary prediction, which took the results of improved LSTM as input.

(c) The Bayesian methods can set the model parameter as a distribution, and infers the posterior distribution of the parameter. The advantage of the Bayesian methods relies on the assigned prior distributions. When the prior distribution is set as Dirichlet distribution, the classical parametric model becomes nonparametric. If the prior distribution is well designed, fewer data are needed to achieve accurate performance.

The commonly used inference methods are the variational inference and Markov Chain Monte Carlo (MCMC). The variational inference sets a variational distribution, and optimizes the distribution to approach the target posterior distribution. The target function of the variational distribution is Evidence Lower Bound (ELBO), which is derived from the Kullback–Leibler (KL) divergence. The MCMC methods are based on simulation, which can generate samples of the target posterior distribution. The advantage of the MCMC is that any posterior distribution can be simulated, while the variational inference can only generate posterior distribution according to variational distribution. The disadvantage of the MCMC is that the computational burden is heavy. Wang et al. [121] proposed a Bayesian multikernel model for multiresolution data, and solve the posterior distributions by variational inference. Liu et al. [122] combined Bayesian methods with deep learning methods, and proposed a variational Bayesian spatiotemporal model. Liu et al. [83] applied Dirichlet Process Mixture Model (DPMM) for probabilistic prediction, which is solved by the MCMC method.

(d) Kalman filter methods can fuse observations and provide reliable estimation according to state equation and measurement equation [123]. The Kalman filter has a feedback mechanism, which can update Kalman gain to minimize estimation error. The time series is not an explicit dynamic system, so the state equation is hard to be inferred. To solve this problem, Liu et al. [85] proposed a state equation estimation method based on the ARIMA, which sets ARIMA coefficients as the state equation parameters. To improve nonlinear fitting capacity, the extended Kalman filter and unscented Kalman filter are applied for prediction. Chen and Yu [124] utilized the SVR to estimate the state equation, and applied the unscented Kalman filter for prediction.

(e) Markov models belong to directed graphical models, which can describe the switch of the states. Song et al. [125] proposed a Markov-switching model to fit evolution of the wind speed, which is solved by Bayesian inference. Wang et al. [126] applied the Markov model to discover the switching behavior of the forecasting residual, and improve the forecasting performance of the Least Squares Support Vector Machine (LSSVM) model. The Hidden Markov Model (HMM) is a commonly used Markov model, which assumes the wind observations are affected by hidden states. In the HMM model, the future observations are independent of the past observations when the present observation is known. Hocaoğlu et al. [127] applied the HMM model to fit the interaction functions of the wind speed and pressure, and generate accurate wind speed prediction results.

With the rapid development of the artificial intelligence technologies, the intelligent wind speed forecasting methods are widely used in recent studies. The intelligent models have strong nonlinear fitting capacity, which can outperform the statistical models. The commonly used intelligent models contain MLP, ELM, Elman Neural Network (ENN), deep learning methods, reinforcement learning methods, etc. The details are presented as follows:

(a) The MLP neural networks are the most typical artificial intelligent methods for wind speed, which are inspired by human neurons. According to the number of the layer, the MLP can be divided into Single-hidden-Layer Feedforward Network (SLFN), Two-hidden-Layer Feedforward Network (TLFN), etc. The SLFN is composed of three layers, including the input layer, hidden layer, and output layer; the TLFN is similar to the SLFN, but has an additional hidden layer. Because the TLFN has more complex structure than the SLFN, the fitting capacity is better [128]. According to the training optimizer, the Back Propagation (BP), Broyden, Flecther, Goldfard, and Shanno Quasi-Newton (BFGS) are frequently used for the MLP [129].

(b) The ELM neural networks have similar structures with the MLP models, but have different training methods. The parameters of the ELM can be calculated through matrix operation without iterative optimization. So, the ELM models have high computational efficiency. Despite this, the ELMs can achieve comparable performance with the MLPs in some cases. Liu et al. [130] applied the Outlier Robust Extreme Learning Machine (ORELM) for multiresolution forecasting, and proved the ORELM is better than the fine-tuned MLP model.

(c) Different from the MLP and ELM models, the ENN models have a different structure. The additional context layer enables the ENN model to discover time dependency within the wind speed series. Liu et al. [131] applied the ENN for wind speed prediction, and proved the ENN is better than the MLP model. Yu et al. [132] applied the ENN model with improved decomposition methods, and achieved excellent performance. Qin et al. [133] applied the ENN model for error correction, and verified forecasting capacity to nonlinear wind speed series.

(d) The deep learning methods have more deep structure than the classical neural networks. In wind speed prediction, the LSTM, GRU, and Convolutional Neural Network (CNN) are widely used. The LSTM and GRU are all recurrent networks, which solve gradient exploration and vanishing problem in deep recurrent networks through specific structure designs. Due to their complexity, the input variables of the LSTM and GRU should be carefully selected. Too many model inputs will result in reduced generalization performance, and too few model inputs will result in underfitting. Memarzadeh and Keynia [134] optimized the LSTM's input variables based on the entropy and mutual information to improve forecasting performance. Niu et al. [135] developed an attention-based GRU model for wind prediction. The attention mechanism can select the most suitable input subset from the raw input variables for the GRU model. The CNN model can take multidimension tensor as input and generate results. Liu et al. [83] used a decomposition algorithm to construct two-dimension matrixes from raw wind speed time series. The causal convolution is a special convolution operation in the CNN model, which can handle time series. Different from classical convolution, the causal convolution only deals with past observations, no future observations are involved. By stacking the dilated causal convolutions, Seriesnet is proposed [136]. The Seriesnet has been proved to have good performance for wind speed prediction [137].

(e) The reinforcement learning methods have strong decision-making capacity. In the wind speed forecasting researches, the reinforcement learning methods can be applied for optimization. Liu et al. [138] applied Q-learning and State-Action-Reward-State-Action (SARSA) methods to calculate the ensemble coefficients. The results proved the Q-learning method outperforms the SARSA and heuristic algorithms, including Grey Wolf Optimizer (GWO), Particle Swarm Optimizer (PSO), etc. Besides, the reinforcement learning methods can be utilized as the forecasting algorithm directly. Sharma et al. proposed a modified fuzzy Q-learning algorithm as the wind speed forecasting model, and verified the superiority over the SVR and KNN.

Based on the statistical and intelligent models, the hybrid models should be proposed to improve performance further. The hybrid models applied preprocessing methods, ensemble methods, and postprocessing methods to improve performance. The details are presented as follows:

(a) The preprocessing methods can make the raw wind speed series more predictable. The decomposition and filtering methods are widely used.

(a-1) The decomposition algorithms can separate raw series into several subseries. In the preprocessing algorithms, the decomposition algorithms have the highest usage frequency [139]. The frequency components of the raw wind speed are mixed, so the single forecasting model can hardly capture the precise autocorrelation function of the series. The decomposition methods can divide the frequency components of the raw series, and produce several subseries with narrow frequency bands. The obtained subseries contain more simple frequency components, so they are more predictable. The commonly used decomposition algorithms contain mode decomposition algorithm family, wavelet decomposition algorithm family, etc. The mode decomposition algorithm family is developed from Empirical Mode Decomposition (EMD). The EMD algorithm can extract Intrinsic Mode Functions (IMFs) by calculating envelopes of the series. The classical EMD has several disadvantages including ending effect, mode mixing, etc. [140]. To improve decomposition performance of the EMD algorithm, a series of decomposition algorithms are proposed, including Ensemble Empirical Mode Decomposition (EEMD), Fast Ensemble Empirical Mode Decomposition (FEEMD), Complementary Ensemble Empirical Mode Decomposition (CEEMD), Complete Ensemble Empirical Mode Decomposition with Adaptive Noise (CEEMDAN), Improved Complete Ensemble Empirical Mode Decomposition with Adaptive Noise (ICEEMDAN), etc. The wavelet decomposition algorithm family is developed from Wavelet Decomposition (WD). The WD algorithm can decompose the wind speed series into detailed and approximate components according to Discrete Wavelet Transform (DWT). One of the disadvantages of the WD is that only approximate components are decomposed at each level, the detailed components keep still. To better decompose the series, the Wavelet Packet Decomposition (WPD) algorithm is proposed, which can decompose both of the detailed and approximate components at each level. Besides, maximal overlap wavelet decomposition algorithms are proposed, including Maximal Overlap Discrete Wavelet Transform (MODWT), Maximal Overlap Discrete Wavelet Packet Transform (MODWPT). The maximal overlap wavelet decomposition algorithms do not need the series length, while the classical wavelet decomposition algorithms fill zeros until the series length meets the requirement. Apart from the above decomposition algorithms, there are many advanced decomposition technologies. One of them is secondary decomposition method. The secondary decomposition method can further decompose subseries to improve performance. How to rationally select the subseries for secondary decomposition is important to ensure performance. Liu et al. [131] applied the FEEMD to decompose high-frequency subseries. Sun et al. [141] used Variational Mode Decomposition (VMD) algorithm to decompose the first IMF of the EEMD. Liu et al. [142] applied sampling entropy to estimate the predictability of the subseries, and decomposed the series with the poorest predictability.

(a-2) Due to the measurement and transmission noise, the raw wind speed series has noise. The contained noise is not predictable, so the noise should be discarded to ensure forecasting performance. The denoizing algorithms are similar to the decomposition algorithms. In the wind speed series, the main informative components are low frequency. The subseries with the highest frequency can be regarded as noise. Guo et al. [143] applied the EMD algorithm to generate IMFs of the raw wind speed series, and predicted all IMFs except the high-frequency IMF. Cheng et al. [144] applied wavelet threshold denoizing method for wind speed series.

(b) The ensemble methods can be divided into optimization methods, boosting methods, stacking methods, etc.

(b-1) The optimization methods can generate optimal ensemble weights to maximize the forecasting accuracy. Ma et al. [145] applied the PSO algorithm to generate ensemble weights for the base models. Yang and Wang [146] used an improved water cycle algorithm based on a quasi-Newton algorithm to optimize No-Negative Constraint theory (NNCT) function. The above optimization methods are single-objective. In the ensemble models, the complexity may lead to overfitting. The single-objective optimization methods only consider the forecasting accuracy. The robustness of the ensemble model is not included. To improve performance, the multi-objective optimization is proposed for ensemble wind speed forecasting models. The commonly used multi-objective optimization algorithms include Multi-Objective Grey Wolf Optimization (MOGWO) [147], Multi-Objective Particle Swarm Optimization (MOPSO) [148], Multi-Objective Multi-Verse Optimization (MOMVO) [130], etc.

(b-2) The ensemble weights can also be calculated with the Boosting algorithms. The Boosting algorithms can assign ensemble weights according to the model's performance. The accurate model has a high ensemble weight to ensure forecasting performance. The commonly used Boosting algorithms contain Adaptive Boosting (AdaBoost), Gradient Boosting (GBoost), Linear Programming Boosting (LPBoost), etc. The AdaBoost and GBoost belong to corrective Boosting algorithms, which can generate ensemble weights of one base model at each iteration. The LPBoost is a total corrective Boosting algorithm, which can change all ensemble weights at each iteration. Li et al. [149] proved these two types of Boosting algorithms have no significant difference.

(b-3) The ensemble weights of the optimization methods can only combine the base models with the linear transformation. The stacking algorithm can combine the base models in a nonlinear way. The stacking methods are composed of two sets of models, i.e., base models and meta model. The meta model can take the forecasting results of the base models as input, and generate ensemble forecasting results. The neural networks are commonly used as the meta models. Chen et al. [150] combined forecasting results of LSTMs with the SVR model. Quresghi et al. [151] proposed a stacking model using a deep belief network as the meta network. The optimization algorithms can also be applied for the meta model. Chen and Liu [152] applied the Butterfly Optimization Algorithm (BOA) to optimize the SVR meta model to improve performance.

(c) The postprocessing methods in the wind speed forecasting contain error correction methods, filtering methods, etc.

(c-1) The error correction methods apply an additional model to forecast the prediction residuals. The static correction methods do not analyze characteristics of the forecasting residuals, and correct them directly. The dynamic error correction methods can assign a suitable correction model for the residuals according to residuals characteristics. Wang and Li [153] matched correction models based on predictability and heteroscedasticity of the residuals. Liu et al. [83] used Ljung-Box Q-test (LBQ-test) to determine whether to perform error correction, and corrected the residuals until the residuals become unpredictable. This design can ensure the predictable components in the residuals can be fully grasped.

(c-2) The filtering methods can be applied for the forecasting results to improve forecasting performance further. The subseries after decomposition have specific frequency bands. However, the forecasting results of the subseries contain components beyond the bands. To improve the forecasting performance, the additional components are filtered. Liu et al. [142] applied the Wavelet Packet Filter (WPF) to process the forecasting results after the WPD. The forecasting results verified the WPF can improve big multistep forecasting performance.

1.4.3 Spatial wind forecasting along railways

The spatial wind speed forecasting methods can produce forecasting results over an area. The spatial wind forecasting technologies can be divided into physical methods, spatial correlation methods, etc. The structure is shown in Fig. 1.4.

The physical methods can simulate spatial meteorological behavior, and generate spatial forecasting results naturally. The NWP methods mainly focus on the mesoscale forecasting with medium- or long-term prediction horizons. To obtain fine results along railways, the downscaling methods should be carried out. The statistical methods can be applied for the output of the NWP models, and generate short-term downscaling forecasting results [154].

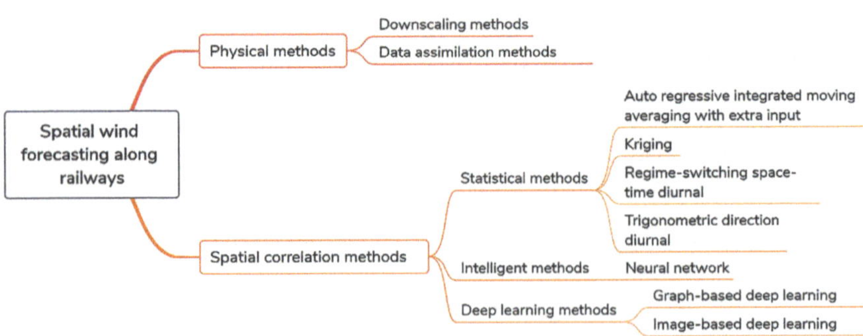

Figure 1.4 Structure of spatial wind forecasting methods.

The initial values of the NWP values are important to ensure accuracy. The data assimilation methods can take meteorological fields with different sources as the input, and generate the best initial field for the NWP model. Cheng et al. [155] applied Real-Time Four-Dimensional Data Assimilation (RTFDDA) system to apply raw wind speed data for the NWP.

The spatial correlation methods can apply statistical, intelligent, or deep learning methods to discover spatial correlation of the wind field, and generate spatial forecasting results.

(a) The commonly used statistical methods contain ARIMAX, Kriging model, Regime-Switching Space-Time Diurnal (RSTD), Trigonometric Direction Diurnal (TDD) model, etc. The ARIMAX model can take the meteorological parameters in near positions as input, fit temporal correlation and spatial correlation, and generate forecasting results. Tascikaraoglu et al. [156] proposed an input parameter selection algorithm based on compressive sensing, then applied a linear model for prediction. Dowell and Pinson [157] applied a sparse vector autocorrelation model for prediction, which can fit sparse spatial correlation. The Kriging method assumes the wind speed of a specific position has a linear correlation with wind speed in nearby stations. The correlation coefficients of the Kriging methods can be calculated to minimize variance. The TDD method is improved from the RSTD model. In the TDD model, the wind speed values obey truncated normal distribution. Zhu et al. [158] used geostrophic wind as the predictor for the TDD to improve forecasting performance.

(b) As for the intelligent spatial correlation models, the neural networks are widely used for their strong fitting capacity. In the neural networks, the input parameters typically are the meteorological parameters. To improve forecasting performance, these parameters should be carefully selected. Li et al. [159] proposed a correlative station selection algorithm considering the wind direction, and applied Radial Basis Function Neural Network (RBFNN) algorithm for prediction. Xu et al. [160] applied a random probe algorithm to select spatial features for neural network prediction.

(c) The deep learning methods have better fitting capacity than the classical neural network models. Graph-based, image-based deep networks are commonly used in the spatial deep learning networks. The graph-based deep networks define a graph convolution operator, to process spatial graph data. Khodayar and Wang [161] combined LSTM and CNN for a scalable Graph Convolutional Deep Learning Architecture (GCDLA), which applied fuzzy system to deal with uncertainty in input data. The image-based deep network can regard the spatial wind speed data as multidimension matrixes, and process them with CNN like images. Chen et al. [162] constructed wind speed data images, and produced forecasting results with LSTM and CNN.

1.5 Scope of this book

This book introduces the wind flow analysis, single-station wind forecasting, and spatial wind forecasting technologies in wind engineering. This book contains eight chapters, which are introduced as follows:

1.5.1 Chapter 1: Introduction

In this chapter, the typical scenarios of the wind engineering are introduced. The key technologies, including wind anemometer layout, single-station wind forecasting, and spatial wind forecasting, are overviewed.

1.5.2 Chapter 2: Analysis of flow field characteristics along railways

In this chapter, the real flow field in 100 Miles Wind Area and the 30 Miles Wind Area are provided as analysis example. The local and global Moran's I indexes are applied to analyze the spatial characteristics of the wind flow field. The Planar Maximally Filtered Graph (PMFG) is applied to extract key spatial correlation structure of the flow field, and the relationship between flow field key correlation and flow streamlines are analyzed. The Fast Fourier Transform (FFT) is applied to analyze frequency spectrum of the flow field, and the main frequencies are discovered. The Bayesian Fuzzy Clustering (BFC) is used to extract key flow field seasonal templates.

1.5.3 Chapter 3: Description of single-point wind time series along railways

In this chapter, wind anemometer layout optimization methods for single-station wind speed measurement are introduced firstly. Then, the seasonal and heteroscedastic characteristics of wind speed and wind direction are analyzed. Finally, the Seasonal Autoregressive Integrated Moving Average (SARIMA), ARCH, and GARCH are utilized for wind description.

1.5.4 Chapter 4: Single-point wind forecasting methods based on deep learning

In this chapter, three advanced deep learning methods, including LSTM, GRU, and Seriesnet, are introduced for wind forecasting. The decomposition methods are applied to further improve performance. At last, the deterministic forecasting performance of the deep learning methods is analyzed.

1.5.5 Chapter 5: Single-point wind forecasting methods based on reinforcement learning

In this chapter, the reinforcement learning methods are introduced for static ensemble weight optimization, feature selection, etc. The Q-learning, Deep Q-Network (DQN), and Deep Deterministic Policy Gradient (DDPG) are investigated. Finally, the advantages and disadvantages of the reinforcement learning methods are summarized.

1.5.6 Chapter 6: Single-point wind forecasting methods based on ensemble modeling

In this chapter, three mainstream ensemble methods for single-station wind forecasting are introduced, including multi-objective ensemble, Stacking ensemble, and Boosting ensemble. The designed ensemble forecasting methods can combine diverse base forecasting models.

1.5.7 Chapter 7: Description methods of spatial wind along railways

In this chapter, the spatial wind correlation characteristics are evaluated by mutual information, Pearson coefficient, Kendall coefficient, and Spearman coefficient. Then, the WRF is built to describe spatial wind. Finally, the performance evaluation indicators of spatial forecasting are introduced.

1.5.8 Chapter 8: Data-driven spatial wind forecasting methods along railways

In this chapter, two statistical spatial forecasting methods are introduced for spatial prediction firstly, which apply mutual information for spatial feature selection. Then, the intelligent spatial forecasting methods are investigated, which are combined with four binary optimization algorithms. Finally, three deep learning methods are applied for spatial prediction, which use sparse autoencoder for spatial feature extraction.

References

[1] J. Liu, F. Schmid, W. Zheng, et al., Understanding railway operational accidents using network theory, Reliab. Eng. Syst. Saf. 189 (2019) 218–231.
[2] M. Mohebbi, M.A. Rezvani, Analysis of the effects of lateral wind on a high speed train on a double routed railway track with porous shelters, J. Wind Eng. Ind. Aerod. 184 (2019) 116–127.
[3] Q. Xie, X. Zhi, Wind tunnel test of an aeroelastic model of a catenary system for a high-speed railway in China, J. Wind Eng. Ind. Aerod. 184 (2019) 23–33.
[4] Y. Tamura, Wind-induced damage to buildings and disaster risk reduction, in: 7th Asia-Pacific Conference on Wind Engineering, APCWE-VII, 2009, pp. 1–11.
[5] Z. Yao, J. Xiao, F. Jiang, Characteristics of daily extreme-wind gusts along the Lanxin Railway in Xinjiang, China, Aeolian Res. 6 (2012) 31–40.
[6] Abc 8news, High Winds Derail 26-Car Train in New Mexico, 2019. https://www.wric.com/news/high-winds-derail-26-car-train-in-new-mexico/. (Accessed 16 October 2020).
[7] U. Hoppmann, S. Koenig, T. Tielkes, et al., A short-term strong wind prediction model for railway application: design and verification, J. Wind Eng. Ind. Aerod. 90 (2002) 1127–1134.
[8] N. Kobayashi, M. Shimamura, Study of a strong wind warning system, Jr. East Tech. Rev. 2 (2003) 61–65.
[9] M. Burlando, A. Freda, C.F. Ratto, et al., A pilot study of the wind speed along the Rome–Naples HS/HC railway line. Part 1—numerical modelling and wind simulations, J. Wind Eng. Ind. Aerod. 98 (2010) 392–403.
[10] D. Delaunay, L.-M. Cléon, C. Sacré, et al., Designing a wind alarm system for the TGV-Méditerranée, in: 11th International Conference on Wind Engineering, Texas, USA, 2003, pp. 1–8.
[11] Y. Deng, X. Xiao, Effect of cross-wind on high-speed vehicle dynamic derailment, in: Logistics: The Emerging Frontiers of Transportation and Development in China, 2009, pp. 2287–2293.
[12] J. Yang, Aerodynamic Effect on Running Safety and Stability of High-Speed Train, Southwest Jiaotong University, 2010.
[13] S. Yi, Dynamic Analysis of High-Speed Railway Alignment: Theory and Practice, Academic Press, 2017.
[14] M.S. Kim, G.Y. Kim, H.T. Kim, et al., Theoretical cross-wind speed against rail vehicle derailment considering the cross-running wind of trains and the dynamic wheel-rail effects, J. Mech. Sci. Technol. 30 (2016) 3487–3498.
[15] B. Chen, X. Pang, X. Cheng, et al., The forecasting method about the rate of wheel load reduction based on NARX neural network, in: 2012 Spring Congress on Engineering and Technology, 2012, pp. 1–4.
[16] D. Liu, T. Wang, X. Liang, et al., High-speed train overturning safety under varying wind speed conditions, J. Wind Eng. Ind. Aerod. 198 (2020) 104111.
[17] J. Xiang, D. He, Q.Y. Zeng, Effect of cross-wind on spatial vibration responses of train and track system, J. Cent. South Univ. Technol. 16 (2009) 520–524.
[18] D. Liu, G. Marita Tomasini, D. Rocchi, et al., Correlation of car-body vibration and train overturning under strong wind conditions, Mech. Syst. Signal Process. 142 (2020) 106743.
[19] L. Zhang, Research on Collection Quality and Operation Security of Pantograph-Catenary under the Action of Strong Wind, Southwest Jiaotong University, 2013.
[20] Y. Song, Z. Liu, H. Wang, et al., Nonlinear analysis of wind-induced vibration of high-speed railway catenary and its influence on pantograph–catenary interaction, Veh. Syst. Dyn. 54 (2016) 723–747.

[21] J. Zhang, Z. Liu, X. Lu, et al., Study on aerodynamics development of high-speed pantograph and catenary, J. China Railw. Soc. 37 (2015) 7–14.
[22] Y. Song, Z. Liu, X. Lu, et al., Study on characteristics of dynamic current collection of high-speed pantograph-catenary considering aerodynamics of catenary, J. China Railw. Soc. 38 (2016) 48–53.
[23] J. Luo, J. Luo, Z. Yang, et al., Numerical research on aerodynamic characteristic optimization of pantograph fixing place on high speed train, in: 2009 2nd International Conference on Power Electronics and Intelligent Transportation System (PEITS), vol. 1, 2009, pp. 94–97.
[24] Y. Song, Z. Liu, H. Ouyang, et al., Sliding mode control with PD sliding surface for high-speed railway pantograph-catenary contact force under strong stochastic wind field, Shock Vib. 2017 (2017) 1–16.
[25] D. Xiang, Study on Correlation of Buffeting Force and Identification of Aerodynamic Admittance Function of Bridge Section under Pulsating Airflow, Zhengzhou University, 2016.
[26] B.-F. Wu, M. Guo, X.-Y. Qiao, Measures study on wind resistance of long-span bridges, Build. Technol. Dev. 47 (2020) 127–128.
[27] Z. Liu, Z. Zhu, W. Chen, et al., Observation of vortex-induced vibration and wind characteristics of cables across the Yangtze river bridge, J. Railw. Sci. Eng. 17 (2020) 1760–1768.
[28] B. Jin, L. Tang, W. Zhou, et al., Parameter optimization of multiple tuned mass damper based on long-span suspension bridge damping control, Highw. Eng. 45 (2020) 98–104.
[29] M. Gu, F. Hai, Flutter and buffeting control of long-span bridges, Bull. Natl. Nat. Sci. Found. China 3 (1999) 3–5.
[30] W. Wang, X. Wang, X. Hua, et al., Vibration control of vortex-induced vibrations of a bridge deck by a single-side pounding tuned mass damper, Eng. Struct. 173 (2018) 61–75.
[31] A. Casalotti, A. Arena, W. Lacarbonara, Mitigation of post-flutter oscillations in suspension bridges by hysteretic tuned mass dampers, Eng. Struct. 69 (2014) 62–71.
[32] J. Yang, H.-L. Zhang, J.-J. Zhou, et al., Air resistance coefficient of hump rolling wagon based on fluent simulation, J. Transp. Syst. Eng. & Inf. Technol. 18 (2018) 168–174.
[33] D. Kun, Research on influence of wind resistance on cars humping, J. China Railw. Soc. 34 (2012) 63–69.
[34] Y. Yu, S. Ma, L. Chen, Optimization of longitudinal section design of railway coal logistics yard in Xinjiang, in: 2010 International Conference on Optoelectronics and Image Processing, vol. 2, 2010, pp. 86–89.
[35] D. Tian, Design research of meshed windbreak used in railway coal storage yard in plateau region, Railw. Stand. Des. 58 (2014) 44–49.
[36] D. Wang, Y.-M. Zhang, C.-Y. Wang, et al., Applying integrated technology of wind-break and dust-control on coal storage yard, Environ. Sci. Technol. 33 (2010) 84–85.
[37] X. Pan, X. Ma, J. Xu, Analysis and evaluation about anti-wind efficiency of wind-break experimental section in Lan-Xin high railway, J. Arid Meteorol. 37 (2019) 496–499.
[38] J. Zhang, K. He, J. Wang, et al., Numerical simulation of flow around a high-speed train subjected to different windbreak walls and yaw angles, J. Appl. Fluid Mech. 12 (2019) 1137–1149.
[39] H. Xiang, Y. Li, Y. Su, et al., Surrogate model optimizations for protective effects of railway wind barriers, J. Southwest Jiaot. Univ. 51 (2016) 1098–1104.

[40] N. Huang, K. Gong, B. Xu, et al., Investigations into the law of sand particle accumulation over railway subgrade with wind-break wall, Eur. Phys. J. E 42 (145) (2019).

[41] X. Guowei, H. Ning, Z. Jie, Wind-tunnel experiment on sand deposition mechanism and optimal measures of wind-break wall along railway in strong wind area, Chin. J. Theor. Appl. Mech. 52 (2020) 635–644.

[42] Y. Zou, Z. Fu, X. He, et al., Wind load characteristics of wind barriers induced by high-speed trains based on field measurements, Appl. Sci. 9 (2019) 4865.

[43] M. Tokunaga, M. Sogabe, T. Santo, et al., Dynamic response evaluation of tall noise barrier on high speed railway structures, J. Sound Vib. 366 (2016) 293–308.

[44] J. Liu, M. Li, J. Zhang, et al., Aerodynamic optimization design of streamline head of high-speed train, Sci. Sin. 43 (2013) 689–698.

[45] S. Yao, D. Guo, Z. Sun, et al., Optimization design for aerodynamic elements of high speed trains, Comput. Fluid 95 (2014) 56–73.

[46] J.-M. Li, Y.-L. Shan, P. Lin, Analysis on aerodynamic performance of anti-ice/snow dome of high speed motor train unit bogie, Comput. Aided Eng. 22 (2013) 20–26.

[47] J. Wang, J. Zhang, F. Xie, et al., A study of snow accumulating on the bogie and the effects of deflectors on the de-icing performance in the bogie region of a high-speed train, Cold Reg. Sci. Technol. 148 (2018) 121–130.

[48] R. Wang, A study on the application rules of high-speed railway wind monitoring system, Railw. Transp. & Econ. 40 (2018) 48–51+57.

[49] R. Wang, R. Chen, Y. Bao, The study on JR-east monitoring technology of strong wind, China Railw. 07 (2018) 96–102.

[50] H. Liu, S. He, J. Chen, Data-driven adaptive adjustment strategy for strong wind alarm in high-speed railway, Acta Autom. Sin. 45 (2019) 2242–2250.

[51] Z. Zhou, Characteristics Analysis and Experimental Research on the Three-Cup Anemometer, Harbin Institute of Technology, 2007.

[52] S. Lang, E. Mckeogh, LIDAR and SODAR measurements of wind speed and direction in upland terrain for wind energy purposes, Rem. Sens. 3 (2011) 1871–1901.

[53] P. Xv, Research on Strong Wind Monitoring and Train Traffic Control System of Qinghai-Tibet Railway, Central South University, 2009.

[54] T. Liu, H. Wei, K. Zhang, Wind power prediction with missing data using Gaussian process regression and multiple imputation, Appl. Soft Comput. 71 (2018) 905–916.

[55] B. Fallah, K.T.W. Ng, H.L. Vu, et al., Application of a multi-stage neural network approach for time-series landfill gas modeling with missing data imputation, Waste Manag. 116 (2020) 66–78.

[56] F. Qu, J. Liu, X. Hong, et al., Data imputation of wind turbine using generative adversarial nets with deep learning models, in: International Conference on Neural Information Processing, 2018, pp. 152–161.

[57] S. Oehmcke, O. Zielinski, O. Kramer, kNN ensembles with penalized DTW for multivariate time series imputation, in: 2016 International Joint Conference on Neural Networks (IJCNN), 2016, pp. 2774–2781.

[58] F.T. Liu, K.M. Ting, Z.-H. Zhou, Isolation forest, in: 2008 Eighth IEEE International Conference on Data Mining, 2008, pp. 413–422.

[59] G. Gan, M.K.-P. Ng, k -means clustering with outlier removal, Pattern Recognit. Lett. 90 (2017) 8–14.

[60] P. Marti-Puig, A. Blanco-M, J.J. Cárdenas, et al., Effects of the pre-processing algorithms in fault diagnosis of wind turbines, Environ. Model. Software 110 (2018) 119–128.

[61] J. Ting, E. Theodorou, S. Schaal, A Kalman filter for robust outlier detection, in: 2007 IEEE/RSJ International Conference on Intelligent Robots and Systems, 2007, pp. 1514–1519.

[62] D. Pradeepkumar, V. Ravi, Forecasting financial time series volatility using particle swarm optimization trained quantile regression neural network, Appl. Soft Comput. 58 (2017) 35–52.
[63] Q. Xu, C. Jiang, Y. He, An exponentially weighted quantile regression via SVM with application to estimating multiperiod VaR, Stat. Methods Appl. 25 (2015) 285–320.
[64] Z. Zhang, H. Qin, Y. Liu, et al., Wind speed forecasting based on quantile regression minimal gated memory network and kernel density estimation, Energy Convers. Manag. 196 (2019) 1395–1409.
[65] R. Cekik, A.K. Uysal, A novel filter feature selection method using rough set for short text data, Expert Syst. Appl. 160 (2020) 113691.
[66] H. Liu, Z. Duan, H. Wu, et al., Wind speed forecasting models based on data decomposition, feature selection and group method of data handling network, Measurement 148 (2019) 106971.
[67] D. Rodrigues, V.H.C. De Albuquerque, J.P. Papa, A multi-objective artificial butterfly optimization approach for feature selection, Appl. Soft Comput. 94 (2020) 106442.
[68] R.C. Thom De Souza, C.A. De Macedo, L. Dos Santos Coelho, et al., Binary coyote optimization algorithm for feature selection, Pattern Recogn. 107 (2020) 107470.
[69] H. Liu, Z. Duan, C. Chen, A hybrid framework for forecasting PM2.5 concentrations using multi-step deterministic and probabilistic strategy, Air Qual. Atmos. & Health 12 (2019) 785–795.
[70] P. Papazek, I. Schicker, C. Plant, et al., Feature selection, ensemble learning, and artificial neural networks for short-range wind speed forecasts, Meteorol. Zogische Zeitschrift 1–17 (2020).
[71] M.C. Alexiadis, P.S. Dokopoulos, H.S. Sahsamanoglou, Wind speed and power forecasting based on spatial correlation models, IEEE Trans. Energy Convers. 14 (1999) 836–842.
[72] T. Filik, Improved spatio-temporal linear models for very short-term wind speed forecasting, Energies 9 (2016) 168.
[73] H. Liu, C. Chen, Spatial air quality index prediction model based on decomposition, adaptive boosting, and three-stage feature selection: a case study in China, J. Clean. Prod. 265 (2020) 121777.
[74] R. Yu, J. Gao, M. Yu, et al., LSTM-EFG for wind power forecasting based on sequential correlation features, Future Generat. Comput. Syst. 93 (2019) 33–42.
[75] Y. Noorollahi, M.A. Jokar, A. Kalhor, Using artificial neural networks for temporal and spatial wind speed forecasting in Iran, Energy Convers. Manag. 115 (2016) 17–25.
[76] X. Shen, C. Zhou, X. Fu, Study of time and meteorological characteristics of wind speed correlation in flat terrains based on operation data, Energies 11 (2018) 219.
[77] Z. Zhang, H. Qin, Y. Liu, et al., Long short-term memory network based on neighborhood gates for processing complex causality in wind speed prediction, Energy Convers. Manag. 192 (2019) 37–51.
[78] T. Ouyang, X. Zha, L. Qin, A combined multivariate model for wind power prediction, Energy Convers. Manag. 144 (2017) 361–373.
[79] M. Jiang, X. Gao, H. An, et al., Reconstructing complex network for characterizing the time-varying causality evolution behavior of multivariate time series, Sci. Rep. 7 (2017) 10486.
[80] N. Huang, C. Yuan, G. Cai, et al., Hybrid short term wind speed forecasting using variational mode decomposition and a weighted regularized extreme learning machine, Energies 9 (2016) 989.

[81] Y. Jiang, S. Liu, L. Peng, et al., A novel wind speed prediction method based on robust local mean decomposition, group method of data handling and conditional kernel density estimation, Energy Convers. Manag. 200 (2019) 112099.

[82] M. Rhif, A. Ben Abbes, I. Farah, et al., Wavelet transform application for/in non-stationary time-series analysis: a review, Appl. Sci. 9 (2019) 1345.

[83] H. Liu, Z. Duan, C. Chen, et al., A novel two-stage deep learning wind speed forecasting method with adaptive multiple error corrections and bivariate Dirichlet process mixture model, Energy Convers. Manag. 199 (2019) 111975.

[84] C.E. Robles-Rodriguez, D. Dochain, Decomposed threshold ARMAX models for short-to medium-term wind power forecasting, IFAC-Pap. OnLine 51 (2018) 49−54.

[85] H. Liu, H.-Q. Tian, Y.-F. Li, Comparison of two new ARIMA-ANN and ARIMA Kalman hybrid methods for wind speed prediction, Appl. Energy 98 (2012) 415−424.

[86] S.S. Soman, H. Zareipour, O. Malik, et al., A review of wind power and wind speed forecasting methods with different time horizons, in: North American Power Symposium (NAPS)., 2010, pp. 1−8.

[87] C. Xv, Research of Disaster Prevention and Safety Monitoring System for High Speed Railway, China Academy of Railway Sciences, 2010.

[88] J. Wang, J. Wang, Design of disaster prevention and safety monitoring system for high-speed railway, China Saf. Sci. J. 28 (2018) 39−45.

[89] L. Liu, Analysis and Protection of Electromagnetic Environment Interference in GSM-R System, Yunnan University, 2019.

[90] Y. Sun, H. Han, R. Sun, et al., Comparison and analysis of GSM-R wireless network coverage schemes, Electron. World 8 (2016) 149+153.

[91] X. Zhou, Modeling Analysis and Verification of Temporary Speed Restriction Server in High Speed Railway Train Control System Based on Timed Automata, Lanzhou Jiaotong University, 2017.

[92] K. Araki, T. Imai, K. Tanemoto, et al., Evaluation of the influence of anemometer position around railway structures on wind observation data, Q. Rep. RTRI 53 (2012) 113−120.

[93] C. Ge, Study on safe train operation in windy weather in Xinjiang railway wind region, Railw. Transp. & Econ. 31 (2009) 32−34+84.

[94] A. Freda, G. Solari, A. Torrielli, et al., Comparison between field measurements and numerical simulations of the wind speed along the HS/HC Rome-Naples railway line, in: Proceedings of the BBAA VI—International Colloquium on Bluff Bodies Aerodynamics and Applications, Milano, Italy, 2008, pp. 1−12.

[95] G.-J. Gao, J. Zhang, X.-H. Xiong, Location of anemometer along Lanzhou-Xinjiang railway, J. Cent. S. Univ. 21 (2014) 3698−3704.

[96] M. Xiujuan, Wind anemometer location determination on railway bridge in valley, J. Railw. Sci. Eng. 13 (2016) 1332−1337.

[97] C.J. Friedland, T.A. Joyner, C. Massarra, et al., Isotropic and anisotropic kriging approaches for interpolating surface-level wind speeds across large, geographically diverse regions, Geomat. Nat. Hazards Risk 8 (2016) 207−224.

[98] G. Ravazzani, A. Ceppi, S. Davolio, Wind speed interpolation for evapotranspiration assessment in complex topography area, Bull. Atmos. Sci. & Technol. 1 (2020) 13−22.

[99] Y. Liu, W. Peng, X. Zhang, et al., Fiber-optic anemometer based on distributed bragg reflector fiber laser technology, IEEE Photon. Technol. Lett. 25 (2013) 1246−1249.

[100] A. Garcia-Ruiz, A. Dominguez-Lopez, J. Pastor-Graells, et al., Long-range distributed optical fiber hot-wire anemometer based on chirped-pulse ΦOTDR, Optic Express 26 (2018) 463−476.

[101] Y. Sakuma, M. Suzuki, A. Ido, et al., Measurement of air velocity and pressure distributions around high-speed trains on board and on the ground, J. Mech. Syst. Transp. & Logist. 3 (2010) 110–118.
[102] L.E. Mitchell, E.T. Crosman, A.A. Jacques, et al., Monitoring of greenhouse gases and pollutants across an urban area using a light-rail public transit platform, Atmos. Environ. 187 (2018) 9–23.
[103] Z. Di, J. Ao, Q. Duan, et al., Improving WRF model turbine-height wind-speed forecasting using a surrogate- based automatic optimization method, Atmos. Res. 226 (2019) 1–16.
[104] H. Wang, S. Han, Y. Liu, et al., Sequence transfer correction algorithm for numerical weather prediction wind speed and its application in a wind power forecasting system, Appl. Energy 237 (2019) 1–10.
[105] J. Zhao, Y. Guo, X. Xiao, et al., Multi-step wind speed and power forecasts based on a WRF simulation and an optimized association method, Appl. Energy 197 (2017) 183–202.
[106] H. Cai, X. Jia, J. Feng, et al., Gaussian process regression for numerical wind speed prediction enhancement, Renew. Energy 146 (2020) 2112–2123.
[107] W. Xu, P. Liu, L. Cheng, et al., Multi-step wind speed prediction by combining a WRF simulation and an error correction strategy, Renew. Energy 163 (2021) 772–782.
[108] T.-H. Yang, C.-C. Tsai, Using numerical weather model outputs to forecast wind gusts during typhoons, J. Wind Eng. Ind. Aerod. 188 (2019) 247–259.
[109] C. Singh, S.K. Singh, P. Chauhan, et al., Simulation of an extreme dust episode using WRF-CHEM based on optimal ensemble approach, Atmos. Res. 249 (2021) 105296.
[110] J. Zhao, Z.-H. Guo, Z.-Y. Su, et al., An improved multi-step forecasting model based on WRF ensembles and creative fuzzy systems for wind speed, Appl. Energy 162 (2016) 808–826.
[111] Aasim, S.N. Singh, A. Mohapatra, Repeated wavelet transform based ARIMA model for very short-term wind speed forecasting, Renew. Energy 136 (2019) 758–768.
[112] H. Liu, H.-Q. Tian, Y.-F. Li, An EMD-recursive ARIMA method to predict wind speed for railway strong wind warning system, J. Wind Eng. Ind. Aerod. 141 (2015) 27–38.
[113] R.G. Kavasseri, K. Seetharaman, Day-ahead wind speed forecasting using f-ARIMA models, Renew. Energy 34 (2009) 1388–1393.
[114] J. Hu, J. Wang, Short-term wind speed prediction using empirical wavelet transform and Gaussian process regression, Energy 93 (2015) 1456–1466.
[115] J. Yu, K. Chen, J. Mori, et al., A Gaussian mixture copula model based localized Gaussian process regression approach for long-term wind speed prediction, Energy 61 (2013) 673–686.
[116] J. Hu, J. Wang, L. Xiao, A hybrid approach based on the Gaussian process with t-observation model for short-term wind speed forecasts, Renew. Energy 114 (2017) 670–685.
[117] S. Fang, H.-D. Chiang, A high-accuracy wind power forecasting model, IEEE Trans. Power Syst. 32 (2017) 1589–1590.
[118] J. Wang, J. Hu, A robust combination approach for short-term wind speed forecasting and analysis – combination of the ARIMA (autoregressive integrated moving average), ELM (extreme learning machine), SVM (support vector machine) and LSSVM (least square SVM) forecasts using a GPR (Gaussian process regression) model, Energy 93 (2015) 41–56.
[119] C. Zhang, H. Wei, X. Zhao, et al., A Gaussian process regression based hybrid approach for short-term wind speed prediction, Energy Convers. Manag. 126 (2016) 1084–1092.

[120] Z. Zhang, L. Ye, H. Qin, et al., Wind speed prediction method using shared weight long short-term memory network and Gaussian process regression, Appl. Energy 247 (2019) 270–284.
[121] Y. Wang, Q. Hu, D. Meng, et al., Deterministic and probabilistic wind power forecasting using a variational Bayesian-based adaptive robust multi-kernel regression model, Appl. Energy 208 (2017) 1097–1112.
[122] Y. Liu, H. Qin, Z. Zhang, et al., Probabilistic spatiotemporal wind speed forecasting based on a variational Bayesian deep learning model, Appl. Energy 260 (2020) 114259.
[123] M.N. Khoshrodi, M. Jannati, T. Sutikno, A review of wind speed estimation for wind turbine systems based on Kalman filter technique, Int. J. Electr. Comput. Eng. 6 (2016) 1406–1411.
[124] K. Chen, J. Yu, Short-term wind speed prediction using an unscented Kalman filter based state-space support vector regression approach, Appl. Energy 113 (2014) 690–705.
[125] Z. Song, Y. Jiang, Z. Zhang, Short-term wind speed forecasting with Markov-switching model, Appl. Energy 130 (2014) 103–112.
[126] Y. Wang, J. Wang, X. Wei, A hybrid wind speed forecasting model based on phase space reconstruction theory and Markov model: a case study of wind farms in Northwest China, Energy 91 (2015) 556–572.
[127] F.O. Hocaoğlu, Ö.N. Gerek, M. Kurban, A novel wind speed modeling approach using atmospheric pressure observations and hidden Markov models, J. Wind Eng. Ind. Aerod. 98 (2010) 472–481.
[128] E. Paluzo-Hidalgo, R. Gonzalez-Diaz, M.A. Gutierrez-Naranjo, Two-hidden-layer feed-forward networks are universal approximators: a constructive approach, Neural Netw. 131 (2020) 29–36.
[129] H. Liu, H.-Q. Tian, Y.-F. Li, et al., Comparison of four Adaboost algorithm based artificial neural networks in wind speed predictions, Energy Convers. Manag. 92 (2015) 67–81.
[130] H. Liu, Z. Duan, C. Chen, Wind speed big data forecasting using time-variant multi-resolution ensemble model with clustering auto-encoder, Appl. Energy 280 (2020) 115975.
[131] H. Liu, H.-Q. Tian, X.-F. Liang, et al., Wind speed forecasting approach using secondary decomposition algorithm and Elman neural networks, Appl. Energy 157 (2015) 183–194.
[132] C. Yu, Y. Li, H. Xiang, et al., Data mining-assisted short-term wind speed forecasting by wavelet packet decomposition and Elman neural network, J. Wind Eng. Ind. Aerod. 175 (2018) 136–143.
[133] S. Qin, J. Wang, J. Wu, et al., A hybrid model based on smooth transition periodic autoregressive and Elman artificial neural network for wind speed forecasting of the Hebei region in China, Int. J. Green Energy 13 (2016) 595–607.
[134] G. Memarzadeh, F. Keynia, A new short-term wind speed forecasting method based on fine-tuned LSTM neural network and optimal input sets, Energy Convers. Manag. 213 (2020) 112824.
[135] Z. Niu, Z. Yu, W. Tang, et al., Wind power forecasting using attention-based gated recurrent unit network, Energy 196 (2020) 117081.
[136] K.P. Seriesnet, A dilated causal convolutional neural network for forecasting, in: Proceedings of the International Conference on Pattern Recognition and Machine Intelligence, Union, NJ, USA, 2018, pp. 1–4.
[137] K. Shivam, J.-C. Tzou, S.-C. Wu, Multi-step short-term wind speed prediction using a residual dilated causal convolutional network with nonlinear attention, Energies 13 (2020) 1772.

[138] H. Liu, C. Yu, C. Yu, et al., A novel axle temperature forecasting method based on decomposition, reinforcement learning optimization and neural network, Adv. Eng. Inf. 44 (2020) 101089.
[139] H. Liu, C. Chen, Data processing strategies in wind energy forecasting models and applications: a comprehensive review, Appl. Energy 249 (2019) 392–408.
[140] Z. Wu, N.E. Huang, Ensemble empirical mode decomposition: a noise-assisted data analysis method, Adv. Adapt. Data Anal. 1 (2009) 1–41.
[141] N. Sun, J. Zhou, L. Chen, et al., An adaptive dynamic short-term wind speed forecasting model using secondary decomposition and an improved regularized extreme learning machine, Energy 165 (2018) 939–957.
[142] H. Liu, Z. Duan, F.-Z. Han, et al., Big multi-step wind speed forecasting model based on secondary decomposition, ensemble method and error correction algorithm, Energy Convers. Manag. 156 (2018) 525–541.
[143] Z. Guo, W. Zhao, H. Lu, et al., Multi-step forecasting for wind speed using a modified EMD-based artificial neural network model, Renew. Energy 37 (2012) 241–249.
[144] L. Cheng, H. Zang, T. Ding, et al., Ensemble recurrent neural network based probabilistic wind speed forecasting approach, Energies 11 (2018) 1958.
[145] T. Ma, C. Wang, J. Wang, et al., Particle-swarm optimization of ensemble neural networks with negative correlation learning for forecasting short-term wind speed of wind farms in western China, Inf. Sci. 505 (2019) 157–182.
[146] Z. Yang, J. Wang, A combination forecasting approach applied in multistep wind speed forecasting based on a data processing strategy and an optimized artificial intelligence algorithm, Appl. Energy 230 (2018) 1108–1125.
[147] J. Wang, S. Wang, W. Yang, A novel non-linear combination system for short-term wind speed forecast, Renew. Energy 143 (2019) 1172–1192.
[148] Z. He, Y. Chen, Z. Shang, et al., A novel wind speed forecasting model based on moving window and multi-objective particle swarm optimization algorithm, Appl. Math. Model. 76 (2019) 717–740.
[149] Y. Li, H. Shi, F. Han, et al., Smart wind speed forecasting approach using various boosting algorithms, big multi-step forecasting strategy, Renew. Energy 135 (2019) 540–553.
[150] J. Chen, G.-Q. Zeng, W. Zhou, et al., Wind speed forecasting using nonlinear-learning ensemble of deep learning time series prediction and extremal optimization, Energy Convers. Manag. 165 (2018) 681–695.
[151] A.S. Qureshi, A. Khan, A. Zameer, et al., Wind power prediction using deep neural network based meta regression and transfer learning, Appl. Soft Comput. 58 (2017) 742–755.
[152] C. Chen, H. Liu, Medium-term wind power forecasting based on multi-resolution multi-learner ensemble and adaptive model selection, Energy Convers. Manag. 206 (2020) 112492.
[153] J. Wang, Y. Li, Multi-step ahead wind speed prediction based on optimal feature extraction, long short term memory neural network and error correction strategy, Appl. Energy 230 (2018) 429–443.
[154] A. Dupré, P. Drobinski, B. Alonzo, et al., Sub-hourly forecasting of wind speed and wind energy, Renew. Energy 145 (2020) 2373–2379.
[155] W.Y. Cheng, Y. Liu, A.J. Bourgeois, et al., Short-term wind forecast of a data assimilation/weather forecasting system with wind turbine anemometer measurement assimilation, Renew. Energy 107 (2017) 340–351.
[156] A. Tascikaraoglu, B.M. Sanandaji, K. Poolla, et al., Exploiting sparsity of interconnections in spatio-temporal wind speed forecasting using Wavelet transform, Appl. Energy 165 (2016) 735–747.

[157] J. Dowell, P. Pinson, Very-short-term probabilistic wind power forecasts by sparse vector autoregression, IEEE Trans. Smart Grid 7 (2015) 763–770.
[158] X. Zhu, K.P. Bowman, M.G. Genton, Incorporating geostrophic wind information for improved space–time short-term wind speed forecasting, Ann. Appl. Stat. 8 (2014) 1782–1799.
[159] W. Li, Z. Wei, G. Sun, et al., Multi-interval wind speed forecast model based on improved spatial correlation and RBF neural network [J], Electr. Power Autom. Equip. 29 (2009) 89–92.
[160] Q. Xu, D. He, N. Zhang, et al., A short-term wind power forecasting approach with adjustment of numerical weather prediction input by data mining, IEEE Trans. Sustain. Energy 6 (2015) 1283–1291.
[161] M. Khodayar, J. Wang, Spatio-temporal graph deep neural network for short-term wind speed forecasting, IEEE Trans. Sustain. Energy 10 (2019) 670–681.
[162] Y. Chen, S. Zhang, W. Zhang, et al., Multifactor spatio-temporal correlation model based on a combination of convolutional neural network and long short-term memory neural network for wind speed forecasting, Energy Convers. Manag. 185 (2019) 783–799. (Accessed 16 October 2020).

CHAPTER 2

Analysis of flow field characteristics along railways

Contents

2.1 Introduction

The difference in temperature and geostrophic motion causes a difference in air pressure, and air flows under pressure to form wind [1]. Strong winds along railways seriously affected the train operation, which can even blow over the train. The wind field along the railway is very long with the length of the railway, but the wind field at every single point has only a few meaningful information of wind speed for moving trains [2]. Once the train passes a certain point, the point of the wind flow field will be useless for the train. So, it is significant to study the fluctuating wind flow field [3]. The characteristics of wind flow field along railways are affected by climate, ground roughness, ground clearance, topography, etc. Studying the characteristics of the wind flow field helps the management department to

Wind Forecasting in Railway Engineering
ISBN 978-0-12-823706-9
https://doi.org/10.1016/B978-0-12-823706-9.00002-8

understand the temporal and spatial evolution of the wind field, find the most dangerous time points and spatial locations, and guide the safe operation of the train.

The space filled with fluid is called flow field. The fluid is composed of infinitely many particles and the fluid particles are microbodies with the smallest volume that make the fluid have macroscopic properties. There are relative motions and mutual motions of fluid particles, and the research of fluid motion needs to analyze the movement changes of each particle [4]. The fluid in the flow field satisfies the continuous medium hypothesis, all the spaces in the flow field are occupied by the corresponding fluid particles without any gaps. Under the continuous medium hypothesis, the macroscopic physical quantities that characterize the fluid state, such as velocity, pressure, density, temperature, etc., are continuously distributed and can be used as continuous functions of time and space [5]. The Lagrangian method and Eulerian method are commonly used to define the flow field.

(a) The Lagrangian method of describing fluid movement can track the entire movement of each fluid particle and record the physical quantities and the changing laws during the movement [6]. After the physical parameters such as the position, velocity, acceleration, pressure, temperature, and density of all fluid particles in the flow space are determined, and then the flow is determined. The Lagrangian method describes the movement of each fluid particle, so it is necessary to distinguish all fluid particles [7]. The change of the position coordinate $\boldsymbol{x}_i(t)$ of the discrete mass point in space with time is the trajectory of the mass point movement. The infinite number of particles in the flow field cannot be represented by discrete coordinates, the initial coordinates of the particle i is $a = x_i(0)$, $b = y_i(0)$, $c = z_i(0)$. Different mass points can be represented by changing the value of (a, b, c). The position coordinates of the mass point at any time can be represented as $\boldsymbol{x}_i(t) = \boldsymbol{x}(a, b, c, t)$, which is called Lagrangian variable [8]. The continuous existence of fluid particles means the continuity of the Lagrangian variable. The spatial position of the fluid particle and other parameters are both a function of its Lagrangian variable and time. The curve formed by the position of a fluid particle at different times is called trace [9]. The Lagrangian method has a clear physical concept and provides a direct description of the detailed time-varying process of each particle. But this method involves complicated calculations when expressing the spatial distribution or mass conservation equations of motion elements.

(b) The Eulerian method observes the changes of motion elements over-time at each spatial point in the flow space, and does not study the time-varying characteristics of a single particle [10]. After the physical parameters of each fluid particle in space at each moment are determined, and then the flow is determined. The rectangular coordinates (x, y, z) are used to represent the spatial coordinates, and the Eulerian variable of the physical quantity β is $\beta(x, y, z, t)$, which represents the spatial and temporal distribution of the physical quantity [11]. The Eulerian method observes passing-by fluid particles in fixed points at different times. The collection of velocity vectors at all spatial points constitutes an instantaneous velocity vector field, and the vector curve describing the flow direction of each point in the velocity field at a certain time is called streamline. The tangent direction of any point on the streamline is consistent with the velocity direction of the fluid at that point [12].

2.2 Analysis of spatial characteristics of railway flow field

2.2.1 Spatial statistical analysis

2.2.1.1 Spatial statistics

Spatial analysis can extract the hidden information in spatial data by characteristic description of geographic objects. It can also study the change of spatial position and attribute data of geographic objects. Spatial statistical analysis is mainly based on Geographic Information System (GIS), and reveals the relationship and laws between research objects through comprehensive analysis of multidimensional information [13].

The core of spatial statistical analysis is spatial dependence, spatial association, or spatial autocorrelation between data of different geographic positions, and to establish statistical relationships [14]. The first law of geography states everything is related to each other spatially [15]. The ubiquitous spatial association or spatial autocorrelation makes the sample independence required by classical statistics is not satisfied, and the spatial changes of geographic phenomena are not random. The second law of geography states that the phenomena in one spatial location are always different from the phenomena in other locations, which is called spatial heterogeneity [16]. Spatial analysis methods include spatial weight matrix, global spatial autocorrelation, and local spatial autocorrelation.

2.2.1.1.1 Spatial weight matrix

Spatial statistical analysis focuses on the spatial multidimensional characteristics and spatiotemporal correlation of data, so the relative position between the data must be included in the analysis process. The neighbor relationship between observation objects is defined by spatial weight matrix. The spatial weight matrix is to measure and test the spatial autocorrelation, and to establish spatial econometric modeling. The number of observation objects is n, the spatial weight matrix W is presented as follows [17]:

$$W = \begin{bmatrix} 0 & w_{12} & \cdots & w_{1n} \\ w_{21} & 0 & \cdots & w_{2n} \\ \vdots & \vdots & \vdots & \vdots \\ w_{n1} & w_{n2} & \cdots & 0 \end{bmatrix} \tag{2.1}$$

where the value on the diagonal represents the connection between object i and itself, and w_{ij} represents the spatial connection information between object i and object j.

Different methods of determining w_{ij} define different spatial weight matrices, the commonly used methods are presented as follows:

(a) Contiguity matrix

According to neighborhood relation, $w_{ij} = 1$ when the objects i and j have a common boundary, otherwise $w_{ij} = 0$. There are Rook contiguity and Queen contiguity to determine common boundary. The Rook contiguity only defines neighborhood relation by common boundary, and the Queen contiguity defines it by common boundary and vertex.

(b) K-nearest neighbor weight matrix

The distance between object i and j is calculated and sorted by size. Set $N_k(i)$ contains the k nearest values to object i. The weight $w_{ij} = 1$ when the distance between object i and j belongs to set $N_k(i)$, otherwise $w_{ij} = 0$.

(c) Distance matrix

The elements of weight matrix are determined by centroid distance or economic distance between object i and j, the form of distance matrix W is as follows [17]:

$$W = \begin{bmatrix} 0 & \frac{1}{d_{12}} & \cdots & \frac{1}{d_{1n}} \\ \frac{1}{d_{21}} & 0 & \cdots & \frac{1}{d_{2n}} \\ \vdots & \vdots & \vdots & \vdots \\ \frac{1}{d_{n1}} & \frac{1}{d_{n2}} & \cdots & 0 \end{bmatrix} \tag{2.2}$$

where $\frac{1}{d_{ij}}$ is generalized distance between object i and j.

2.2.1.1.2 Global spatial autocorrelation

Global autocorrelation analysis is overall description of spatial attribute characteristics through the estimation of global autocorrelation statistics. In spatial statistics, Moran's I index is generally used to evaluate spatial autocorrelation, it represents the similarity of same attribute value in spatial neighborhood area. The global Moran's I index is calculated as follows [18]:

$$I = \frac{n \sum_{i=1}^{n} \sum_{j=1}^{n} w_{ij}(x_i - \overline{x})(x_j - \overline{x})}{\sum_{i=1}^{n} \sum_{j=1}^{n} w_{ij} \sum_{i=1}^{n} (x_i - \overline{x})^2} \tag{2.3}$$

where n is number of objects, x_i and x_j are variables of object i and j, respectively, and $\overline{x} = (1/n) \sum_{i=1}^{n} x_i$.

The Z-statistic is constructed for two-sided hypothesis testing [18]:

$$Z = \frac{I - E(I)}{\sqrt{Var(I)}} \tag{2.4}$$

where $E(I)$ and $Var(I)$ are expectation and variance of the Moran's I index.

The null hypothesis H_0 and alternative hypothesis H_1 are shown as follows:

H_0: There is no spatial autocorrelation between objects in overall space.

H_1: There is spatial autocorrelation between objects in overall space.

Spatial autocorrelation is determined by P-value corresponding to Z-statistic and significance level. If the Z-statistic passes the significance test, the null hypothesis is rejected, and there is spatial autocorrelation in spatial distribution. Otherwise, there is no spatial autocorrelation.

2.2.1.1.3 Local spatial autocorrelation

The global spatial autocorrelation cannot fully explain the spatial relationship between each object and its neighborhood. Therefore, local spatial autocorrelation is introduced to explore spatial correlation between each object and its neighborhood. Then, whether different observation values are generated due to different spatial observation positions can be observed. The main research methods of local spatial autocorrelation include Moran scatter plot, local Moran's I_i index, and Getis-Ord G_i^* index.

The local Moran's I_i index is calculated as follows [18]:

$$I_i = \frac{n(x_i - \overline{x}) \sum_{j=1}^{n} w_{ij}\left(x_j - \overline{x}\right)}{\sum_{i=1}^{n} (x_i - \overline{x})^2} \tag{2.5}$$

At a significance level, $I_i > 0$ means there is a spatial cluster of attribute values surrounding the object i, which means a positive local spatial autocorrelation. Otherwise, there is a negative local spatial autocorrelation. The significance test of local Moran's I_i index is the same as that of global Moran's I index.

2.2.1.2 Spatial statistical analysis of wind field along railways

Strong wind will seriously affect train scheduling and safety along railways. To analyze the spatial characteristics of windy weather, the areas along Southern Xinjiang Railway and Lanzhou-Xinjiang Railway in Xinjiang, China are taken as an example. The selected geographic area of collection data area contains the Hundred Miles Wind Area and the Thirty Miles Wind Area, which are severely affected by wind sand disaster [19]. The data comes from physical sciences laboratory (https://www.psl.noaa.gov/data/index.html) of National Oceanic and Atmospheric Administration (NOAA). The spatial coverage is 41N−44N, 84E-94E and temporal coverage is 2011-01-01 00:00:00 UTC to 2015-12-31 21:00:00 UTC. The time resolution is 3 h and spatial resolution is 1.0-degree latitude × 1.0-degree longitude global grid. Fig. 2.1 is the topographic map of wind velocity data collection area.

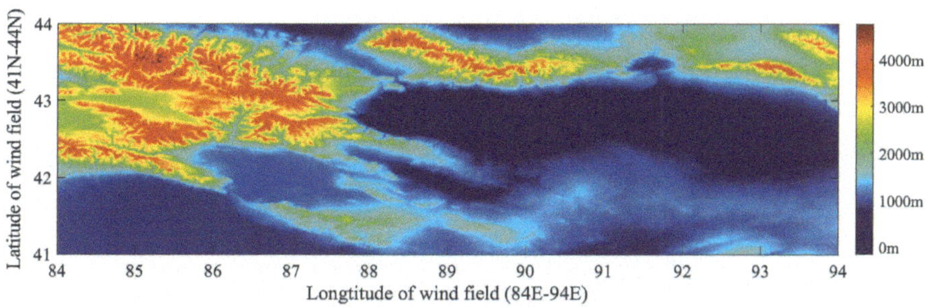

Figure 2.1 Topographic map of wind velocity data collection area.

Fig. 2.2 is the flow field distribution in the selected data collection area, the selected sampling times are 2011-03-31 06:00:00 UTC, 2011-06-30 06:00:00 UTC, 2011-09-30 06:00:00 UTC, and 2011-12-31 06:00:00 UTC. When the cold air flow moves from the north, the air pressure difference between the north and south of Tianshan Mountains increases. When the pressure difference increases, the cold air flow over the Tianshan Mountains causes northerly gale along the Tianshan Mountains.

Take the four sampling points (2011-03-31 06:00:00 UTC, 2011-06-30 06:00:00 UTC, 2011-09-30 06:00:00 UTC, and 2011-12-31 06:00:00 UTC) as examples to analyze spatial statistical characteristics. Fig. 2.3 is the standardized spatial weight matrix of data collection area, the spatial weight matrix is a distance matrix, the distance is calculated on the basis of the latitude and longitude of the coordinates.

In Fig. 2.3, the horizontal and vertical coordinates indicate 44 points in the studied area. The sort of the points is shown in Table 2.1.

Table 2.2 is the global Moran's I index of wind speed and direction, where all global Moran's I indices of data attributes are greater than 0. There are positive spatial correlations under *P*-value of significance level is .05, the larger index value indicates more significant spatial correlation.

Fig. 2.4 is the local Moran's I_i index of wind speed under significance level of 0.05. There is spatial agglomeration between 41N89E, 41N90E, and 41N94E at 2011-03-31 06:00:00 UTC, 43N93E and 43N94E at 2011-06-30 06:00:00 UTC, 41N90E and 41N91E at 2011-09-30 06:00:00 UTC, 44N93E and 44N94E at 2011-12-31 06:00:00 UTC.

Fig. 2.5 is the local Moran's I_i index of wind direction under significance level of 0.05. There is spatial agglomeration between 44N90E, 44N91E, 44N92E, 44N93E, and 44N94E at 2011-03-31 06:00:00 UTC, 43N84E, 44N85E, and 43N85E at 2011-09-30 06:00:00 UTC, 42N84E, 43N86E, 41N87E, 42N88E, 41N88E, and 41N89E at 2011-12-31 06:00:00 UTC.

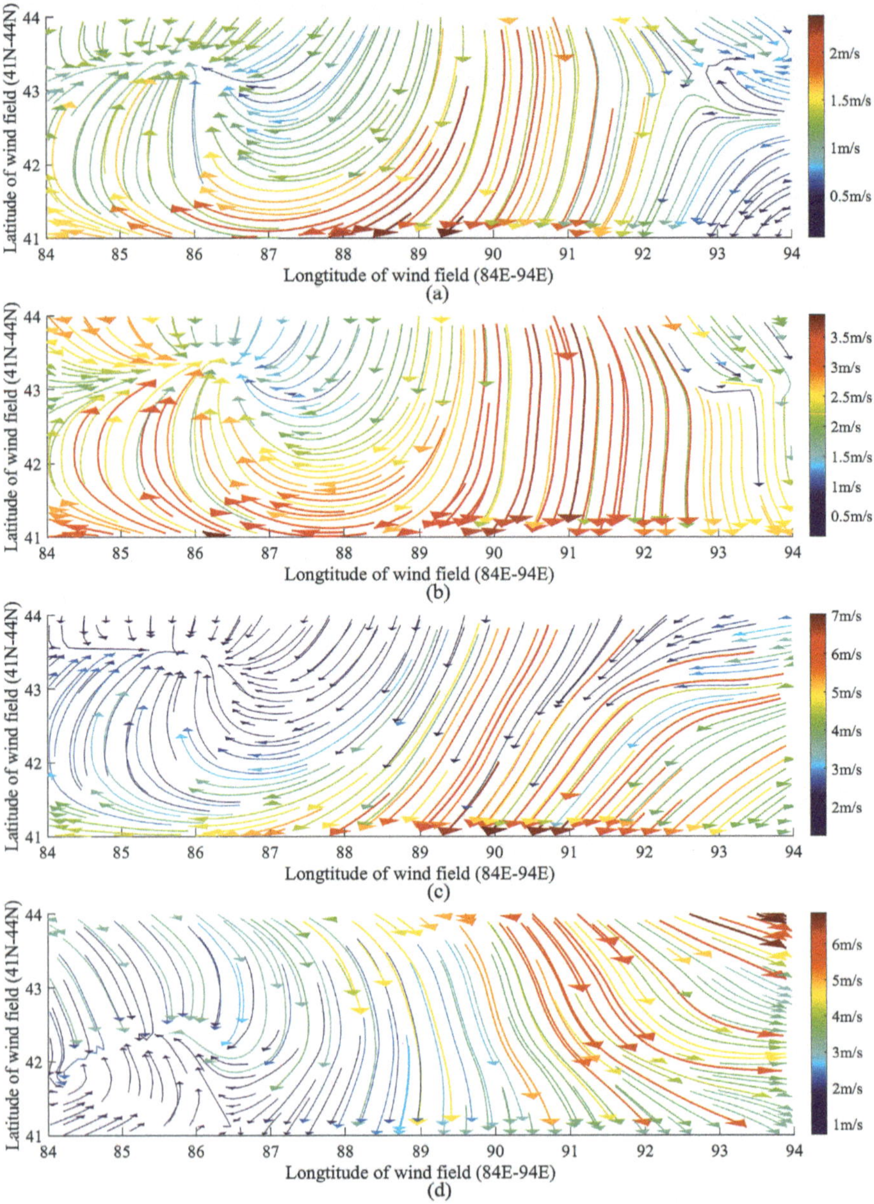

Figure 2.2 Wind field distribution at 41N–44N, 84E-94E. (A) 2011-03-31 06:00:00 UTC, (B) 2011-06-30 06:00:00 UTC, (C) 2011-09-30 06:00:00 UTC, (D) 2011-12-31 06:00:00 UTC.

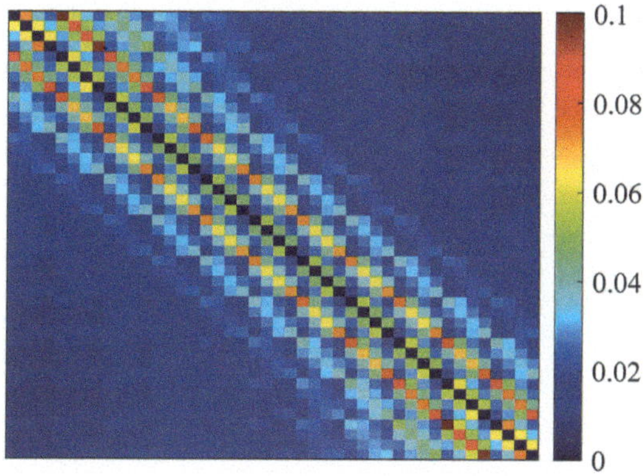

Figure 2.3 Spatial weight matrix of data collection area.

Table 2.1 The sort of 44 points in the studied area.

Sorted index	Sample points
1–4	44N84E ~ 41N84E
5–8	44N85E ~ 41N85E
9–12	44N86E ~ 41N86E
13–16	44N87E ~ 41N87E
17–20	44N88E ~ 41N88E
21–24	44N89E ~ 41N89E
25–28	44N90E ~ 41N90E
29–32	44N91E ~ 41N91E
33–36	44N92E ~ 41N92E
37–40	44N93E ~ 41N93E
41–44	44N94E ~ 41N94E

Table 2.2 Global Moran's I index of wind speed and direction.

	Global Moran's I index	
Time	**Wind speed**	**Wind direction**
2011-03-31 06:00:00 UTC	0.1161	0.0798
2011-06-30 06:00:00 UTC	0.0820	0.0775
2011-09-30 06:00:00 UTC	0.1772	0.1286
2011-12-31 06:00:00 UTC	0.2437	0.0928

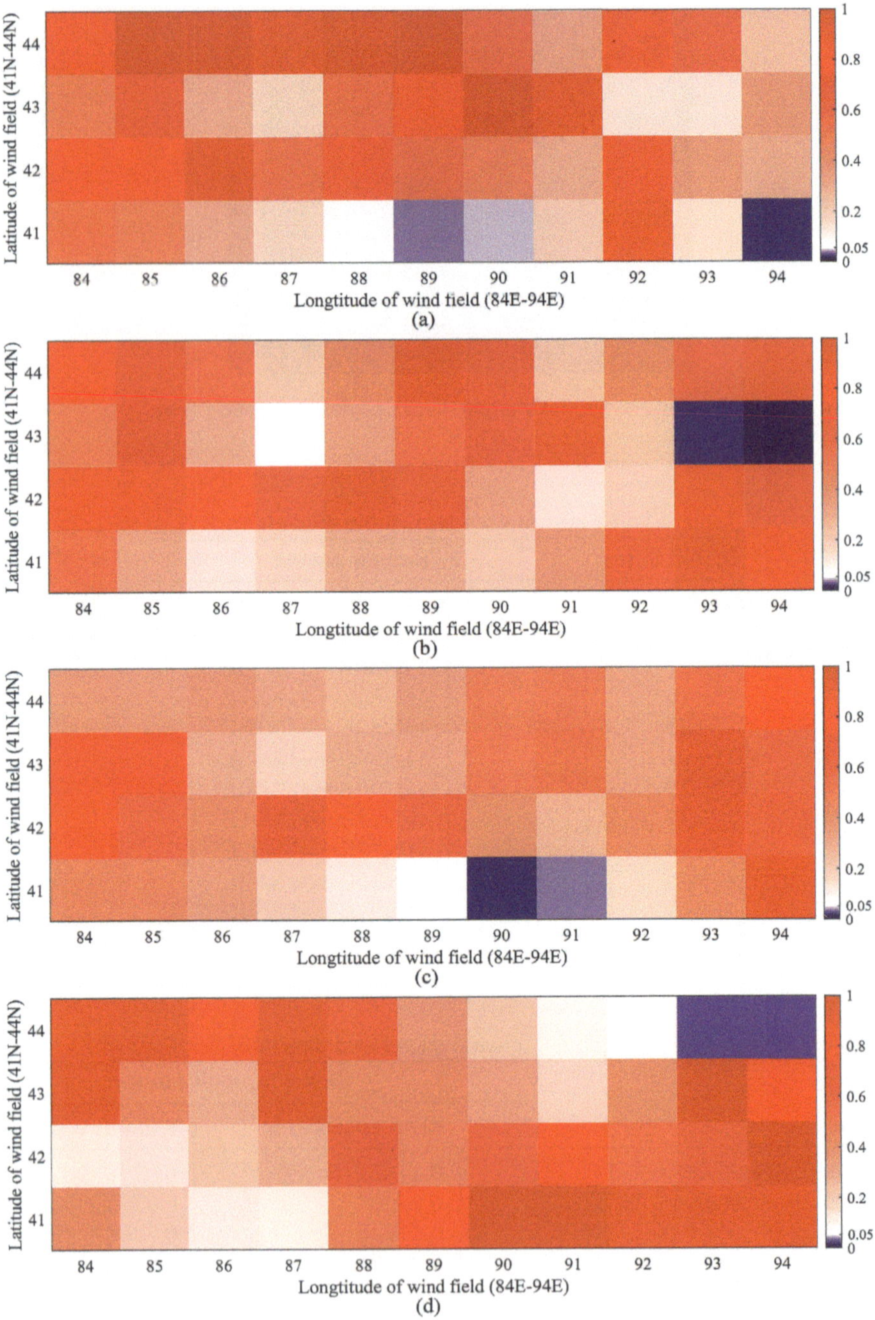

Figure 2.4 *P*-values of local Moran's I_i Z-test for wind speed. (A) 2011-03-31 06:00:00 UTC, (B) 2011-06-30 06:00:00 UTC, (C) 2011-09-30 06:00:00 UTC, (D) 2011-12-31 06:00:00 UTC.

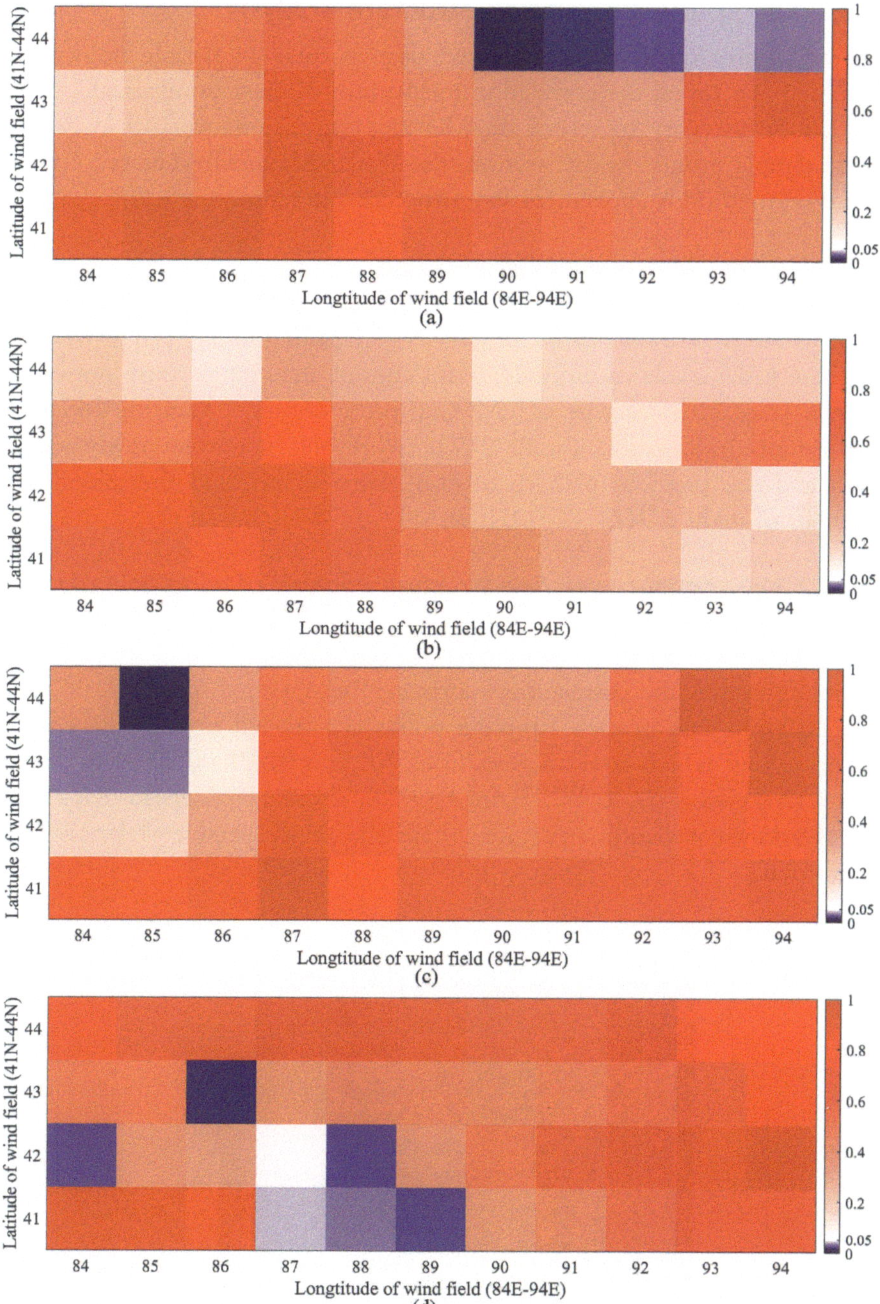

Figure 2.5 *P*-values of local Moran's I_i Z-test for wind direction. (A) 2011-03-31 06:00:00 UTC, (B) 2011-06-30 06:00:00 UTC, (C) 2011-09-30 06:00:00 UTC, (D) 2011-12-31 06:00:00 UTC.

2.2.2 Key spatial correlation structure analysis

In the studied area, the wind speed data in total 44 sample points are included. To reveal the correlation mechanism of these points in the flow field, it is necessary to screen out the key correlation structure between these sample points. In this section, the Planar Maximally Filtered Graph (PMFG) method is applied for key structure extraction.

2.2.2.1 Planar Maximally Filtered Graph

The PMFG is an extension of the Minimum Spanning Tree (MST), the MST is used to connect N nodes with $N - 1$ paths and the sum of weights of these paths is minimum, these paths are not allowed to form loops and cliques [20]. The structure of PMFG is more complex than the MST to retain more information, the PMFG network only needs the newly added path and the previous path to form a planar graph, and the maximum number of paths is $3(N-2)$ [21].

2.2.2.2 Key spatial correlation structure analysis of wind field along railways

By calculating the Spearman correlation coefficients of wind speed in 44 sampling points, the correlation matrix can be obtained in Fig. 2.6. Total 14,608 wind data in each sampling points are applied to calculate correlation values, which cover 5 years from 2011 to 01-01 00:00:00 UTC to 2015-12-31 21:00:00 UTC. From Fig. 2.6, it can be observed that there exists a key correlation structure where the wind speed correlations are significant.

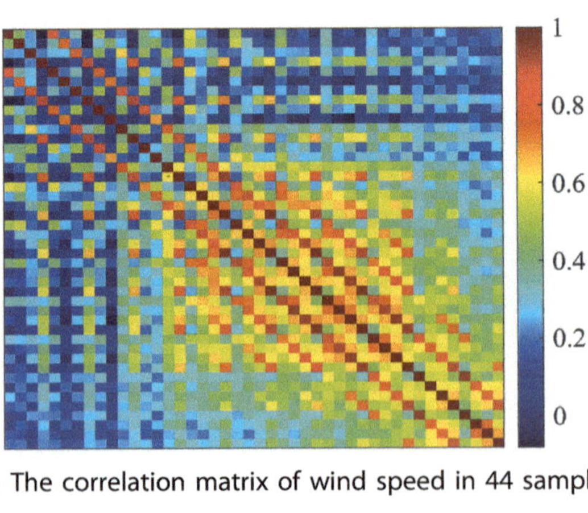

Figure 2.6 The correlation matrix of wind speed in 44 sampling points.

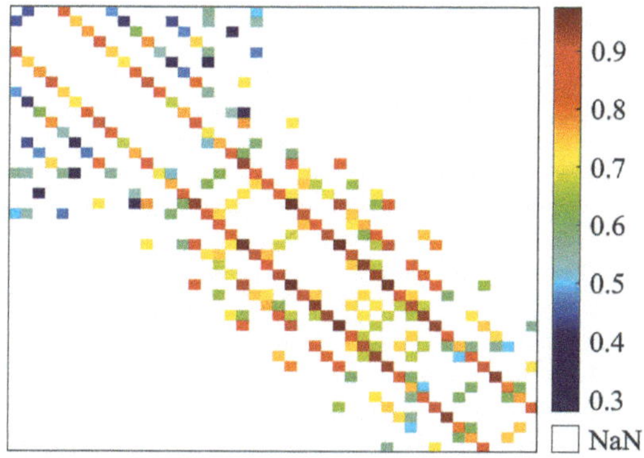

Figure 2.7 The correlation matrix after PMFG calculation.

The PMFG method is applied to extract the key correlations. The correlation matrix after the calculation is shown in Fig. 2.7, where auto-correlations between the same points are discarded. From Fig. 2.7, it can be seen that the strong correlations in the original matrix are kept, and several weak correlations are kept to complete the information.

The histogram of the correlation values before and after the PMFG calculation is shown in Fig. 2.8. From Fig. 2.8, it can be seen that almost all strong correlation values are kept. This phenomenon indicates the PMFG is effective to discover the key correlation structure.

By mapping extracted correlation into the real position, the correlation structure can be visualized in Fig. 2.9A. To explain the phenomenon in the correlation structure, the averaging flow field from 2011 to 2015 is shown

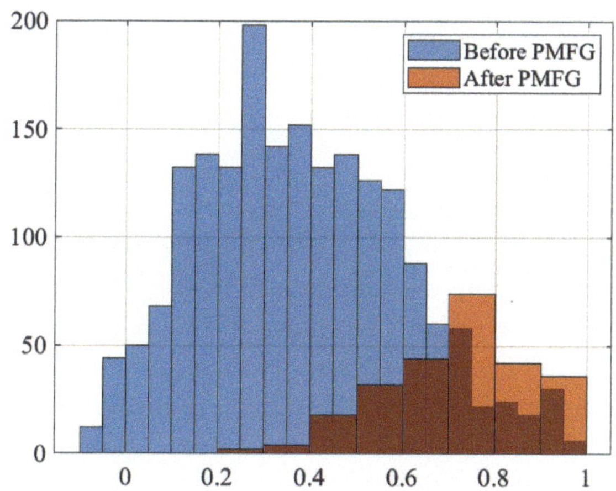

Figure 2.8 The histograms of the correlations before and after PMFG calculation.

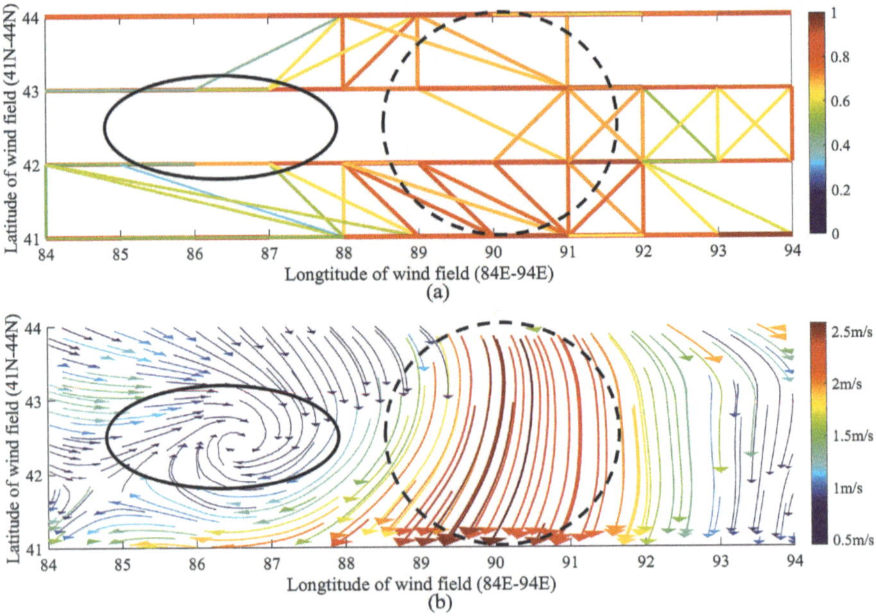

Figure 2.9 The comparison between the key correlation structure and flow field: (A) key correlation structure, (B) flow field.

in Fig. 2.9B. In the dotted line curve, there are serval strong streamlines in Fig. 2.9B. So, there are strong correlations in Fig. 2.9B. In the solid line curve of Fig. 2.9B, there is a significantly rotational flow. So, the correlations in the solid line curve are weak as shown in Fig. 2.9A.

2.3 Analysis of seasonal characteristics of railway flow field

2.3.1 Frequency analysis

2.3.1.1 Fast Fourier transform

Fourier transform can represent any continuous-time sequence signal as an infinite superposition of sine signals of different frequencies, which decompose complex time-domain superimposed signals into frequency-domain signals. Fast Fourier Transform (FFT) decomposes the original sequence of length N into a series of short sequences [22]. The computational complexity of the Discrete Fourier Transform (DFT) for a sequence of length N is $O(N^2)$, the FFT reduces repeated calculations as $O(N \log 2\, N)$ [23].

2.3.1.2 Frequency analysis of wind field along railways

To visualize the frequency characteristics of the wind speed in the wind field, the correlation of the frequency components and sampling points is shown in Fig. 2.10. In Fig. 2.10, the static term is not concluded to shed light on the seasonal terms. From Fig. 2.10, it can be observed that there exist several significant components in the frequency spectrum. Almost all sampling points have the same main frequencies, although the amplitudes are different.

Taking the sampling point #1 as an example, the frequency spectrum is shown in Fig. 2.11. It can be observed that the wind speed has six main frequencies. The corresponding periods are 365-day, 183-day, 1-day, 12-h, 8-h, and 6-h, respectively. This phenomenon indicates the wind field has yearly and daily seasonality.

Taking the yearly seasonality as an example, the yearly averaged wind speed data are shown in Fig. 2.12. As can be seen from Fig. 2.12, April to August is the peak season for strong winds. Some sampling points have much higher amplitude than others.

To further study the relationship between sampling point location and yearly wind speed amplitude, the relationship between yearly wind speed amplitude and location is shown in Fig. 2.13. From Fig. 2.13, it can be seen that the area with 42N and 90E-92E has the highest yearly wind speed components.

Taking the daily seasonality as an example, the daily averaged wind speed data are shown in Fig. 2.14. As can be seen from Fig. 2.14, the strong wind is concentrated from 6:00 to 18:00 approximately.

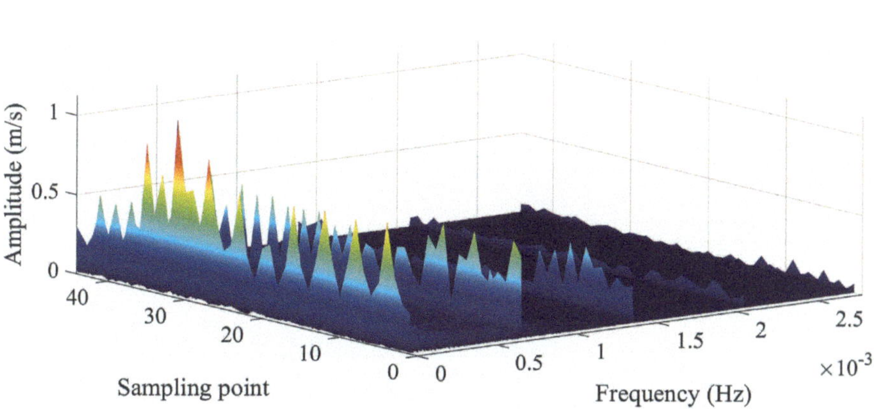

Figure 2.10 The correlation of the frequency components and sampling points.

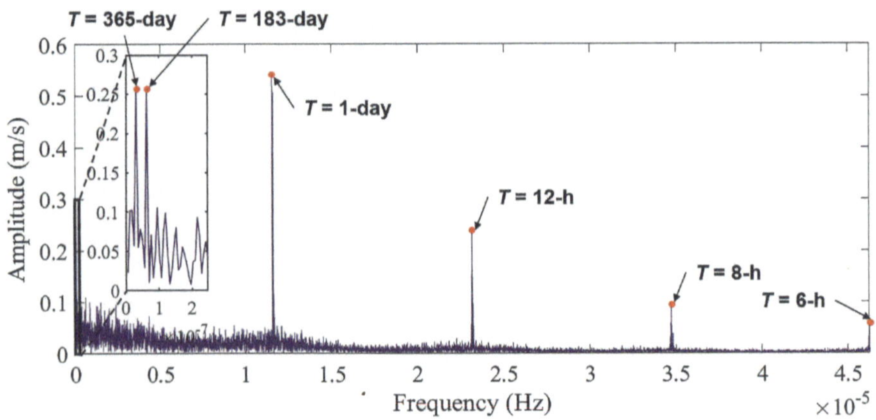

Figure 2.11 Frequency spectrum of sampling point #1.

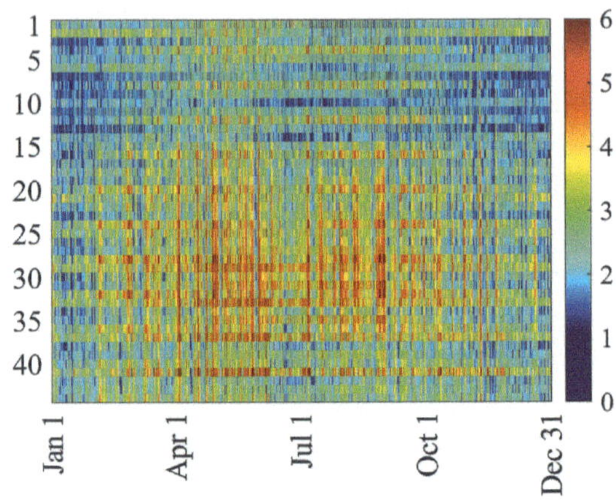

Figure 2.12 The yearly averaged wind speed data.

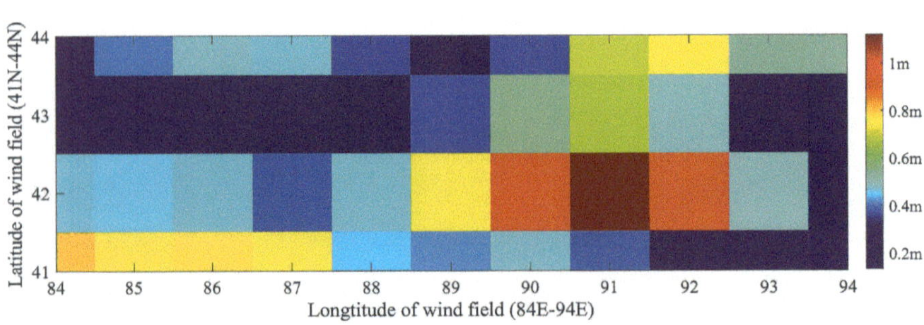

Figure 2.13 The amplitudes of yearly wind speed components over the studied area.

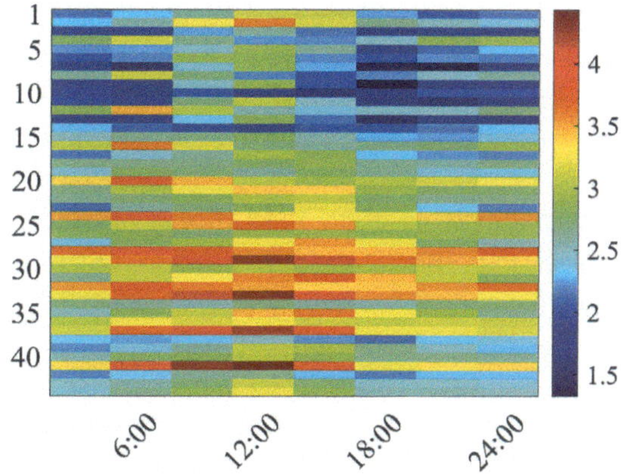

Figure 2.14 The daily averaged wind speed data.

To further study the relationship between sampling point location and daily wind speed amplitude, the relationship between daily wind speed amplitude and location is shown in Fig. 2.15. From Fig. 2.15, it can be seen that the area with 44N and 93E-94E has the highest daily wind speed amplitude components.

2.3.2 Clustering analysis

2.3.2.1 Bayesian Fuzzy Clustering

Bayesian Fuzzy Clustering (BFC) is a clustering model that combines fuzzy clustering and probabilistic clustering. Based on prior knowledge and Bayesian theory, the Maximum A Posteriori (MAP) is used to process fuzzy clustering, and the ability to adaptively learn clustering number is further obtained, and the method can obtain the global optimal solution [24]. The BFC model includes Fuzzy Data Likelihood (FDL), Fuzzy Cluster Prior (FCP), and Gaussian prior distribution on cluster prototypes [25]. The

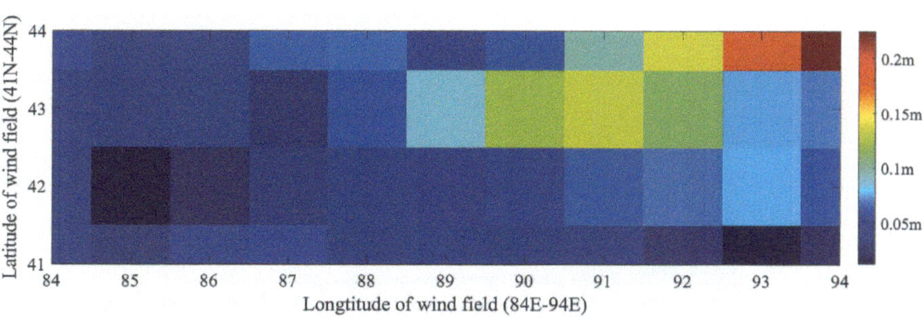

Figure 2.15 The amplitudes of daily wind speed components over the studied area.

objective function of the joint likelihood of data and parameters can be calculated based on the BFC model, then use the MAP inference to iteratively solve the global optimum of membership degree and clustering center parameters through the Metropolis–Hastings algorithm.

2.3.2.2 Clustering analysis of wind field along railways

The raw wind field data have 44 dimensions. To avoid the curse of dimension for clustering, the dimension of the wind field data should be reduced. The Principal Components Analysis (PCA) is applied to extract key information. The number of principal components of the PCA should be carefully selected. The Pareto diagram of the latent variables is shown in Fig. 2.16. Setting the 90% as the threshold, the number of features is selected as 8.

With the extracted features, the BFC can be applied for clustering. The number of clusters is an important parameter for clustering. If the number of clusters is not suitable, the natural difference between the data cannot be well described, so the information quality decreases. In this section, the likelihood and Davies Bouldin score are applied to select the number of clusters. The likelihood represents the matching degree between the data and the fitted model. In the BFC algorithm, the likelihood is an important indicator to evaluate whether the data are well clustered. The larger the likelihood, the better the clustering performance. The Davies Bouldin score measures the ratio between distances within clusters and between clusters. The small Davies Bouldin score indicates good clustering performance. Setting the searching range from 3 to 20, the likelihoods and Davies Bouldin scores are shown in Fig. 2.17. From Fig. 2.17, it can be observed that the likelihood increases with a large number of clusters. In extreme cases, the number of clusters is equal to the number of data. The likelihood becomes the best. However, the fully clustered results are not useful for seasonality analysis. To compromise the likelihood and information quality,

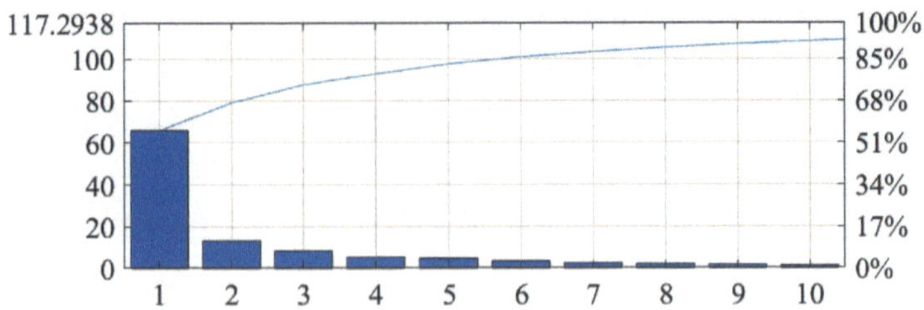

Figure 2.16 The Pareto diagram of the principal components.

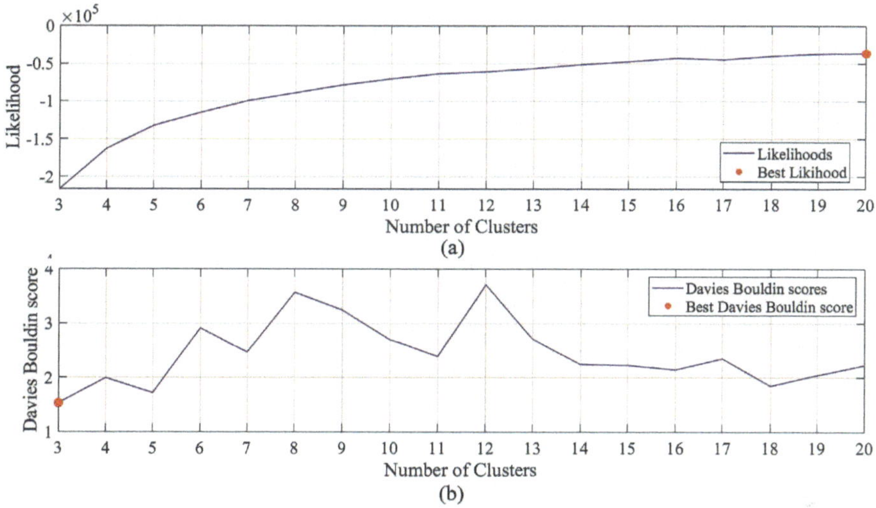

Figure 2.17 The likelihoods and Davies Bouldin scores with different numbers of clusters.

the number of clusters should be selected as the point where the increase of likelihood gets slow. The Davies Bouldin score is the best when the number of clusters is equal to 3 and 5 approximately. When the number of clusters is larger than 5, the likelihood curve increases slowly. So, the number of clusters is set as 5.

Given the best number of the clusters as 5, the likelihood curve of the BFC algorithm when clustering is shown in Fig. 2.18. From Fig. 2.18, it can be observed that the loss decreases sharply at the beginning, and then keep stable. This phenomenon indicates the BFC algorithm can well converge.

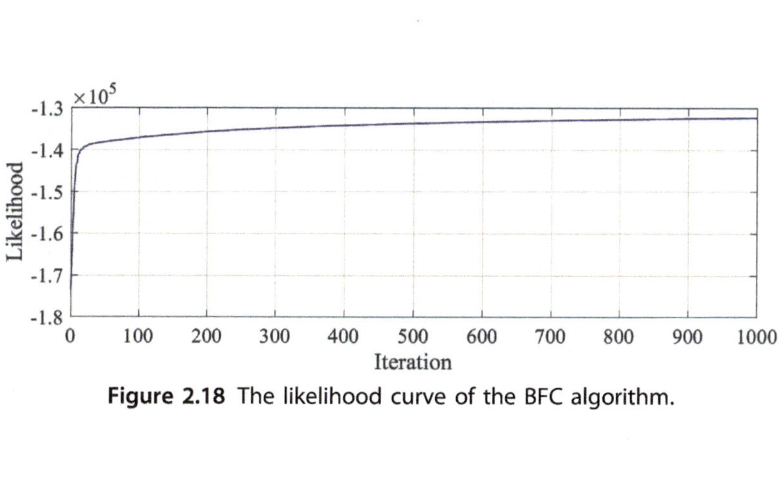

Figure 2.18 The likelihood curve of the BFC algorithm.

After the clustering computation, the wind field of the studied area can be clustered into clusters. To compare the difference of the clusters, the averaged flow fields of the clusters are shown in Fig. 2.19. It can be observed that the flow fields of different clusters are significantly different, which indicates the effectiveness of the clustering algorithm.

These clustering results can reflect the typical wind speed patterns. To analyze the seasonality of these patterns, the distributions of the clusters over 1 year are shown in Fig. 2.20. From Fig. 2.20, it can be observed that the wind field cluster #1 is evenly distributed over 1 year; the wind field cluster #2 and #5 mainly occur in Apr to Oct; the wind field cluster #3 mainly occurs in Jan and Dec; the wind field cluster #4 mainly occurs in Mar and Oct.

2.4 Summary and outlook

This chapter describes and analyzes the flow field characteristics along railways. Strong winds have a direct impact on train safety and operation, to study the influence of wind flow field on the train, the description methods (Lagrangian method and Eulerian method) of the flow field are introduced.

To calculate spatial correlation, this chapter introduces the spatial weight matrix, global spatial autocorrelation, and local spatial autocorrelation. And these three statistical analysis methods are used to analyze the spatial characteristics of the wind field of the Hundred Miles Wind Area and Thirty Miles Wind Area in China, which indicates the spatial correlation and spatial agglomeration of wind field attributes along railways. Then, the key spatial correlation structure analysis is calculated by PMFG to indicate the most relevant area. In temporal analysis based on statistics data, this chapter uses 5 years' wind velocity datum along railways for frequency and temporal clustering analysis. The frequency characteristics of wind speed show that it has six main frequencies. After clustering calculation, the clustering area are indicated and the flow fields of different clusters are significantly different in data attributes.

In the future research, the flow field simulation needs to study and build the physics-based prediction model. For the spatial statistical analysis of large-scale flow fields, it is necessary to use flow field data with higher accuracy and larger coverage to analyze the correlation and seasonal characteristics between each unit. Then, some spatial models can be established, such as spatial interaction model, time-space prediction model, and dynamic spatial panel data model.

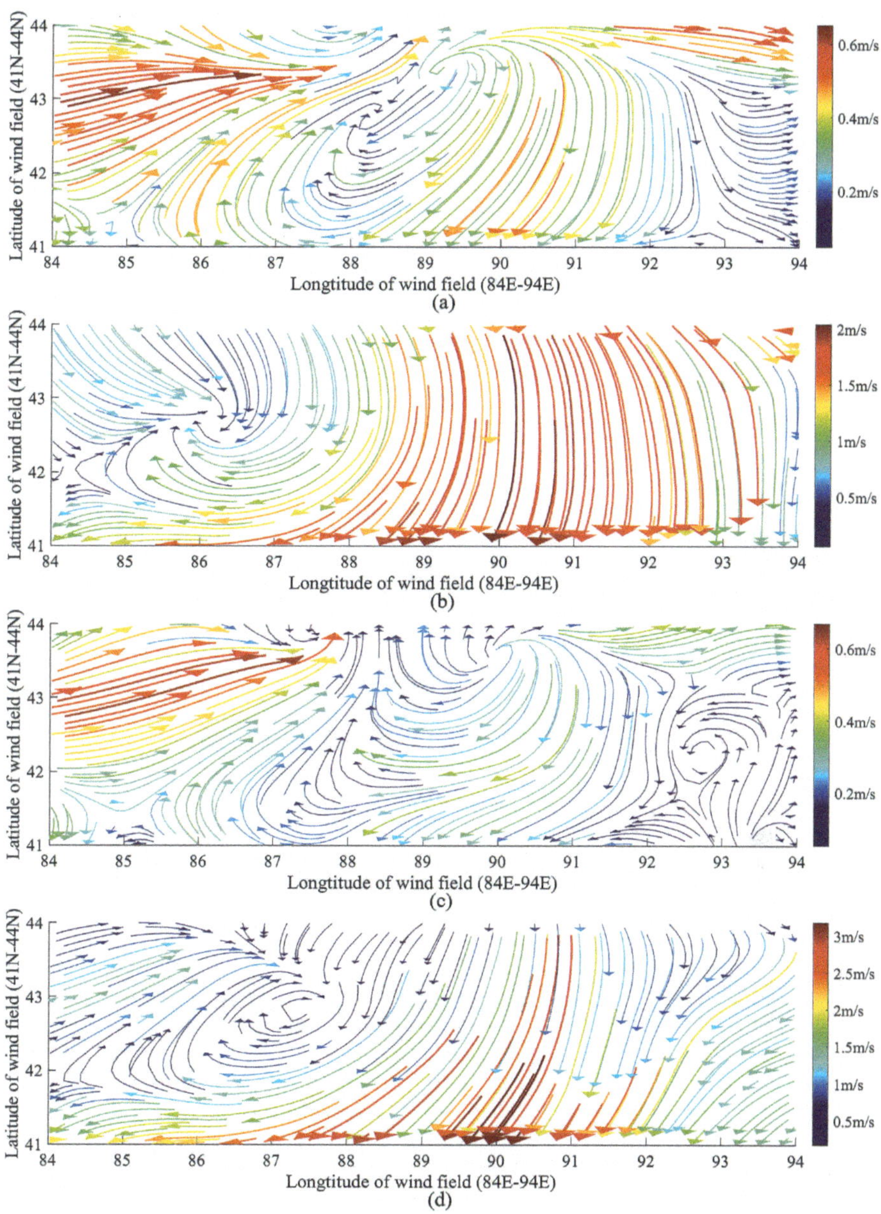

Figure 2.19 The averaged flow fields of the clusters: (A) cluster #1, (B) cluster #2, (C) cluster #3, (D) cluster #4, (E) cluster #5.

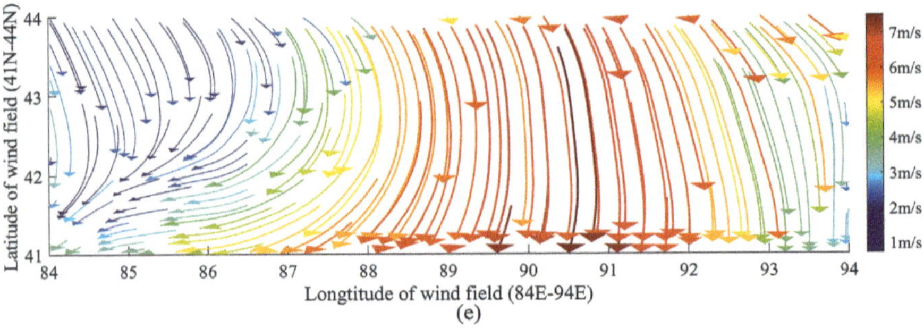

Figure 2.19 Cont'd

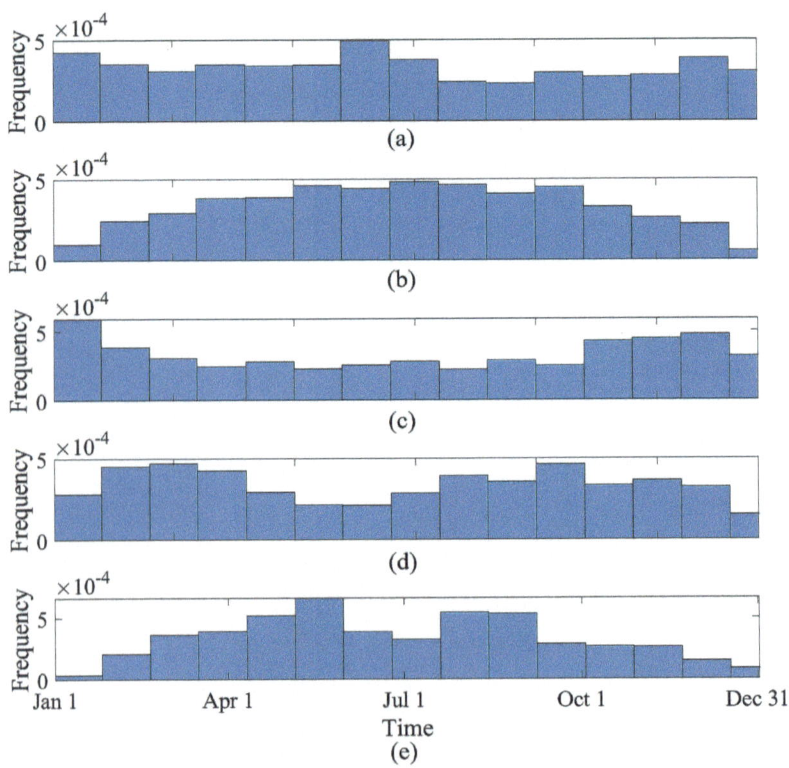

Figure 2.20 The distributions of the wind field clusters over 1 year: (A) cluster #1, (B) cluster #2, (C) cluster #3, (D) cluster #4, (E) cluster #5.

References

[1] P. Beaucage, M.C. Brower, J. Tensen, Evaluation of four numerical wind flow models for wind resource mapping, Wind Energy 17 (2014) 197–208.
[2] R. Li, N. Zhou, W. Zhang, Fluctuating wind field and wind-induced vibration response of catenary based on AR model, J. Traffic Transport. Eng. 13 (2013) 56–62.
[3] X. Li, J. Xiao, D. Liu, et al., An analytical model for the fluctuating wind velocity spectra of a moving vehicle, J. Wind Eng. Ind. Aerod. 164 (2017) 34–43.
[4] X. Pan, G. Wang, Z. Lu, Flow field simulation and a flow model of servo-valve spool valve orifice, Energy Convers. Manag. 52 (2011) 3249–3256.
[5] A. Furman, C. Breitsamter, Turbulent and unsteady flow characteristics of delta wing vortex systems, Aero. Sci. Technol. 24 (2013) 32–44.
[6] K. Trachenko, Lagrangian formulation and symmetrical description of liquid dynamics, Phys. Rev. 96 (2017) 062134.
[7] C. Meneveau, Lagrangian dynamics and models of the velocity gradient tensor in turbulent flows, Annu. Rev. Fluid Mech. 43 (2011) 219–245.
[8] A. Constantin, Some three-dimensional nonlinear equatorial flows, J. Phys. Oceanogr. 43 (2013) 165–175.
[9] C. Baker, The flow around high speed trains, J. Wind Eng. Ind. Aerod. 98 (2010) 277–298.
[10] S. Subramaniam, Lagrangian–Eulerian methods for multiphase flows, Prog. Energy Combust. Sci. 39 (2013) 215–245.
[11] S. Ii, K. Sugiyama, S. Takeuchi, et al., An implicit full Eulerian method for the fluid–structure interaction problem, Int. J. Numer. Methods Fluid. 65 (2011) 150–165.
[12] T.-Y. Lee, O. Mishchenko, H.-W. Shen, et al., View point evaluation and streamline filtering for flow visualization, in: 2011 IEEE Pacific Visualization Symposium, 2011, pp. 83–90.
[13] L. Anselin, I. Syabri, Y. Kho, GeoDa: an introduction to spatial data analysis, in: Handbook of Applied Spatial Analysis, Springer, 2010, pp. 73–89.
[14] P.J. Diggle, Statistical Analysis of Spatial and Spatio-Temporal Point Patterns, CRC press, 2013.
[15] A. Klippel, F. Hardisty, R. Li, Interpreting spatial patterns: an inquiry into formal and cognitive aspects of Tobler's first law of geography, Ann. Assoc. Am. Geogr. 101 (2011) 1011–1031.
[16] T. Foresman, R. Luscombe, The second law of geography for a spatially enabled economy, Int. J. Digit. Earth 10 (2017) 979–995.
[17] Y. Weng, P. Gong, Modeling spatial and temporal dependencies among global stock markets, Expert Syst. Appl. 43 (2016) 175–185.
[18] W. Musakwa, A. Van Niekerk, Monitoring urban sprawl and sustainable urban development using the Moran index: a case study of Stellenbosch, South Africa, Int. J. Appl. Geospatial Res. (IJAGR) 5 (2014) 1–20.
[19] Z. Yao, J. Xiao, F. Jiang, Characteristics of daily extreme-wind gusts along the Lanxin Railway in Xinjiang, China, Aeolian Res. 6 (2012) 31–40.
[20] M. Tumminello, T. Aste, T. Di Matteo, et al., A tool for filtering information in complex systems, Proc. Natl. Acad. Sci. U. S. A. 102 (2005) 10421–10426.
[21] G.P. Massara, T. Di Matteo, T. Aste, Network filtering for big data: triangulated maximally filtered graph, J. Complex Netw. 5 (2016) 161–178.
[22] K.R. Rao, D.N. Kim, J.J. Hwang, Fast Fourier Transform-Algorithms and Applications, Springer Science & Business Media, 2011.

[23] M. Garrido, F. Qureshi, J. Takala, et al., Hardware architectures for the fast Fourier transform, in: Handbook of Signal Processing Systems, Springer, 2019, pp. 613–647.
[24] T.C. Glenn, A. Zare, P.D. Gader, Bayesian fuzzy clustering, IEEE Trans. Fuzzy Syst. 23 (2014) 1545–1561.
[25] X. Gu, F. Chung, H. Ishibuchi, et al., Imbalanced TSK fuzzy classifier by cross-class Bayesian fuzzy clustering and imbalance learning, IEEE Trans. Syst. Man & Cybern. Syst. 47 (2016) 2005–2020.

CHAPTER 3

Description of single-point wind time series along railways

Contents

Wind Forecasting in Railway Engineering
ISBN 978-0-12-823706-9
https://doi.org/10.1016/B978-0-12-823706-9.00003-X

3.1 Introduction

Train accidents caused by strong crosswinds can cause great losses to traffic and personnel safety [1]. To ensure the safety of the train in the wind area, it is very necessary to set up wind speed anemometers along railways to monitor the change of wind speed. It is one of the key factors to ensure the safety of train operation to reflect the wind environment along the railway by wind anemometer layout optimization.

Based on wind anemometers along railways, a detailed description of single-point wind time series along railways is helpful to provide reasonable speed control information for the railway traffic command and control system. It can reduce the impact of strong winds on the high-speed railway trains, and improve train operation efficiency.

The railway wind time series presents an obvious seasonal effect, and the variation of the meteorological environment along the railways will lead to heteroscedasticity of wind speed and direction variance. Many scholars have paid attention to the analysis of the seasonality and heteroscedasticity of the wind time series. Yao et al. described the time series of strong winds along the Lanzhou–Xinjiang railway line in Xinjiang, China. By analyzing the characteristics of the series, they found that the series is seasonally distributed, with most of the wind occurring between April and August [2]. Jiang et al. carried out a simulation study of wind behavior along the Qinghai–Tibet railway, and found that most of the wind erosion occurred in the unseasonal months from December to April [3]. Ziel et al. used a model framework including conditional heteroscedasticity to predict wind power, which reflects the periodic fluctuation and heteroscedasticity distribution [4].

This chapter will firstly introduce the wind anemometer layout along railways. Then, this chapter will mainly describe the wind speed and wind direction time series along the railways and analyze its seasonality and heteroscedasticity. The description algorithms applied in this chapter are introduced and their performance is deeply analyzed.

3.2 Wind anemometer layout optimization methods along railways

3.2.1 Development progress

Strong crosswind has a great impact on the operational safety of the train. To improve the safety of train operation, scholars in various countries have done a lot of research on the aerodynamic performance of trains in windy conditions.

In the United Kingdom, Copley et al. proposed a numerical method and programmed the calculation program, which could predict the aerodynamic force of the train under the average wind [5]. Chui et al. led to the development of a three-dimensional source-disaster face-element method that could predict the aerodynamic loads acting on the train's surface in a crosswind environment [6]. Baker et al. conducted in-depth research on wind tunnel simulation and real vehicle test, and obtained excellent research results, which greatly promoted the development of train aerodynamics [7]. Fauchier et al. studied the train on the embankment subjected to crosswind, the significance of the embankment retaining wall and the acoustic isolation [8]. The aerodynamic characteristics of the train were calculated by changing the height and location of the retaining wall and the angle of natural wind.

In Sweden, Iedrichs et al. studied the flow field around a traction vehicle running on a high embankment through two methods of experimental research and numerical simulation [9]. Then the results of the experiment and numerical simulation are compared. The experimental results show that the data obtained by these two methods are in good agreement. This indicates that the trains running on the windward side have better aerodynamic characteristics than those running on the leeward side.

In Japan, Suzuki used numerical simulation to study the outflow field distribution of high-speed trains running on high embankments and bridges [10]. The results show that the transverse aerodynamic force of the train grows with the increase of the height of the main beam, and the distribution of the ground boundary layer has a significant influence on the aerodynamic characteristics of the train on the high embankment. Suzuki also compared the aerodynamic force of trains running on the high embankment and low embankment, and the results show that the former has greater aerodynamic force.

Besides, wind warning systems have been developed in Germany, Japan, France, China, and other countries. The system can synchronously adjust the train speed according to the change of wind speed and minimize the adverse effect of lateral wind on the train. At present, many pieces of research have been carried out on the aerodynamic characteristics of trains under a strong crosswind.

3.2.2 Numerical simulation methods

The embankment, mat, and bridge structures and their windbreaks on high-speed railways have great impact on flow field along railways [11]. The irregular flow field along railways makes it difficult to select suitable measurement points. At present, the Computational Fluid Dynamics (CFD) methods are widely used for railway aerodynamics research. Compared with the wind tunnel test, the CFD methods have unique advantages [12]. First, the parameter setting range is wide, which can simulate a variety of external environments. The calculation area can be set large enough, and the model size can also be set to the same as the research object. Therefore, the cost of the test can be greatly reduced. In addition, the flow analysis is not limited by the flow field properties and the direction of the incoming flow. Second, more information is available than the wind tunnel test. In this way, the complex flow corresponding to the test can be calculated and observed, and the physical quantities difficult to be measured in the test can also be obtained. Based on the above advantages, the CFD numerical simulation methods can systematically analyze the physical phenomena of fluid flow. Therefore, the CFD methods have become one of the important techniques in wind engineering research.

Through the fundamental equations of hydrodynamics, the differential equations that control fluid flow are solved, and the physical quantities in equations are calculated to describe the complex flow phenomena along railways.

3.2.2.1 Hydrodynamic equations

3.2.2.1.1 Continuity equation

According to the law of conservation of mass, the reduction of fluid mass in the control volume per unit time is equal to the mass flux passing through the control surface. The fluid continuity equation that can be expressed as a differential form is as follows [13]:

$$\frac{\partial \rho}{\partial t} + \nabla \cdot (\rho U) = 0 \tag{3.1}$$

where ρ is fluid density, U is fluid velocity, ∇ is Hamilton operator, and $\nabla = i\frac{\partial}{\partial x} + j\frac{\partial}{\partial y} + k\frac{\partial}{\partial z}$.

3.2.2.1.2 Momentum equation

The momentum equation expresses the law of conservation of momentum for moving fluid. The rate of change of total momentum of any micro unit in flow field is equal to the resultant force of all external forces acting on the micro unit. The expression of fluid momentum equation is as follows [14]:

$$\begin{cases} \dfrac{\partial(\rho u)}{\partial t} + \nabla \cdot (\rho u U) = -\dfrac{\partial p}{\partial x} + \dfrac{\partial \tau_{xx}}{\partial x} + \dfrac{\partial \tau_{yx}}{\partial y} + \dfrac{\partial \tau_{zx}}{\partial z} + \rho f_x \\ \dfrac{\partial(\rho v)}{\partial t} + \nabla \cdot (\rho v U) = -\dfrac{\partial p}{\partial y} + \dfrac{\partial \tau_{xy}}{\partial x} + \dfrac{\partial \tau_{yy}}{\partial y} + \dfrac{\partial \tau_{zy}}{\partial z} + \rho f_y \\ \dfrac{\partial(\rho w)}{\partial t} + \nabla \cdot (\rho w U) = -\dfrac{\partial p}{\partial z} + \dfrac{\partial \tau_{xz}}{\partial x} + \dfrac{\partial \tau_{yz}}{\partial y} + \dfrac{\partial \tau_{zz}}{\partial z} + \rho f_z \end{cases} \tag{3.2}$$

where $U = ui + vj + wk$ is velocity of micro unit, τ_{ij} is viscosity stress component of micro unit, and f_i is body force of micro unit.

3.2.2.1.3 Energy equation

According to the law of conservation of energy, the energy change is equal to the sum of heat and working by the external force (including mass force and surface force) in the fluid control volume, which can be expressed as follows [15]:

$$\frac{\partial(\rho e)}{\partial t}+\nabla\cdot(\rho e U)=\frac{\partial}{\partial x}\left(k\frac{\partial T}{\partial x}\right)+\frac{\partial}{\partial y}\left(k\frac{\partial T}{\partial y}\right)+\frac{\partial}{\partial z}\left(k\frac{\partial T}{\partial z}\right)+S_T \quad (3.3)$$

where e is internal energy per unit mass, T is fluid temperature, k is heat transfer coefficient of fluid, and S_T is viscous dissipation.

3.2.2.2 Numerical methods in CFD

The CFD model is a series of partial differential equations, and it is difficult to obtain analytical solutions. The basic idea of the numerical method to solve the CFD model is to replace the field of continuous physical quantities in space-time with a collection of values on finite discrete nodes [16]. The CFD model has a variety of different solving methods. According to the different discretization methods of control equations, the methods can be divided into Finite Difference Method (FDM), Finite Element Method (FEM), Finite Volume Method (FVM), particle method, and Lattice Boltzmann Method (LBM).

3.2.2.2.1 Finite difference method

The computational domain is discretized into difference grids, and the differences formed by the value on nodes are used to approximately replace the derivatives of the partial differential equations [17]. The partial differential equations are discretized into finite algebraic equations, and the continuous computational domain is replaced by finite grid nodes. The difference scheme of the FDM is mainly constructed by Taylor expansion, and the difference quotient on the grid is used to approximately replace the spatial derivative, to transform the differential equations into difference equations with the field value on grid node as unknown [18].

3.2.2.2.2 Finite element method

The computational domain is discretized into nonoverlapping and interconnected units, and an interpolation function is constructed on each unit to transform the control equations into finite element equations. All finite element equations of each unit are combined to represent the entire computational domain.

3.2.2.2.3 Finite volume method

The computational domain is discretized into nonoverlapping control volumes, and the difference equation is integrated into each control volume [19]. The calculation of volume integral must assume the changing law of variable values between grids [20]. The physical meaning of the FVM is clear, and the conserved variable satisfies the conservation law in all control volumes of any size.

3.2.2.2.4 Particle method

The particle method describes the continuous fluid into interacting particles, the motion characteristics of the system are a combination of each particle. The particle method can be divided into Smoothed Particle Hydrodynamics (SPH), Moving Particle Semiimplicit (MPS), and Finite Volume Particle (FVP).

The idea of the SPH is to discretize the uninterrupted problem domain into some interacting particles, and then use the properties of kernel function to convert the control equations into particle expressions [21]. According to the initial information and boundary conditions of the particles, the change of the particle information in each step can be obtained to solve flow fields in different forms.

The MPS obtains the fluid pressure field according to the pressure Poisson equation, and corrects the predicted fluid velocity through the pressure gradient. The MPS uses the Lagrangian method to describe the fluid movement, and it does not require discrete convection terms to avoid numerical diffusions [22]. By introducing the weight function, the differential operators in the fluid control equations are transformed into interaction model between particles. In MPS, the incompressibility of fluid is reflected by keeping the particle number density in the flow field constant. Free surface particles and solid boundary particles are also determined by the particle number density [23]. The MPS uses semi-implicit calculation process to solve the momentum equation, all terms of the equation are solved explicitly, except the pressure term which is solved implicitly. The essence of the FVP is the same as the MPS, the difference compared to the MPS is that the control equations are integrated in the imaginary particle volume space to obtain the new gradient and Laplacian operator [24].

3.2.2.2.5 Lattice Boltzmann method

The LBM is a CFD method based on mesoscopic simulation scale. The basic assumption of the LBM is that the movement of fluid is the statistical average result of the movement of microscopic particles. The kernel of the LBM is to construct a simplified kinetic model containing the microscopic physical mechanism of the system, so that the overall average movement law follows the macroscopic motion control equation [25]. The LBM model consists of lattice structure, local equilibrium distribution function, and kinetic evolution equation.

3.2.2.3 Turbulence model

Atmospheric turbulence is an irregular, multiscale, and structured flow. Turbulent movement is formed by the superimposition of eddies of various scales. The size and direction of the rotation axis of these eddies are random. The pressure, velocity, temperature, and other physical characteristics at each point of the turbulence fluctuate randomly [26]. The occurrence of turbulence requires certain dynamic and thermodynamic conditions. The dynamic condition is that the air layer has obvious wind velocity shear, the thermodynamic condition is that the air layer has a certain degree of instability, and the most favorable condition is the convection condition that the air temperature in the upper layer is lower than the lower layer [27].

The average Navier–Stokes equations are often used to describe the turbulent movement. After averaging the Navier–Stokes equations of three-dimensional unsteady random and irregular turbulence, the Reynolds stress term of unknown turbulence fluctuation is added to the average equations [28]. To close the equations, it is necessary to establish the expression of Reynolds stress or introduce new turbulence equations. The turbulence control equations based on these model assumptions are called turbulence models.

The currently used turbulence simulation methods mainly include Direct Numerical Simulation (DNS) and indirect numerical simulation. Indirect numerical simulation methods do not directly calculate the fluctuation characteristics of turbulence, but approximate and simplify the turbulence [29]. According to different approximation and simplification methods, indirect numerical simulation methods include statistic average method, Large Eddy Simulation (LES) method, and Reynolds Average Navier–Stokes (RANS) method.

The DNS method does not do any simplification or approximation to the turbulent movement equations, and completely solves the instantaneous

Navier–Stokes equations. It can obtain turbulent movements of all scales, so it can solve highly complex eddy structures and turbulence characteristics under severe fluctuation [30]. However, the DNS method requires high grid accuracy and small-time step. And it requires excellent processing capabilities and large storage space for computer, which is difficult in practical engineering applications.

The statistic averaging method directly uses the average equation of turbulence, the most commonly used averaging methods include time average method, spatial average method, and probabilistic average method. Time average method is for steady turbulence. When the sample time is long, according to the ergodic theorem, it can be considered that the sampled time series include all the possible samples of random variables. The time series samples are equivalent to statistical samples, and the time series sample average can be used instead of probability statistics average.

The spatial average method is for uniform turbulent, the statistical average properties of all points in the flow field space are the same. When flow field space is large enough, according to the ergodic theorem, it can be considered that each sample in the flow field space contains all the possible samples of random variables [31]. The spatial sample average is equivalent to the statistical sample average, and spatial sample average can be used instead of probability statistics sample average.

The probabilistic average method is for unsteady and nonuniform turbulence, to perform weighted integration of random variables by the joint probability density function. However, the joint probability density function of turbulent physical quantities is unknown, and the sample statistical average is used instead of the probability statistics average.

For high grid accuracy in the DNS method, the LES method does not perform full-scale eddy simulation, and decomposes turbulence into two-scale flows through filtering. One is solvable scale turbulence (large-scale eddies), the other is unsolvable scale turbulence (small-scale eddies) [32]. The large-scale eddies can be directly simulated by numerical flow field, and the influence of small-scale eddies on the momentum and power transport in large-scale eddies movement is represented by sub-grid models. This method of turbulence decomposition into two scales can simulate higher Reynolds number and complex turbulent movement on a larger grid scale [33]. In turbulent energy transmission, almost all the energy is carried by large-scale movement, and small-scale eddies play a role in dissipating kinetic energy. The basic idea of the LES method is universal and very consistent with physical law of energy transmission in turbulence.

The LES needs to determine the filter function and cutoff width and filter out small-scale eddies by spatial-filtering, to decompose the movement equations describing the large eddies field. The influence of small eddies on the movement of large eddies is reflected by an additional stress term in movement equations of the large eddies field, which is called sub-grid-scale stress (SGS). The SGS indicates the influence of the modeled velocity component on the analytical velocity component. The SGS can be divided into three parts: Leonard stress, cross-stress, and LES Reynolds stress [34]. Furthermore, the LES Reynolds stress can be divided into deviator stress term and normal stress term. The SGS model is needed to close the LES movement equations, the types of SGS models include Smagorinsky model, Smagorinsky-Lilly model, dynamic SGS model, etc. [35,36].

The simulation principle of the RANS method is to split the physical quantities in turbulent into fluctuation and average, and then use time-averaged unsteady Navier–Stokes equations for calculation [37]. The unsteady Navier–Stokes equations are averaged over time to obtain a set of unclosed equations with time-average physical quantities as unknown parameters. Other equations are added to describe the time-average value of the product of fluctuation, and form closed equations with the Navier–Stokes equations to describe turbulent movement.

Common RANS turbulence models include Eddy Viscosity Model (EVM) method and Reynolds Stress Model (RSM) method. The EVM does not directly deal with the Reynolds stress but introduces eddy viscosity to express the turbulence stress as a function of the eddy viscosity coefficient [38]. According to the number of differential equations to determine the eddy viscosity coefficient, the EVM can be divided into zero-equation models, one-equation models, and two-equation models.

The zero-equation model uses algebraic relations to link the eddy viscosity with time-average value. The model only uses the eddy time-average continuity equations and Reynolds equation to form equations, and Reynolds stress in equations is expressed by local velocity gradient of the average velocity field. The zero-equation model only applies to turbulence in local equilibrium state, ignoring the effects of convection and diffusion. It is not suitable for processing complex flows with separation and backflow phenomena [39].

The one-equation model establishes a kinetic energy transport equation based on eddy continuity equations and Reynolds equation, and expresses eddy viscosity as a function of eddy kinetic energy to close the equations [40]. The one-equation model considers the convection and diffusion of turbulence, which is more reasonable than the zero-equation model. But to make the one-equation model closed, an algebraic expression of the length scale must be given in advance. The eddy length scale is also related to specific problem, and needs a differential equation to determine it. The addition of this differential equation generates the two-equation model, which uses the dissipation scale as the characteristic length scale.

The EVM assumes that the eddy viscosity coefficient is isotropic, and uses the product of effective viscosity coefficient and average velocity gradient to simulate Reynolds stress. It is difficult for the EVM to reflect the effects of rotating flow and surface curvature changes in the direction of flow. It is necessary to directly establish the differential equation to solve the Reynolds stress. The RSM simulation methods include the RSM and Algebraic Stress Model (ASM). The RSM directly establishes a differential equation with Reynolds stress as dependent variable based on anisotropy. The RSM directly solves the time-average equation of the product of two stress fluctuation values, and calculates the time-average value of the product of three stress fluctuation values by simulation.

The ASM simplifies the differential equation of Reynolds stress into algebraic expression to reduce complexity, while retaining the characteristics of turbulent anisotropy. When simplifying the Reynolds stress equation in the RSM, it is necessary to deal with the convection term and diffusion term. A simplified scheme is to adopt the local equilibrium assumption, which considers the difference between convection term and diffusion term to be zero. Another simplified scheme is to assume that the difference between convection term and diffusion term of Reynolds stress is proportional to the difference between convection term and diffusion term of turbulent kinetic energy [41]. Compared with the RSM, the ASM greatly reduces calculation, and the requirements for initial conditions and boundary conditions are not strict, but there are difficulties in convergence in three-dimensional calculations.

With further studies on turbulence phenomena and turbulence models, some hybrid turbulence models that combine LES and RANS have gradually emerged, such as Detached-Eddy Simulation (DES), Scale-Adaptive Simulation (SAS), Partially Integrated Transport Model (PITM), Partially-Averaged Navier—Stokes (PANS), etc.

The original DES method was implemented based on the Spalart-Allmaras one-equation turbulence model. When approaching the wall, the DES model returns to the Spalart-Allmaras RANS model. When it is far away from the wall, the wall distance in the DES model is modified to increase the dissipation term, and resolve the eddy movement larger than the cutoff width [42]. However, when calculating the flow separation of smooth surface, since the separation position is sensitive to total Reynolds stress, the switching of the computational domain is not necessarily smooth. Moreover, in some regions of the grid, the grid is not dense enough to be suitable for switching to LES, resulting in lower total Reynolds stress and the separation occurs earlier. At this time, the model performance of the DES is worse than that of the RANS. Therefore, Spalart et al. proposed a further modification, which is Delayed Detached-Eddy Simulation (DDES), to avoid the switch from RANS to LES too close to the wall, and prevent the modeled stress depletion caused by the model entering the LES region too early [43].

After proposing the DDES method, Spalart et al. used a new sub-grid length scale that depends on the wall distance to solve the logarithmic layer mismatch problem of the DES. This Improved Delayed Detached-Eddy Simulation (IDDES) method has achieved better results than DDES when dealing with complex flow problems involving walls [44]. The IDDES combines the DDES and Wall-Modeled Large Eddy Simulation (WMLES) method, and can be converted to the WMLES method when the boundary layer contains turbulence fluctuation. Compared with the LES and (D)DES, the IDDES adopts a new sub-lattice length-scale definition, which depends on the grid size and wall distance [44]. Compared with the LES, the DES methods do not consider small-scale fluctuation movements in turbulent boundary layer, and the number of grids required is greatly reduced. It can not only take advantage of small calculation of the RANS method in the boundary layer but can also simulate large-scale eddies in flow separation region [45].

Inspired by the length-scale idea in DES, Mentor et al. introduced the von Karman length scale into turbulence equations as a second characteristic length scale based on the traditional RANS model, and proposed the SAS [46]. The von Karman length scale can cover all turbulent fluctuation scales in the inertial subregion, and can distinguish eddy dynamics according to known flow field in the unsteady region, and adjust the length scale in turbulence model in real time. The SAS uses a processing method similar to

LES to solve unstable region of the flow field, and the near-wall region is solved by RANS. The length scale is automatically adjusted according to local flow field topology, which greatly reduces the dependence on grid. The von Karman length scale can adaptively change with the size of local turbulent fluctuation to reduce eddy viscosity and dissipation in small-scale high-frequency fluctuation region [47]. Therefore, although the SAS method is based on the RANS equation, it can show performance similar to the LES in large separation region. Because the construction of the von Karman length scale does not include the grid scale, the SAS model can achieve a smooth transition from the RANS to the LES, avoiding the problem of grid sensitivity [48].

The crux of hybrid RANS/LES simulations is the transition between the RANS and LES regions, the PITM can realize this transition with seamless coupling [49]. For the LES using coarse grids, Schiestel and Dejoan proposed modeled transport equations to compatible with direct numerical simulation and full statistical modeling. In this model, the characteristic length scale of sub-grid turbulence is given by the dissipation equation, not by the spatial discrete step size [50]. To realize the continuous change between RANS and LES regions, a new epsilon equation for LES is derived, which is formally consistent with RANS when the filter width is very large. The PITM provides a general formulation based on accurate energy spectrum function valid in both large and small eddy ranges [51].

The PANS model is proposed by Girimaji et al. based on the standard k-epsilon model and variable-filter model [52,53]. The PANS is a turbulence numerical model that can smoothly transition from RANS to DNS. It mainly uses model parameters to modify closed equations. The turbulence model can be solved at any filter width by changing unresolved dissipation and total dissipation [54]. The PANS model solves turbulent flow by local averaging. The control equations are corrected by model parameters to realize the numerical calculation of turbulent movement within different filter widths. The PANS model resolves the velocity field based on turbulence energy, which is equivalent to directly solving the velocity field distribution [55]. The PANS model judges the decomposed and undecomposed velocity fields based on the implicit difference format instead of the explicit difference format. The simulation does not need filtering during the iteration solution process, which improves the computational efficiency.

3.2.3 Anemometer layout optimization

Layout problem means that under certain constraints (such as position constraints, budget constraints, coupling constraints, and space constraints), several objects to be distributed are placed in a specific container or region, and the desired optimization target is optimized as much as possible [56]. Anemometer layout problem is a complex combinatorial optimization problem, which involves many subjects and fields, and is closely related to logic, geometry, computer graphics, operations research, etc. The result of the layout will directly affect the security, efficiency, and rationality of the application mentioned above. In recent years, researchers from various countries have conducted extensive and in-depth studies on layout optimization and proposed a variety of optimization algorithms, such as evolutionary algorithms, deterministic algorithms, heuristic algorithms, and meta-heuristic algorithms.

The layout optimization problem is hard to solve. With the increasing scale of the problem, the traditional exhaustive algorithm relying on the computing speed of the computer has been difficult to solve this problem [57]. Although scholars have put forward many methods to solve the layout optimization problem, they have their advantages and disadvantages. In solving layout problems, deterministic algorithms can generally obtain very accurate solutions, and the commonly used algorithms are nonlinear programming, branch and bound methods, mixed-integer, etc. But the deterministic algorithm is usually only suitable for solving small-scale layout problems. Among these optimization algorithms, heuristics are algorithms designed by summarizing general rules inspired by nature or human social phenomena. It is easy to understand, fast in searching and solving, and has strong universality for different problem fields, which is favored by researchers [57]. However, these algorithms also lack universalities, such as genetic algorithms, simulated annealing algorithms, and other global search algorithms. Due to the lack of effective local search mechanisms, their convergence speed is usually slow. Even in the late stage of algorithm iteration, the search convergence stops, and the precision of the solution still needs to be further improved. Similarly, the ant colony algorithm is also slow in convergence and prone to fall into local minima.

In the existing literature, scholars have conducted extensive research on the optimal microcosmic location of anemometer stations along the railway. By using a better optimization algorithm, a more accurate wake model and a better grasp of wind speed characteristics, the layout optimization of anemometers can be realized. And the hidden danger of high-speed driving caused by strong crosswind can be reduced [58].

The layout optimization of anemometers is a highly complex optimization problem. Mosetti et al. introduced the Genetic Algorithm (GA) for layout optimization [59]. In this study, they hypothesized that a 2 km by 2 km wind field along the route was divided into square grids and set up three typical wind conditions for numerical testing. Case 1: the wind direction and wind speed are unchanged at 12 m/s, and the wind direction is 0 degrees. Case 2: the wind speed is unchanged at 12 m/s, and the wind direction is uniformly distributed within the range of 0–360 degrees. Case 3: The wind speed is 8 m/s, 12 m/s, and 17 m/s, and the wind direction is not evenly distributed.

As the basis of the research project of wind field layout optimization, the work of Mosetti et al. has been continuously improved. By using different optimization algorithms, researchers strived to get more accurate data with the same number of anemometers installed [60]. On this basis, a genetic algorithm with special mutation and selection operators is applied for layout optimization in the wind field [61].

Huang et al. proposed a hybrid distributed genetic algorithm, which is the Distributed Genetic Algorithm (DGA). This was followed by the heuristic method of Hill-Climbing (HC). The DGA divides the population into small ranges and improves the performance of the GA by maintaining diversity. A similar variant, the Multi-Population Genetic Algorithm (MPGA) was used to optimize layout in three typical scenarios. The effectiveness of this method in dealing with real conditions is demonstrated through relevant experimental studies. Ali M et al.'s work in 2018 focused on the use of Binary Real-Coded Genetic Algorithm (BRCGA)-based Local Search (LS) to obtain the optimal layout. The model adopts appropriate wake interaction modeling to collect a stable single wake model. In this model, the binary part of the GA is used to represent the locations. The results were compared with earlier studies using genetic algorithms and random search algorithms. Experiments show that the method is superior in finding the best solution.

3.3 Single-point wind speed–wind direction seasonal analysis

3.3.1 Seasonal analysis

Due to the periodic changes of climate and atmosphere variables, the wind speed and wind direction have significant seasonality. By observing the series waveform diagram of wind speed and wind direction along railways, the seasonal regularity of wind speed and wind direction can be intuitively observed.

In this section, several methods are proposed to analyze the seasonality of wind speed and wind direction data series. Firstly, an Augmented Dickey Fuller (ADF) method is utilized to test the stationarity of the wind data series. Secondly, a Hurst exponent is proposed to verify the seasonality of the data series. Thirdly, a Fast Fourier Transformation (FFT) is utilized to find the seasonal cycle of the data series. Finally, the Autocorrelation Function (ACF) and Partial Autocorrelation Function (PACF) methods are utilized to determine the type of the model. Besides, the Bayesian Information Criterion (BIC) is proposed to find the best parameter settings of the corresponding model types. The specific steps of seasonal analysis are given in Fig. 3.1 in detail.

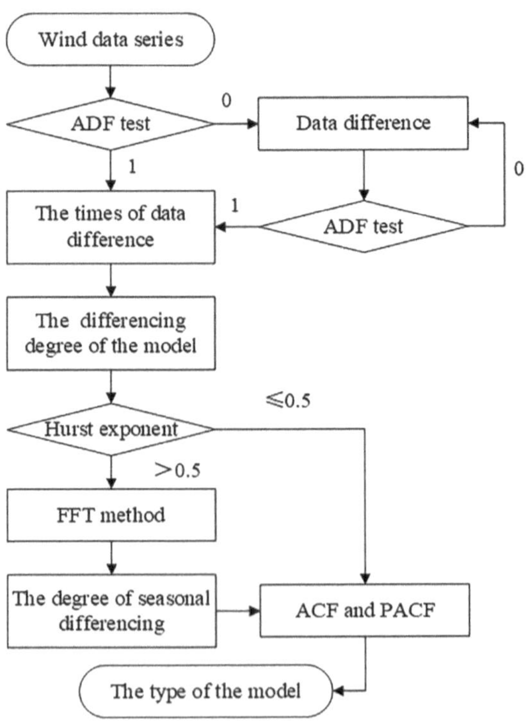

Figure 3.1 The steps of seasonal analysis.

3.3.1.1 Augmented Dickey Fuller test

In the process of modeling nonstationary time series, it is very necessary to transform the sequence into the stationary time series. Cointegration and difference are the basic methods to remove the nonstationarity. Cointegration is a situation in which two variables are nonstationary, but the two variables keep the same trend, and the linear combination of them is stationary. The difference is a situation in which the time series is stationary by subtracting the first-order lag term and reducing the order of the function. The difference method is widely used in the processing and analysis of time series.

ADF test can determine the existence of unit roots in a sequence. A drift d and several additional lags are added [62]:

$$\Delta y_t = b + \alpha^* y_{t-1} + \sum_{i=1}^{n-1} \phi_i \Delta y_{t-i} + \varepsilon_t \tag{3.4}$$

If the null hypothesis $a^* = 0$, the data are nonstationary. If the null hypothesis $a^* = 1$, the data are stationary.

3.3.1.2 Hurst exponent

In this section, the Hurst exponent is utilized to verify the seasonality of the wind speed and wind direction data series. Hurst exponent is defined and calculated according to the R/S analysis method [63]:

$$E\left[\frac{R(n)}{S(n)}\right] = An^H \tag{3.5}$$

where $R(n)$ is the range of data series, $S(n)$ is the deviation, and A is a constant value. The value of Hurst exponent corresponds to different meanings of time series:

(a) If $0 < H < 0.5$, it shows that the time series is of long-term relevance, but the overall trend in the future is the opposite of the past, which is called anti-persistence. If a time series has an upward trend in the first period, then it is likely to go down in the next period, and vice versa. The strength of this anti-persistence behavior depends on how close the H value is to 0, and the closer it is to 0, the more negatively correlated it is.

(b) If $H = 0.5$, it shows that the time series is random and unrelated, and the present does not affect future development.

(c) If $0.5 < H < 1$, then the time series has long-term relevance, that is, the persistence of a process. The closer H is to 1, the stronger the correlation is. If a sequence go up in the previous period, then it will continue to go up in the next period, and vice versa. Therefore, a certain range of records can last a long time and form a large cycle. But these cycles do not have a fixed cycle, and it is difficult to rely on past data to predict future changes.

3.3.1.3 Autocorrelation and partial autocorrelation functions

In this section, the *ACF* and *PACF* methods are utilized to determine the type of models. The *ACF* is calculated by the autocovariance of x_t and x_{t-n} [64]:

$$ACF(n) = \frac{Cov(x_t, x_{t-n})}{Var(x_t)} \tag{3.6}$$

The *PACF* is calculated by the simple correlation between x_t and x_{t-n} [64]:

$$PACF(n) = Corr\left[x_t - E^*\left(x_t | x_{t-1}, \ldots, x_{t-n+1}\right), x_{t-n}\right] \tag{3.7}$$

According to the *ACF* results and *PACF* results of the wind data series, the optimal solution model can be obtained. The type of models can be determined by the *ACF* results and *PACF* results, as shown in Table 3.1.

3.3.1.4 Bayesian information criterion

The Akaike Information Criterion (AIC) is a criterion to measure the performance of statistical model fitting. It is based on the principle of entropy and provides a standard for balancing the model complexity and fitting data. Compared with AIC, the BIC method can deal with the problem of compatible estimation by introducing a larger penalty term.

The BIC method consists of two components. The first component reflects the size of the model parameters. It increases as the order increases. The second component reflects the model fitting effect. It decreases as the order increases. The *BIC* function is as follows [65]:

Table 3.1 The characteristics of ACF and PACF results.

Models	AR	MA	ARMA
ACF	Tails off	Cuts off	Tails off
PACF	Cuts off	Tails off	Tails off

$$BIC(n) = k \bullet ln(n) - 2ln(L_m) \tag{3.8}$$

where k is the number of estimated parameters of the model, n is the number of observation points in the series, and L_m is the maximum likelihood function value of the model.

In this way, when n exceeds the order n_0 of the sequence optimal model, increasing the value of n can reduce the accuracy. The n_{best} value at the minimum value of BIC is the optimal value of the model, which can be expressed as follows [65]:

$$n_{best} = \underset{0 \le n \le n_h}{argminBIC}(n) \tag{3.9}$$

3.3.2 Single-point wind speed seasonal analysis

3.3.2.1 Data description

In this section, the real wind speed data are utilized in the wind speed seasonal analysis. The data are collected from the wind anemometer stations along the strong wind railway line. The wind speed data series contains 1500 data samples, where the former 1000 data points are utilized to train the wind speed description models, and the later 500 data points are utilized to verify the accuracy and performance of different models.

As shown in Fig. 3.2, the time interval of wind speed data is 1 min. The diagram of wind speed data samples is given in Fig. 3.2.

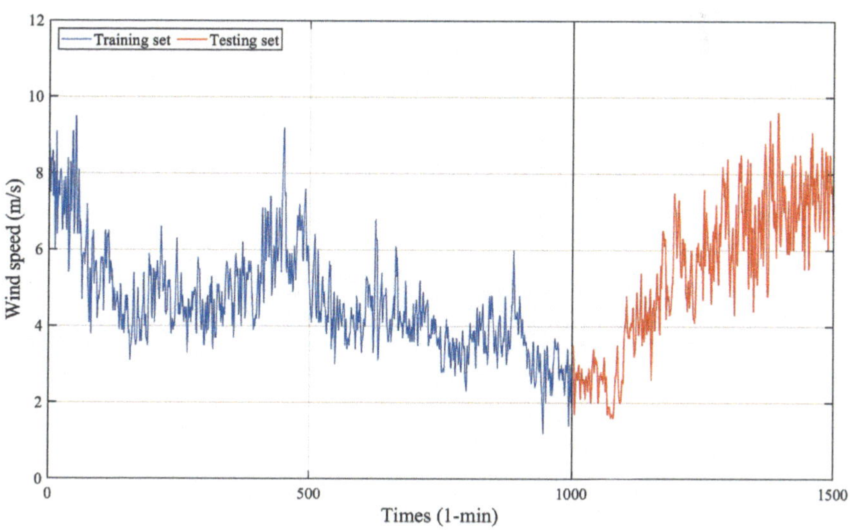

Figure 3.2 The wind speed data series.

3.3.2.2 Data difference

After the ADF test, the result of the original wind speed data is 0. Therefore, the original wind speed data series is unstable. After the first-order difference, the wind speed data are shown in Fig. 3.3.

After the difference, the ADF test result of the data is 1, and the first-difference result of wind speed data is stable. Therefore, the degree of differencing of the Autoregressive Integrated Moving Average (ARIMA) model and Seasonal Autoregressive Integrated Moving Average (SARIMA) model is 1.

3.3.2.3 Seasonal analysis

After calculation, the Hurst exponent of the original wind speed data series H_S is 0.99. The H_S value is greater than 0.5, which means that the wind speed data series is a long-term dependent process. The wind speed data series has a strong trend that a high value following the previous high value. It shows that the wind speed data series has seasonality.

To determine the seasonal cycle of the wind speed data series, the FFT method is utilized. The FFT result of the wind speed data series is shown in Fig. 3.4.

As shown in the amplitude spectrum, the wind speed data series has a peak power around $f = 7.78 \times 10^{-5} HZ$. The frequency is corresponding to the period of 214 min, which is the seasonal pattern of the wind speed data series. Since the time interval of the wind speed data series is 1min, the degree of the seasonal differencing polynomial is 214.

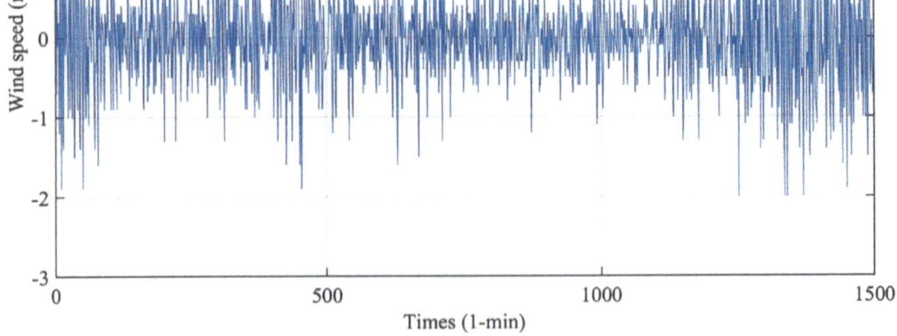

Figure 3.3 The first-difference result of wind speed data.

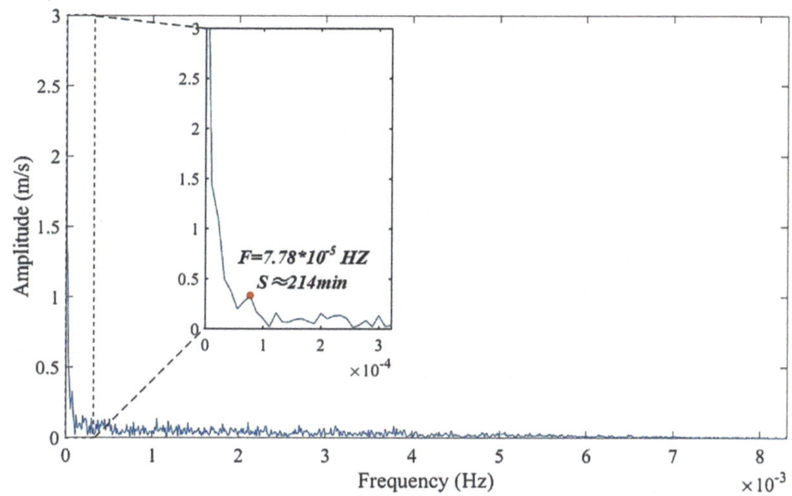

Figure 3.4 The FFT result of wind speed data series.

3.3.2.4 ACF and PACF analysis

After calculating the autocorrelation function, the autocorrelation of the wind speed data series is shown in Fig. 3.5.

After calculating the partial autocorrelation function, the partial autocorrelation of the wind speed data series is shown in Fig. 3.6.

It can be seen from Figs. 3.5 and 3.6 that the ACF of the original wind speed data series shows tails off and the PACF shows cuts off. Therefore, an Autoregressive Integrated (ARI) or Seasonal Autoregressive Integrated (SARI) model should be used to fit the wind speed data. However, the specific autoregressive polynomial degree should be further determined by the BIC method to achieve more precise arguments.

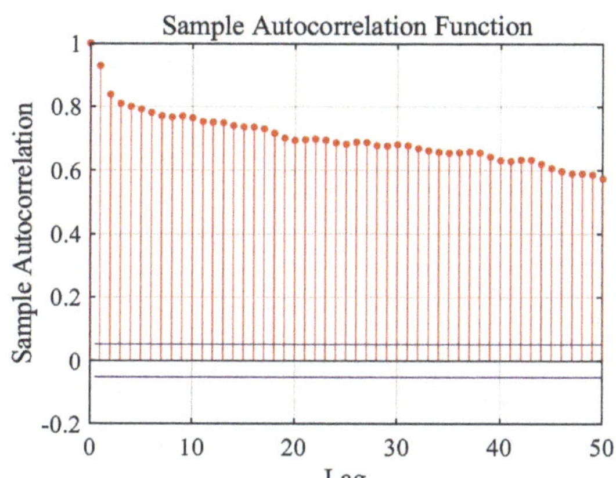

Figure 3.5 The autocorrelation of the original wind speed data series.

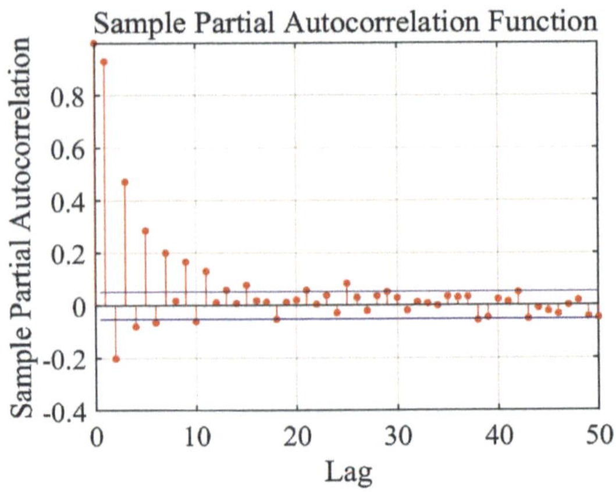

Figure 3.6 The partial autocorrelation of the original wind speed data series.

3.3.3 Single-point wind direction seasonal analysis

3.3.3.1 Data description

The real wind direction data are utilized in the wind direction seasonal analysis. The data are also collected from the wind anemometer stations along the strong wind railway line. The wind direction data series contains 1500 data samples, where the former 1000 data points are training set and the later 500 data points are testing set. In this section, the training set is utilized to train the wind direction description models and the testing set is utilized to verify the accuracy and performance of different models.

As shown in Fig. 3.7, the time interval of wind direction data is also 1 min. The diagram of wind direction data samples is given in Fig. 3.7.

It can also be seen from Fig. 3.7 that the wind direction data show seasonality more obviously, comparing to the wind speed data in Fig. 3.2. The outline of each cycle is also clearer, because the changing range of wind direction data is larger and the change is more obvious.

3.3.3.2 Data difference

After the ADF test, the result of the original wind direction data is 0. Therefore, the original wind direction data series is unstable. After the first-order difference, the wind direction data are shown in Fig. 3.8.

After the difference, the ADF test result of the data is 1, and the first-difference result of wind direction data is stable. Therefore, the degree of differencing of the ARIMA model and SARIMA model is 1.

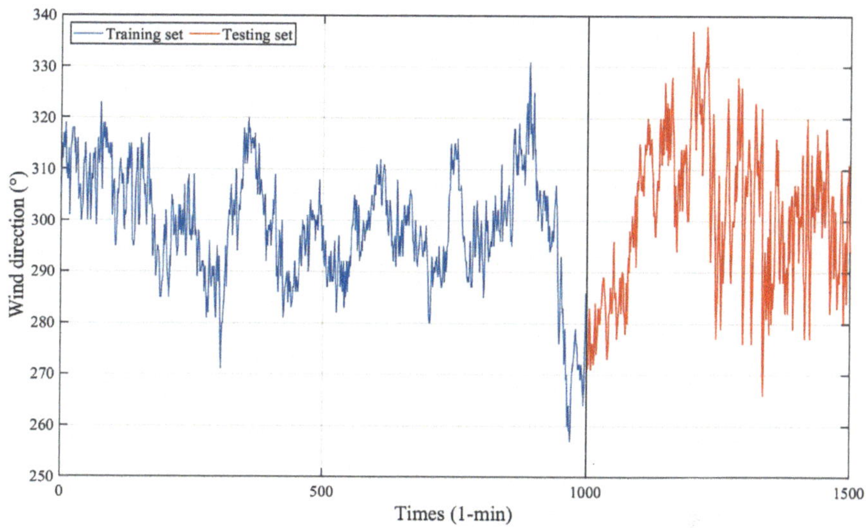

Figure 3.7 The wind direction data series.

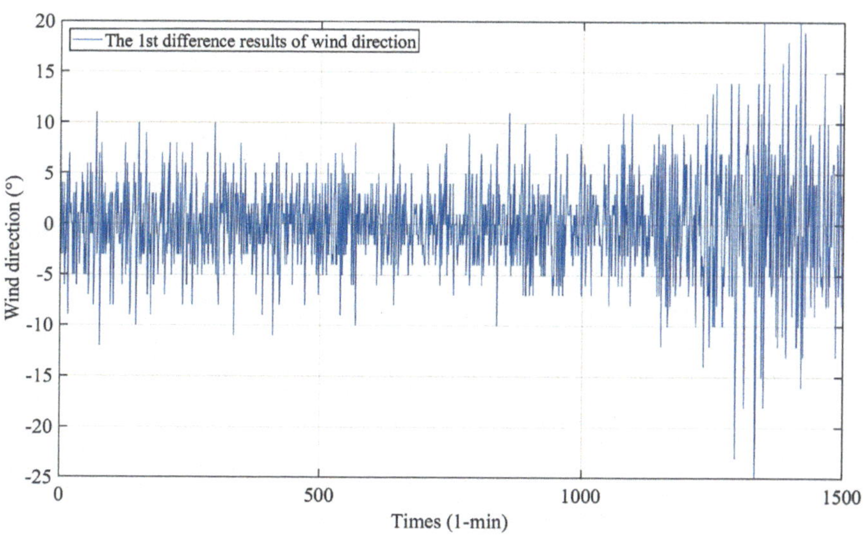

Figure 3.8 The first-difference result of wind direction data.

3.3.3.3 Seasonal analysis

After calculation, the Hurst exponent of the original wind direction data series H_D is 0.89. The H_D value is greater than 0.5, which means that the wind direction data series is also a long-term dependent process. The wind

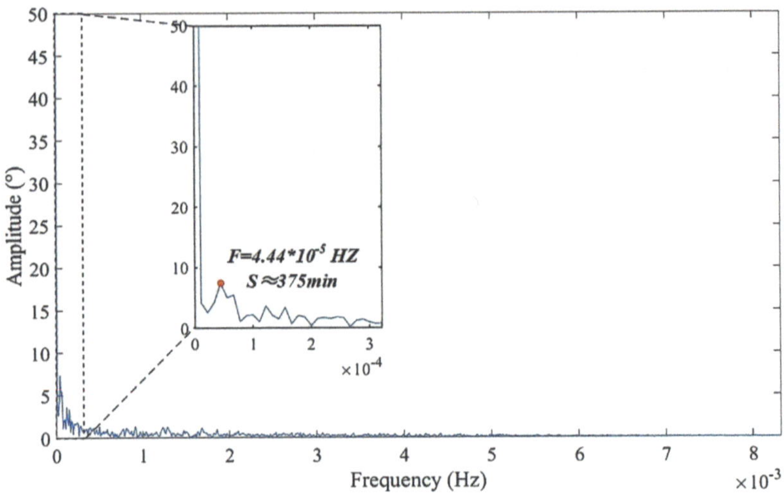

Figure 3.9 The FFT result of wind direction data series.

direction data series has a strong trend that a high value following the previous high value. It shows that the wind direction data series has seasonality.

To determine the seasonal cycle of the wind direction data series, the FFT method is utilized. The FFT result of wind direction data series is shown in Fig. 3.9.

As shown in the amplitude spectrum, the wind direction data series has a peak power around $f = 4.44 \times 10^{-5}$Hz. The frequency is corresponding to the period of 375 min, which is the seasonal pattern of the wind direction data series. Since the time interval of the wind direction data series is 1min, the degree of the seasonal differencing polynomial is 375.

3.3.3.4 ACF and PACF analysis

After calculating the autocorrelation function, the autocorrelation of the wind direction data series is shown in Fig. 3.10.

After calculating the partial autocorrelation function, the partial autocorrelation of the wind direction data series is shown in Fig. 3.11.

It can be seen from Figs. 3.10 and 3.11 that the ACF of the original wind direction data series shows tails off and the PACF shows cuts off. Therefore, an ARI or SARI model should be used to fit the wind direction data. However, the specific autoregressive polynomial degree should be further determined by the BIC method to achieve more precise arguments.

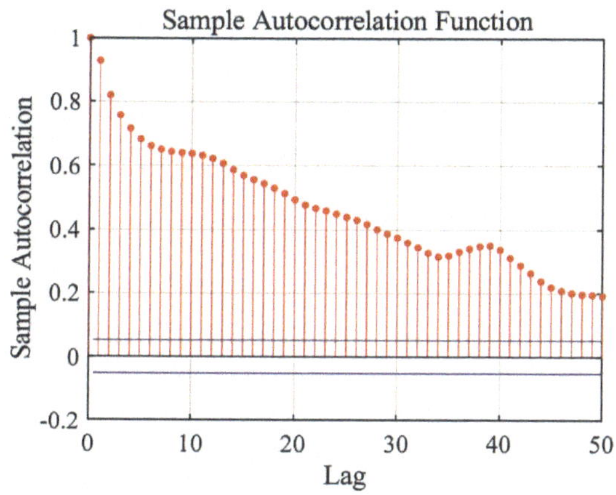

Figure 3.10 The autocorrelation of the original wind direction data series.

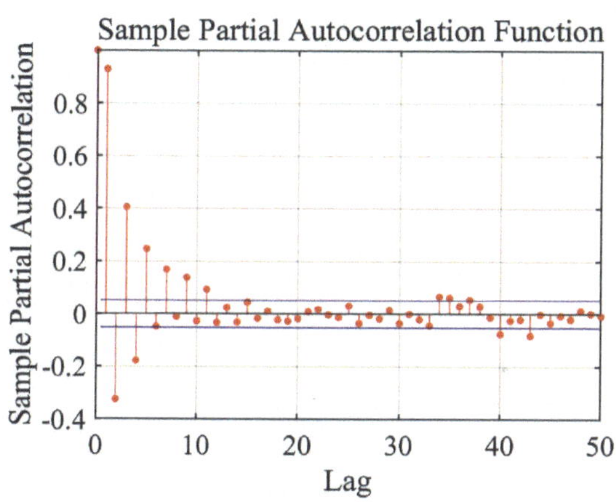

Figure 3.11 The partial autocorrelation of the original wind direction data series.

3.4 Single-point wind speed—wind direction heteroscedasticity analysis

3.4.1 Heteroscedasticity analysis

The concept of heteroscedasticity is based on homoscedasticity. The definition of the heteroscedasticity model is given as follows:

$$Y_i = \beta_0 + \beta_1 X_{1i} + \beta_2 X_{2i} + \ldots + \beta_k X_{ki} + \varepsilon_i \quad (3.10)$$

where $Var(\varepsilon_i) = \delta_i^2$. It means that the variance changes as i changes.

Heteroscedasticity can generally be divided into three types: monotonically increasing type, which means that δ_i^2 increases as X_i increases; monotonically decreasing type, which means that δ_i^2 decreases as X_i increases; complex type, which means that the changes of δ_i^2 and X_i change in complex forms.

The occurrence of heteroscedasticity is affected by many reasons. It generally comes from four aspects: (A) the data is of cross-sectional data; (B) some factors that affect the explanatory variables in the model are omitted; (C) measurement error; (D) use grouped data to estimate the model.

The methods for testing heteroscedasticity contain graphical test and hypothesis tests. They can be used to test the existence of heteroscedasticity. These inspection methods are, respectively, introduced below.

3.4.1.1 *Graphical test*

The heteroscedasticity of the time series appears on the uncertainty. The graphical test can estimate the heteroscedasticity of the series according to the data distribution visually. Because the variance of the heteroscedastic time series varies over time, the distribution of the data changes. To visualize the variance of the time series, it is necessary to calculate the probabilistic components. The probabilistic components of the time series can be calculated as the residuals. The heteroscedasticity indicates the variance of the data is related to the dependent variables. By visualizing the correlation between probabilistic data and variable variables, the heteroscedasticity can be found.

Multi-variate Kernel Density Estimation (MKDE) can be applied to improve visualizing performance [66]. The MKDE algorithm can fit joint distributions between the innovations and dependent variables. The conditional distribution of innovation can be calculated. Then, the conditional variance can be explicitly visualized.

3.4.1.2 *Hypothesis tests*

3.4.1.2.1 Goldfeld–Quandt test

The Goldfeld–Quandt test can divide the dataset into two parts and check the difference between them for the heteroscedasticity test [67]. The Goldfeld–Quandt test can generate the residual sums of squares in the separated data groups. The statistic is calculated as the ratio between these sums. The F-test can be used to validate the hypothesis.

3.4.1.2.2 Breusch–Pagan test

The Breusch–Pagan test assumes the time series obeys function in Eq. (3.12). The null hypothesis is shown as follows [68]:

$$H_o: E\left(\varepsilon_i^2 \middle| X_{1i}, X_{2i}, \ldots, X_{ki}\right) = \sigma^2 \tag{3.11}$$

where ε_i is the innovation of the time series, $X_{1i}, X_{2i}, \ldots, X_{ki}$ are the dependent variables, and σ^2 is the assumed variance.

The Breusch–Pagan test assumes the innovations obey linear function with the dependent variables as follows [69]:

$$\varepsilon_i^2 = \delta_0 + \delta_1 X_{1i} + \delta_2 X_{2i} + \ldots + \delta_k X_{ki} + error \tag{3.12}$$

where δ is regression weight, X is the dependent variable, and *error* is the regression error term.

Then, the null hypothesis becomes $H_o : \delta_1 = \delta_2 = \ldots = \delta_k = 0$. Because the innovation cannot be observed, the δ should be calculated according to the description residuals. This hypothesis test can be calculated by the chi-square test, F-test, and Lagrange Multiplier (LM) test. These test methods are consistent with a large data amount.

3.4.1.2.3 White test

The disadvantage of the Breusch–Pagan test is that only the linear correlation between the innovations and dependent variables is described [70]. The White test adds quadratic terms for the heteroscedasticity test. Assuming there are two dependent variables X_{1i} and X_{2i}, the regression equation is presented as follows:

$$\varepsilon_i^2 = \delta_0 + \delta_1 X_{1i} + \delta_2 X_{2i} + \delta_3 X_{1i}^2 + \delta_4 X_{2i}^2 + \delta_5 X_{2i} X_{2i} + error \tag{3.13}$$

where δ is regression weight, X is the dependent variable, and *error* is the regression error term.

Then the null hypothesis becomes $H_0 : \delta_0 = \delta_1 = \delta_2 = \delta_3 = \delta_4 = \delta_5 = 0$. This hypothesis can be validated like the Breusch–Pagan test. The advantage of the White test is that the quadratic equation can approximate any functions with high precision according to the Taylor expression.

3.4.1.2.4 Park test

The Park test assumes the variance obeys lagged linear correlation with the dependent variables as follows [71]:

$$\ln \varepsilon_i = \delta_0 + \sum \delta_k \ln X_{ki} + error \tag{3.14}$$

where δ is regression weight, X is the dependent variable, and *error* is the regression error term.

Like the Breusch–Pagan test and White test, the Park test can validate the heteroscedasticity of the time series according to the significance of the nonzero γ_k.

3.4.1.2.5 Glejser test

The Glejser test provides three different formula forms between the absolute innovations and dependent variables as follows [72]:

$$\begin{cases} |\varepsilon_i| = \delta_0 + \sum \delta_k X_{ki} + error \\ |\varepsilon_i| = \delta_0 + \sum \delta_k \sqrt{X_{ki}} + error \\ |\varepsilon_i| = \delta_0 + \sum \delta_k / X_{ki} + error \end{cases} \tag{3.15}$$

where δ is regression weight, X is the dependent variable, and *error* is the regression error term.

The Glejser test can compare the goodness of fit over these formulas, and select the most suitable formula for the heteroscedasticity test.

3.4.2 Single-point wind speed heteroscedasticity analysis

The wind data in Section 3.3 is utilized for analysis. The heteroscedasticity should be validated according to the innovations. In this section, the innovations of the wind speed data are estimated as the ARIMA residuals. The estimated innovations of the wind speed are shown in Fig. 3.12.

3.4.2.1 Graphical test

The wind speed and wind direction are applied as the dependent variables. To visualize the correlation between the innovation and dependent variables, the scatter plots and joint distributions are shown in Fig. 3.13.

By calculating the conditional distribution, the variance can be calculated as Fig. 3.14. From Fig. 3.14, it can be observed that the conditional variances of the innovations increase with large wind speed, and the conditional variances are the largest in certain directions. This phenomenon indicates the wind speed data have significant heteroscedasticity.

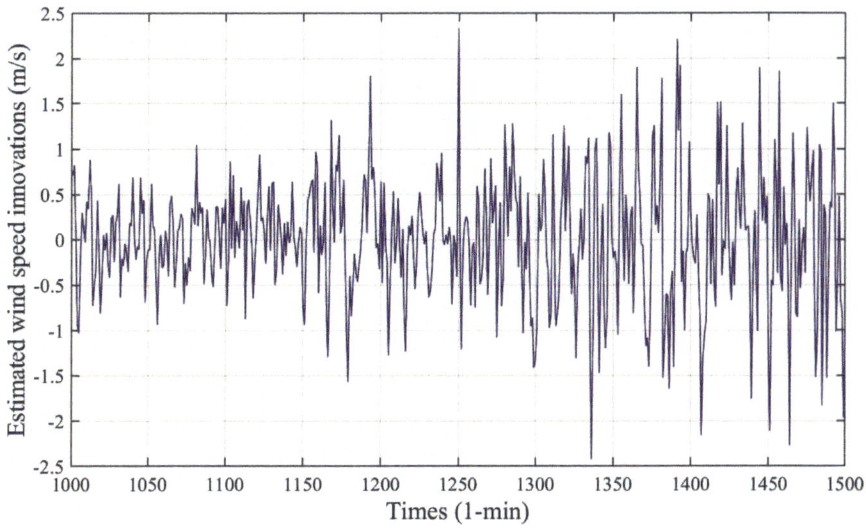

Figure 3.12 The estimated innovations of the wind speed.

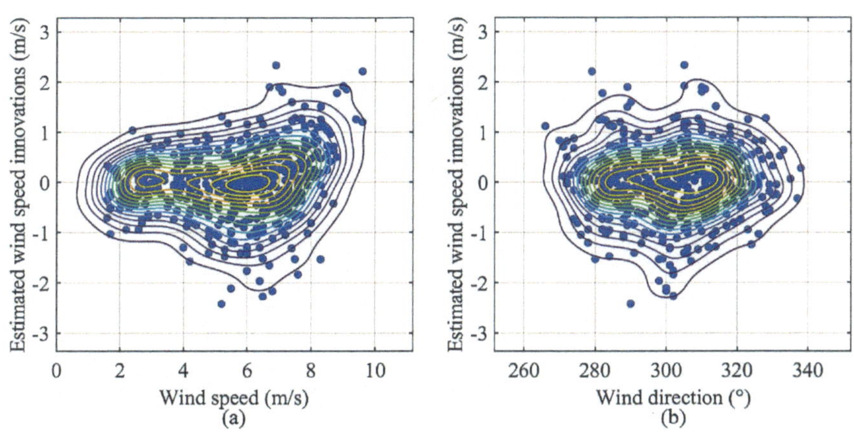

Figure 3.13 The scatter plots between the wind speed innovations and dependent variables (A) when the dependent variable is wind speed and (B) when the dependent variable is wind direction.

3.4.2.2 Hypothesis tests

The Goldfeld–Quandt test, Breusch–Pagan test, White test, Park test, and Glejser test are applied to test the heteroscedasticity of the wind direction innovations. The results are shown in Fig. 3.15. From Fig. 3.15, it can be seen that wind direction innovations have significant heteroscedasticity with 95% confidence level.

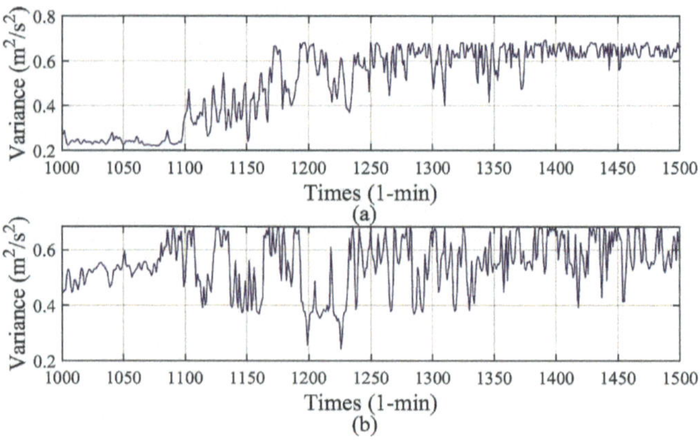

Figure 3.14 The conditional variances of the wind speed innovations (A) when the dependent variable is wind speed and (B) when the dependent variable is wind direction.

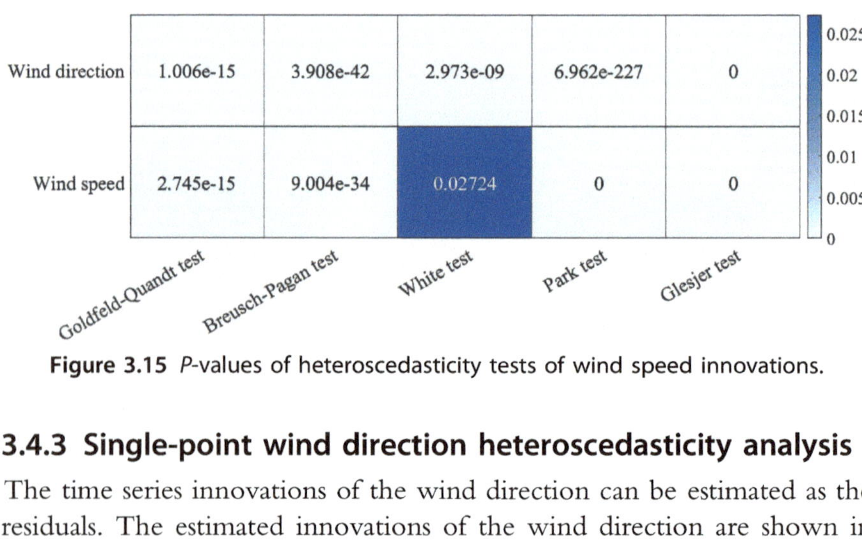

Figure 3.15 *P*-values of heteroscedasticity tests of wind speed innovations.

3.4.3 Single-point wind direction heteroscedasticity analysis

The time series innovations of the wind direction can be estimated as the residuals. The estimated innovations of the wind direction are shown in Fig. 3.16. From Fig. 3.16, it can be seen that the divergence of wind direction innovations varies over time.

3.4.3.1 Graphical test

Taking the wind speed and wind direction as the dependent variables, the correlations between the innovations and dependent variables are visualized in Fig. 3.17. From Fig. 3.17, it can be found that the conditional distributions of the wind direction vary over the dependent variables.

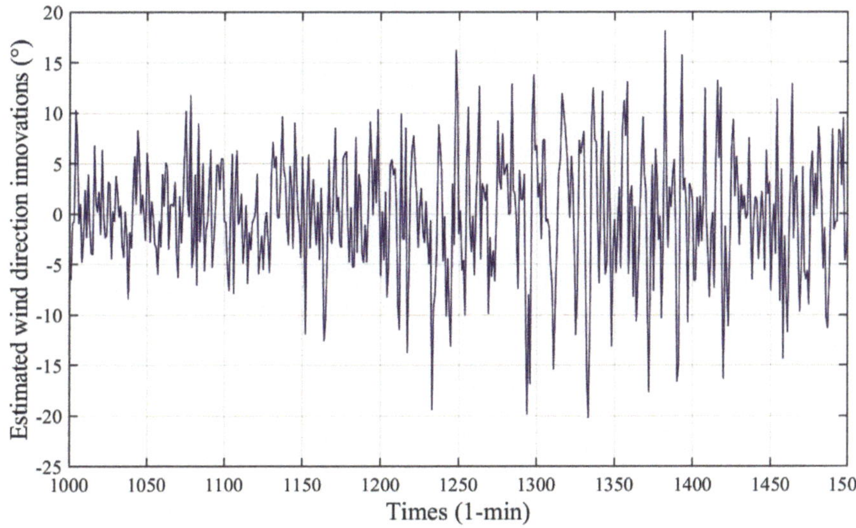

Figure 3.16 The estimated innovations of the wind speed.

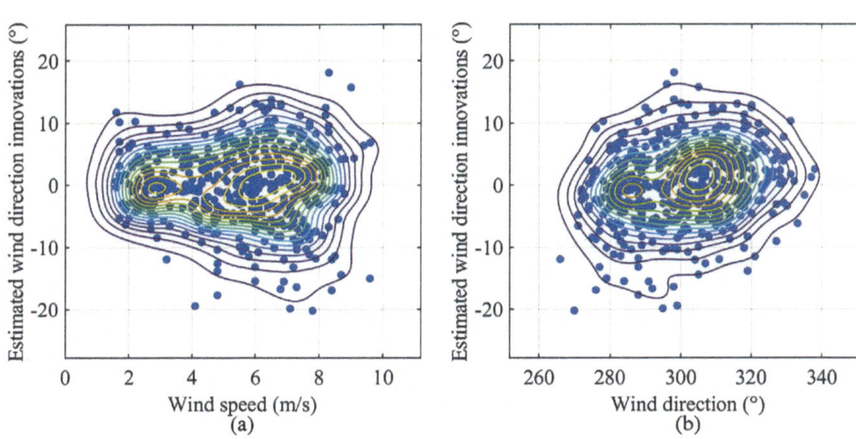

Figure 3.17 The scatter plots between the wind direction innovations and dependent variables (A) when the dependent variable is wind speed and (B) when the dependent variable is wind direction.

By calculating the conditional distribution, the variance of the wind direction can be calculated as Fig. 3.18. From Fig. 3.18, it can be observed that the conditional variances of the innovations increase with large wind speed, and the conditional variances are the largest in certain directions. This phenomenon indicates the wind direction time series have significant heteroscedasticity.

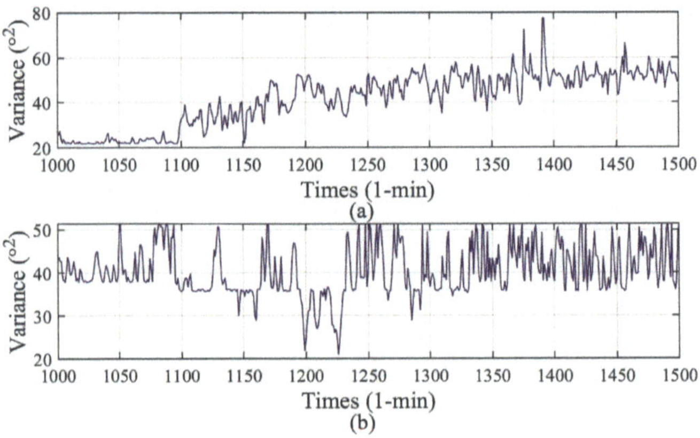

Figure 3.18 The conditional variances of the wind direction innovations (A) when the dependent variable is wind speed and (B) when the dependent variable is wind direction.

3.4.3.2 Hypothesis tests

The Goldfeld–Quandt test, Breusch–Pagan test, White test, Park test, and Glejser test are applied to test the heteroscedasticity of the wind direction innovations. The results are shown in Fig. 3.19. From Fig. 3.19, it can be observed that all of the *P*-values are below 0.05. So, wind direction innovations have significant heteroscedasticity with 95% confidence level.

3.5 Various single-point wind time series description algorithms

3.5.1 Autoregressive Integrated moving average

Traditional time series analysis includes stationary time series analysis and nonstationary time series analysis. For stationary time series, analysis

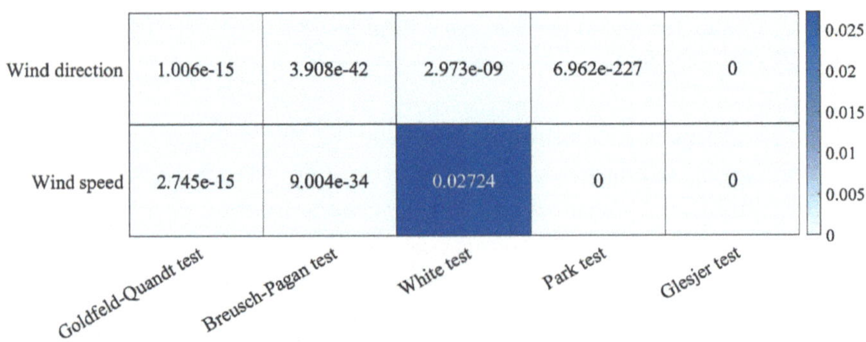

Figure 3.19 *P*-values of heteroscedasticity tests of wind direction innovations.

methods mainly include the Autoregressive (AR) model, the Moving Average (MA), and their combination-the Autoregressive Moving Average (ARMA). For nonstationary time series, make the series stationary through difference, and then use stationary time series analysis methods. The ARIMA is one of the nonstationary time series analysis methods.

The ARIMA can transform the nonstationary time series into the stationary time series [73]. The ARIMA regresses the lag value of the variable, the present value of the random error term, and the lag value of the random error term. Its basic idea is to treat the data sequence as a random sequence, and then use the corresponding mathematical model to describe this sequence. The ARIMA includes the AR process, the MA process, the ARMA process, and the ARIMA process [74].

The ARIMA model is convenient and quick to calculate, and it is widely used in wind time series description along railways. Some scholars have also proposed some new algorithms based on the classic ARIMA algorithm for wind speed. Kavasseri and Seetharaman used fractional-ARIMA and f-ARIMA models to predict day-ahead wind speeds [75]. Cadenas and Rivera combined the ARIMA and the Artificial Neural Network (ANN) model for wind speed in Mexico [76]. Liu et al. used the ARIMA to extract the features of the sequence and applied it to initialize the Kalman filter state equation and observation equation parameters [77]. Experimental results showed that the above methods can improve the accuracy of the ARIMA.

3.5.1.1 Theoretical basis

3.5.1.1.1 The autoregressive model

The AR model uses the linear combination of random variables in the previous period to describe the linear regression model of random variables in the future. It is the simplest linear time series analysis model. The AR (p) model can be expressed as follows [78]:

$$x_t = \varphi_0 + \varphi_1 x_{t-1} + \varphi_2 x_{t-2} + \ldots + \varphi_p x_{p-1} + \varepsilon_t \tag{3.16}$$

where x_t is the value that needs to be predicted at the current moment, φ_0 is a constant, $\varphi_1, \varphi_2, \varphi_3, \ldots, \varphi_p$ are the autoregressive coefficients that need to be calculated using the regression algorithm, p represents for p time points before selecting the current time point to compute, and ε_t is the current random disturbance and represents the zero-mean white noise series.

3.5.1.1.2 The moving average model

The MA model can be obtained by the way of weighting the white noise sequence in the time series. It uses the innovations of the past q period to linearly express the description value. The MA (q) model can be expressed as follows [78]:

$$x_t = \overline{x} + \varepsilon_t + \phi_1\varepsilon_{t-1} + \phi_2\varepsilon_{t-2} + \ \ldots \ + \phi_q\varepsilon_{t-q} \tag{3.17}$$

where x_t is the value that needs to be predicted at the current moment, $\overline{x}$ is the mean of the sequence, ε_t is the current random disturbance, and $\phi_1, \phi_2, \phi_3, \ldots, \phi_q$ are the moving average regression coefficients that need to be calculated using the regression algorithm.

3.5.1.1.3 The autoregressive moving average model

The ARMA model is a combination of the AR and MA. The value of the random variable is not only related to the previous sequence value but also related to the previous random disturbance. The ARMA (p, q) model can be expressed as follows [78]:

$$x_t = \varphi_0 + \varphi_1 x_{t-1} + \varphi_2 x_{t-2} + \ldots + \varphi_p x_{t-p} + \varepsilon_t - \phi_1\varepsilon_{t-1} - \phi_2\varepsilon_{t-2} - \ldots - \phi_q\varepsilon_{t-q} \tag{3.18}$$

where φ_0 is a constant, $\varphi_1, \varphi_2, \varphi_3, \ldots, \varphi_p$ are the autoregressive coefficients that need to be calculated using the regression algorithm, $\phi_1, \phi_2, \phi_3, \ldots, \phi_q$ are the moving average regression coefficients that need to be calculated using the regression algorithm, and $\varepsilon_t, \varepsilon_{t-1}, \varepsilon_{t-2}, \ldots, \varepsilon_{t-q}$ are the random disturbance in the corresponding period.

3.5.1.1.4 The autoregressive integrated moving average model

The above models can just deal with stationary time series, but most time series are not stationary. As for the nonstationary time series, the time series should be smoothed by difference before using the above models. First, the ARIMA performs d—order difference on the time series, and then applies the ARMA model. The ARIMA (p, d, q) model can be expressed as follows [78]:

$$\left\{ \begin{array}{l} \Phi(B)\nabla^d X_t = \Theta(B)\varepsilon_t \\ \Phi(B) = 1 - f_1 B - L - f_p B^P \\ \Theta(B) = 1 - \theta_1 B - L - \theta_q B^q \end{array} \right\} \tag{3.19}$$

where $\nabla X_t = (1 - B)^d$ represents the ARIMA as a d-order difference model, $\Phi(B)$ represents the p-order autoregressive coefficient polynomial of the ARIMA (p, d, q), $\Theta(B)$ represents the q-order moving average coefficient polynomial of the ARIMA (p, d, q), and ε_t represents independent disturbance or random error.

3.5.1.2 Modeling steps

After the ACF and PACF in Section 3.3, the ARI model is chosen for wind speed and wind direction description. In the modeling steps of the ARIMA model, the specific autoregressive polynomial degree is further determined by the BIC method. The modeling steps are given in Fig. 3.20 in detail.

3.5.1.2.1 Wind speed ARIMA description model

Since the degree of difference d has been determined as 1 and the model has been determined as an *ARI* model in Section 3.3, the wind speed ARIMA model is set as $ARIMA(p, 1, 0)$. The autoregressive polynomial degree p is determined by BIC from 1 to 4. Different BIC results of the corresponding autoregressive polynomial degree p are given in Table 3.2. The best BIC results is marked in bold.

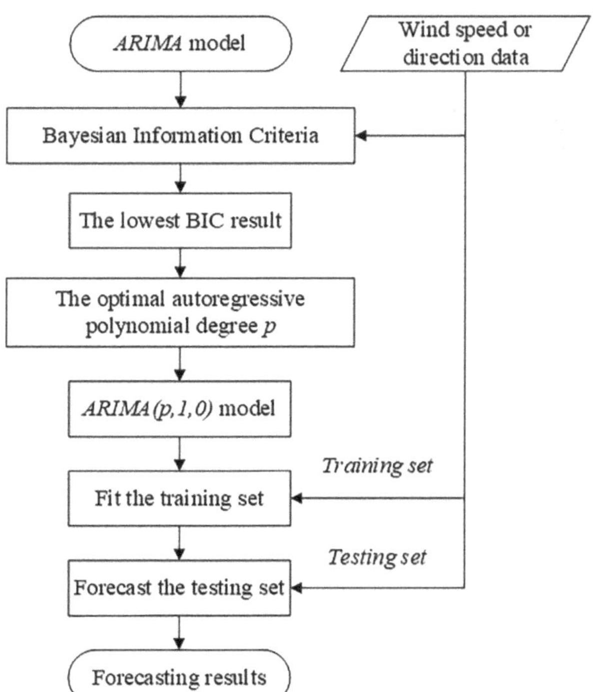

Figure 3.20 The modeling steps of ARIMA models.

Table 3.2 BIC results of different wind speed ARIMA description models.

The autoregressive polynomial degree p	The ARIMA model expression	BIC results
1	$ARIMA(1,1,0)$	**2624.56**
2	$ARIMA(2,1,0)$	3815.35
3	$ARIMA(3,1,0)$	4031.49
4	$ARIMA(4,1,0)$	4084.77

It can be seen from Table 3.2 that the ARIMA model gets the lowest BIC result when the autoregressive polynomial degree is 1. Therefore, an $ARIMA(1,1,0)$ model is built for wind speed description. The estimated formula of the wind speed $ARIMA(1,1,0)$ model is shown as follows:

$$x_t = -0.0057 + x_{t-1} + 0.1585(x_{t-1} - x_{t-2}) + \varepsilon_t \tag{3.20}$$

where ε_t is a series of Gaussian random variables with mean 0 and variance 0.2751.

3.5.1.2.2 Wind direction ARIMA description model

Similarly, the BIC method is used to determine the autoregressive polynomial degree p of wind direction ARIMA description model. Different BIC results of the corresponding autoregressive polynomial degree p are given in Table 3.3. The best BIC results is marked in bold.

It can be seen from Table 3.3 that the ARIMA model gets the lowest BIC result when the autoregressive polynomial degree is 1. Therefore, an $ARIMA(1,1,0)$ model is also built for wind direction description. Meanwhile, it can also be seen that the BIC results of wind direction are much larger than those of the wind speed. This is caused by different changing

Table 3.3 BIC results of different wind direction ARIMA description models.

The autoregressive polynomial degree p	The ARIMA model expression	BIC results
1	$ARIMA(1,1,0)$	**8854.61**
2	$ARIMA(2,1,0)$	10171.39
3	$ARIMA(3,1,0)$	10564.24
4	$ARIMA(4,1,0)$	10764.34

ranges of wind speed data and wind direction data. The estimated formula of the wind direction $ARIMA(1,1,0)$ model is shown as follows:

$$x_t = -0.0295 + x_{t-1} + 0.1876(x_{t-1} - x_{t-2}) + \varepsilon_t \tag{3.21}$$

where ε_t is a series of Gaussian random variables with mean 0 and variance 12.5473.

3.5.1.3 Description results

3.5.1.3.1 Description results of wind speed ARIMA model

Description results of the wind speed ARIMA model are all given in Fig. 3.21.

As shown in Fig. 3.21, the $ARIMA(1,1,0)$ model can fit the wind speed data series accurately. However, the description data points of the model are usually hysteretic when comparing to the corresponding real data points.

3.5.1.3.2 Description results of wind direction ARIMA model

Description results of the wind direction ARIMA model are all given in Fig. 3.22.

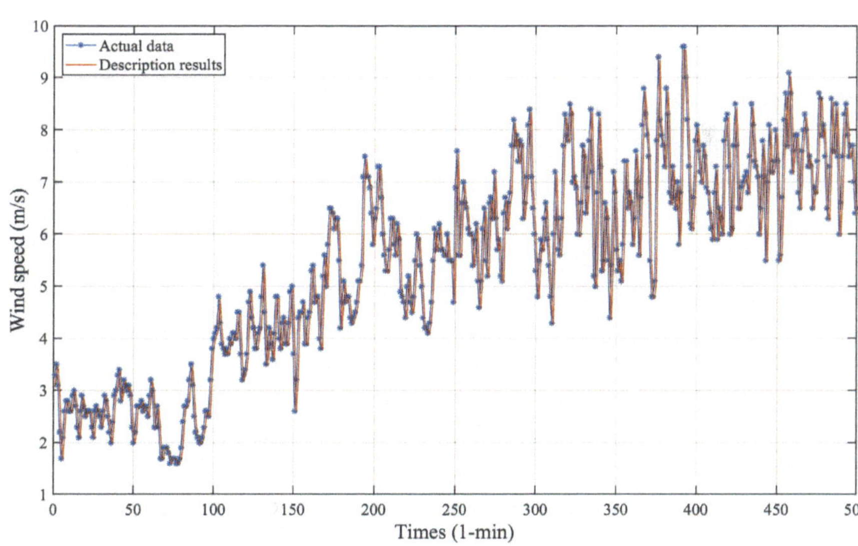

Figure 3.21 Description results of wind speed ARIMA model.

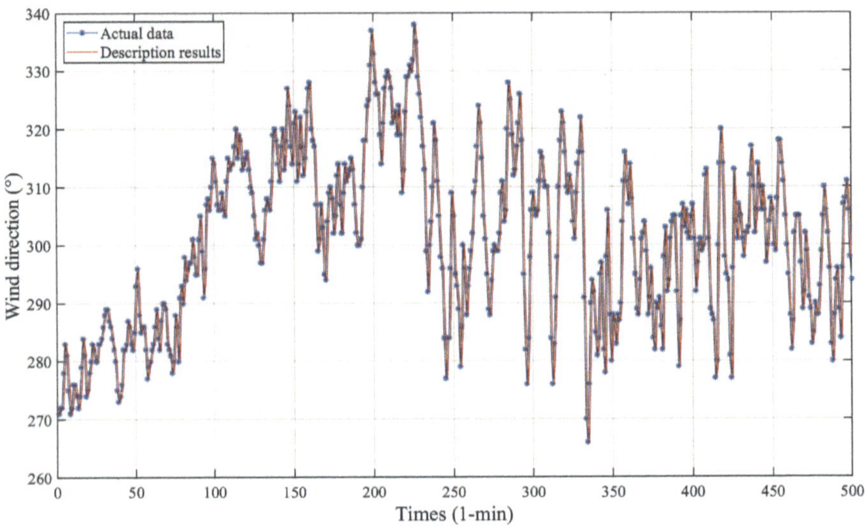

Figure 3.22 Description results of wind direction ARIMA model.

Similarly, the $ARIMA(1,1,0)$ model can fit the wind direction data series accurately. However, the description data points of the model are usually hysteretic when comparing to the corresponding real wind direction data points.

3.5.2 Seasonal autoregressive integrated moving average

The SARIMA is an improvement on the classical ARIMA, which introduces the seasonal item [79,80]. The SARIMA model includes seasonal items and nonseasonal items, where seasonal items describe the periodic seasonal performance in the sequence and nonseasonal items describe the random fluctuation performance in the sequence. According to the different difficulty of extracting seasonal effects of time series, there are the simple seasonal ARIMA model and the multiplicative seasonal ARIMA model. The SARIMA generally refers to the multiplicative seasonal ARIMA model, in which the seasonal and nonseasonal items of the model have a product relationship.

For the time series with a complex correlation between seasonal effects and other effects, the multiplicative seasonal ARIMA model is more suitable than the simple seasonal ARIMA model. It is because the simple additive seasonal model cannot effectively and fully extract the seasonal effects of the time series.

The SARIMA has also been widely used in wind description and has achieved good results. Alencar et al. combined the SARIMA and neural networks for wind speed. This method can learn the behavior of both linear and nonlinear systems [81]. Guo et al. proposed a hybrid model combining the SARIMA and the Least Square Support Vector Machine (LSSVM) algorithm to predict the monthly wind speed, and compared its effectiveness with that of the ARIMA [82]. Wang et al. used the Extreme Learning Machine (ELM) algorithm combined with the Ljung–Box Q-test (LBQ) and the SARIMA to predict the daily and monthly wind speed in northwestern China. The experimental results showed that the performance of this method is better than that of a single model [83].

3.5.2.1 Theoretical basis

The SARIMA uses the product method to deal with the complex relationship between different effect components, so the essence of the SARIMA model fitting time series is the product of $ARIMA(p, d, q)$ and $ARIMA(P, D, Q)_s$. The simple representation of the SARIMA is as follows [84]:

$$\mathrm{SARIMA}(p, d, q) \times (P, D, Q)_s \tag{3.22}$$

The complete structure of the SARIMA model is as follows [84]:

$$\begin{cases} \Phi(B)\Phi_S(B)\nabla_S^D\nabla^d x_i = \Theta(B)\Theta_S(B)\varepsilon_t \\ \Phi(B) = 1 - \phi_1 B - \ldots\phi_p B^p \\ \Phi(B) = 1 - \theta_1 B - \ldots\theta_p B^q \\ \Phi_S(B) = 1 - \phi_1 B^S - \ldots\phi_P B^{PS} \\ \Theta_S(B) = 1 - \theta_1 B^S - \ldots\theta_Q B^{QS} \end{cases} \tag{3.23}$$

where S is the period step used to extract seasonal effect information, d represents $d-$order difference of the ARIMA (p, d, q), $\Phi(B)$ represents the $p-$order autoregressive coefficient polynomial of the ARIMA (p, d, q), $\Theta(B)$ represents the $q-$order moving average coefficient polynomial of the ARIMA (p, d, q), $\Phi_S(B)$ is the $P-$order seasonal autoregressive coefficient polynomial of the ARIMA $(P, D, Q)_S$, $\Theta_S(B)$ is the $Q-$order seasonal moving average coefficient polynomial of the ARIMA $(P, D, Q)_S$, ε_t represents independent interference or random error, ∇^d represents $d-$order nonseasonal difference operator and its expression is $\nabla^d = (1 - B)^d$, and

∇_D^S is the $D-$order seasonal difference operator with S as the period and its expression is $\nabla_S^D = (1 - B^S)^D$.

3.5.2.2 Modeling steps

After the ACF and PACF in Section 3.3, the SARI model is chosen for wind speed and wind direction description. In the modeling steps of the SARIMA model, the specific autoregressive polynomial degree p and seasonal autoregressive polynomial degree P are further determined by the BIC method. The modeling steps are given in Fig. 3.23 in detail.

3.5.2.2.1 Wind speed SARIMA description model

Since the degree of difference d has been determined as 1 and the seasonal cycle of wind speed data has been determined as 214 in Section 3.3, the wind speed SARIMA model is set as $SARIMA(p, 1, 0) \times (P, 1, 0)_{214}$. The autoregressive polynomial degree p and seasonal autoregressive polynomial degree P are both determined by the BIC from 1 to 4. Different BIC results are given in Table 3.4. The best BIC results is marked in bold.

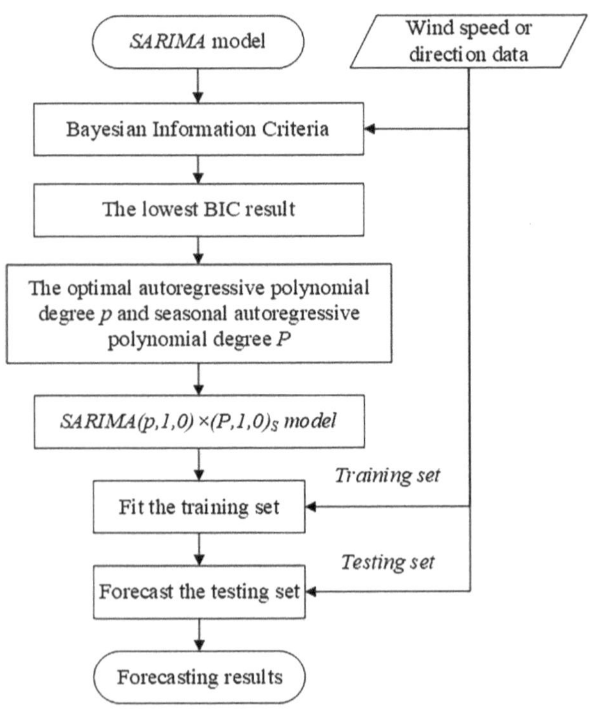

Figure 3.23 The modeling steps of SARIMA models.

Table 3.4 BIC results of different wind speed SARIMA description models.

The autoregressive polynomial degree *p*	The seasonal autoregressive polynomial degree *P*	The SARIMA model expression	BIC results
1	1	$SARIMA(1,1,0) \times (1,1,0)_{214}$	3206.97
	2	$SARIMA(1,1,0) \times (2,1,0)_{214}$	2847.08
	3	$SARIMA(1,1,0) \times (3,1,0)_{214}$	3215.44
	4	$SARIMA(1,1,0) \times (4,1,0)_{214}$	3222.37
2	1	$SARIMA(2,1,0) \times (1,1,0)_{214}$	2847.08
	2	$SARIMA(2,1,0) \times (2,1,0)_{214}$	2941.82
	3	$SARIMA(2,1,0) \times (3,1,0)_{214}$	2961.87
	4	$SARIMA(2,1,0) \times (4,1,0)_{214}$	**2829.24**
3	1	$SARIMA(3,1,0) \times (1,1,0)_{214}$	3215.44
	2	$SARIMA(3,1,0) \times (2,1,0)_{214}$	2961.87
	3	$SARIMA(3,1,0) \times (3,1,0)_{214}$	3243.98
	4	$SARIMA(3,1,0) \times (4,1,0)_{214}$	3243.89
4	1	$SARIMA(4,1,0) \times (1,1,0)_{214}$	3222.37
	2	$SARIMA(4,1,0) \times (2,1,0)_{214}$	2829.24
	3	$SARIMA(4,1,0) \times (3,1,0)_{214}$	3243.89
	4	$SARIMA(4,1,0) \times (4,1,0)_{214}$	3271.88

It can be seen from Table 3.4 that the SARIMA model gets the lowest BIC result at the autoregressive polynomial degree 2 and seasonal autoregressive polynomial degree 4. Therefore, a $SARIMA(2,1,0) \times (4,1,0)_{214}$ model is built for seasonal wind speed description. The estimated formula of the seasonal wind speed model is shown as follows:

$$x_t^* = 0.006 - 0.5573x_{t-2}^* - 0.3411x_{t-217}^* - 0.1901x_{t-219}^* + \varepsilon_t \quad (3.24)$$

where ε_t is a series of Gaussian random variables with mean 0 and variance 0.3338, and x_t^* is derived from x_t by taking the first difference and then 214th seasonal differences.

3.5.2.2.2 Wind direction SARIMA description model

Since the degree of difference d has been determined as 1 and the seasonal cycle of wind direction data has been determined as 375 in Section 3.3, the wind direction SARIMA model is set as $SARIMA(p,1,0) \times (P,1,0)_{375}$. The autoregressive polynomial degree p and seasonal autoregressive polynomial degree P are both determined by the BIC from 1 to 4. Different BIC results are given in Table 3.5. The best BIC results is marked in bold.

Table 3.5 BIC results of different wind direction SARIMA description models.

The autoregressive polynomial degree *p*	The seasonal autoregressive polynomial degree *P*	The SARIMA model expression	BIC results
1	1	$SARIMA(1,1,0) \times (1,1,0)_{375}$	9097.75
	2	$SARIMA(1,1,0) \times (2,1,0)_{375}$	**8786.47**
	3	$SARIMA(1,1,0) \times (3,1,0)_{375}$	9120.15
	4	$SARIMA(1,1,0) \times (4,1,0)_{375}$	9120.37
2	1	$SARIMA(2,1,0) \times (1,1,0)_{375}$	8786.47
	2	$SARIMA(2,1,0) \times (2,1,0)_{375}$	9001.62
	3	$SARIMA(2,1,0) \times (3,1,0)_{375}$	9009.78
	4	$SARIMA(2,1,0) \times (4,1,0)_{375}$	8944.37
3	1	$SARIMA(3,1,0) \times (1,1,0)_{375}$	9120.15
	2	$SARIMA(3,1,0) \times (2,1,0)_{375}$	9009.78
	3	$SARIMA(3,1,0) \times (3,1,0)_{375}$	9191.92
	4	$SARIMA(3,1,0) \times (4,1,0)_{375}$	9192.61
4	1	$SARIMA(4,1,0) \times (1,1,0)_{375}$	9120.37
	2	$SARIMA(4,1,0) \times (2,1,0)_{375}$	8944.37
	3	$SARIMA(4,1,0) \times (3,1,0)_{375}$	9192.61
	4	$SARIMA(4,1,0) \times (4,1,0)_{375}$	9207.99

It can be seen from Table 3.5 that the SARIMA model gets the lowest BIC result at the autoregressive polynomial degree 1 and seasonal autoregressive polynomial degree 2. Therefore, a $SARIMA(1,1,0) \times (2,1,0)_{375}$ model is built for seasonal wind direction description. The estimated formula of the seasonal wind direction model is shown as follows:

$$x_t^* = 0.0307 + 0.2259x_{t-1}^* - 0.4981x_{t-376}^* + 0.1125x_{t-377}^* + \varepsilon_t \quad (3.25)$$

where ε_t is a series of Gaussian random variables with mean 0 and variance 8.1457, and x_t^* is derived from x_t by taking the first difference and then 375th seasonal differences.

3.5.2.3 Description results

3.5.2.3.1 Description results of wind speed SARIMA model

Description results of wind speed SARIMA model are all given in Fig. 3.24.

As shown in Fig. 3.24, the $SARIMA(2,1,0) \times (4,1,0)_{214}$ can describe the basic trend of wind speed data series. It can also be seen that the $SARIMA(2,1,0) \times (4,1,0)_{214}$ cannot fit the data series as accurately as the

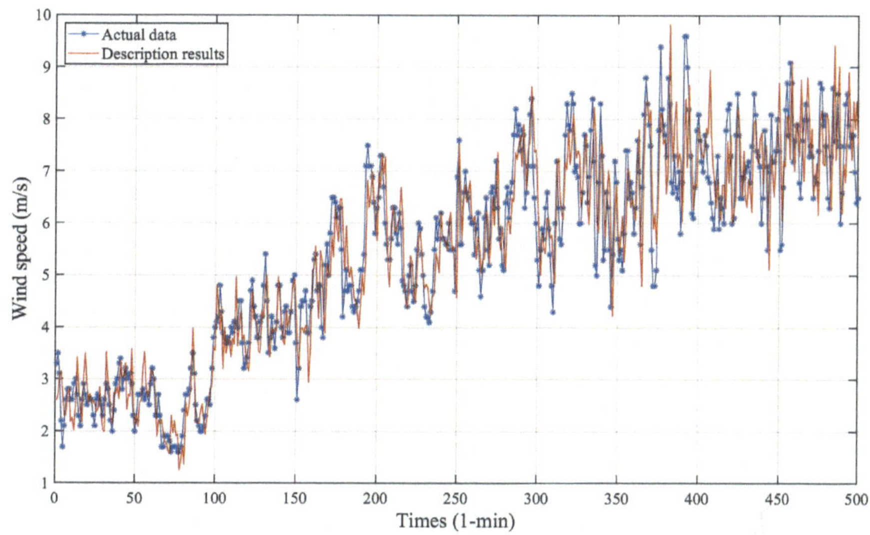

Figure 3.24 Description results of wind speed SARIMA model.

$ARIMA(1,1,0)$. However, the hysteresis is solved in the $SARIMA(2,1,0) \times (4,1,0)_{214}$ model. To analyze the performance of the models more quantitatively, several accuracy evaluation indicators are proposed in Section 3.6.

3.5.2.3.2 Description results of wind direction SARIMA model

Description results of wind direction SARIMA model are all given in Fig. 3.25.

Similarly, it can be seen from Fig. 3.25 that the $SARIMA(1,1,0)\times(2,1,0)_{375}$ can describe the basic trend of wind direction data series. The SARIMA models have satisfactory performance. Besides, several accuracy evaluation indicators are proposed in Section 3.6.1 to analyze the performance of the models quantitatively.

3.5.3 Autoregressive conditional heteroscedasticity model

The Autoregressive Conditionally Heteroscedastic (ARCH) model was first proposed by Engel, which was first applied in the field of economics [85]. For some time series, the data show the characteristics of volatility clustering. In the long term, the data are stable. It means that the long-term variance is constant, which is also called the unconditional variance. But in the short run, the variance is unstable, which is called conditional

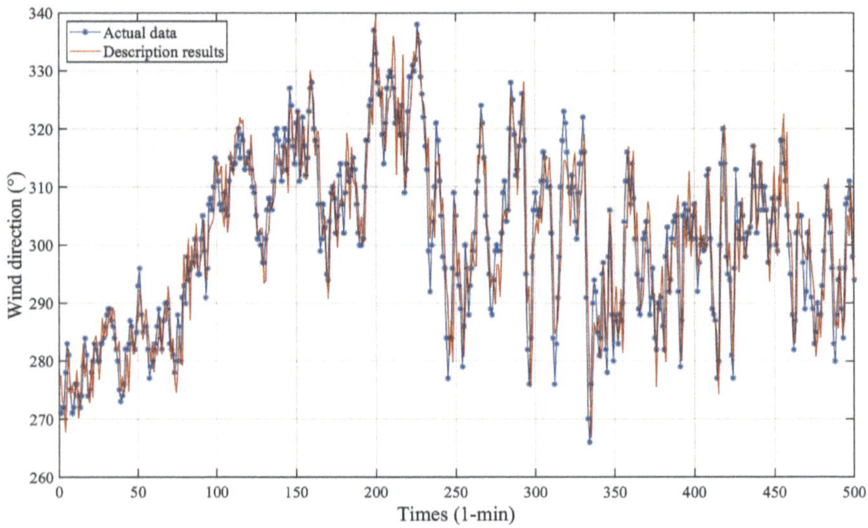

Figure 3.25 Description results of wind direction SARIMA model.

heteroscedasticity. Traditional time models such as the ARMA cannot recognize this feature well, the ARCH can solve this problem.

The ARCH describes the variation of the sequence using autocorrelation between variances. It determines that the series value fluctuation of the model is related to the historical fluctuation and uses the variance of the current data to reflect the future fluctuation [86].

Some scholars have applied the ARCH to model random disturbance of time series models. Gao et al. combined the ARMA with the ARCH model for wind speed. Compared with the single ARMA model, they concluded that the performance of the model combined with the ARCH was better [87]. Lv et al. proposed a hybrid model combining wavelet, the ARIMA, and the ARCH. The hybrid model overcomes the deficiency of the traditional ARIMA model and greatly improves the accuracy [88]. Wang et al. also considered the heteroscedasticity effect of wind speed and combined the ARIMA and ARCH for achieving good performance [89].

3.5.3.1 Theoretical basis

The conditional variance of the ARCH model follows the law of the moving average. It can derive different conditional variances based on historical moment information. The ARCH model can be expressed as follows [90]:

$$\begin{cases} x_t = f(t, x_{t-1}, x_{t-2}, \ldots) + \varepsilon_t \\ \varepsilon_t = \sigma_t e_t \\ \sigma_t^2 = w + \sum_{j=1}^{q} \lambda_j \varepsilon_{t-j}^2 \end{cases} \tag{3.26}$$

where σ_t represents the conditional variance, e_t represents an independently distributed sequence of white noise, and ε_t is the predicted residual.

3.5.3.2 Modeling steps

The modeling of the ARCH model is based on the description residuals of the ARIMA and SARIMA. The wind speed and wind direction description residuals of the ARIMA and SARIMA are shown in Fig. 3.26.

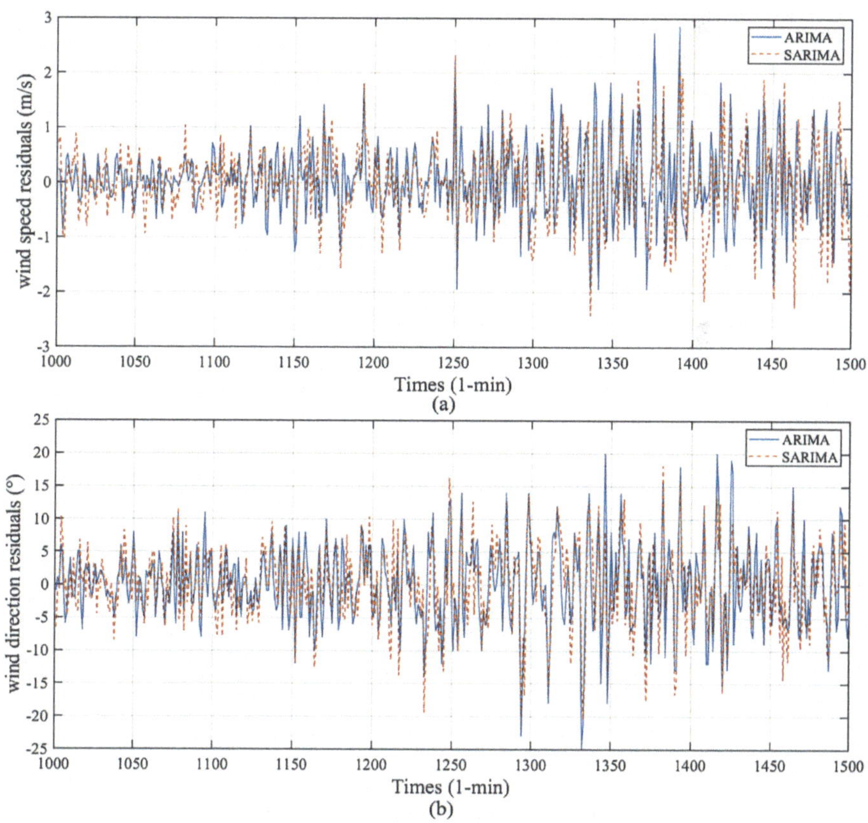

Figure 3.26 The wind speed and wind direction description residuals of the ARIMA and SARIMA: (A) wind speed residuals and (B) wind direction residuals.

By computing the unconditional variances, the description interval of the wind speed and wind direction can be calculated. Applying the Kernel Density Estimation (KDE) to fit the distribution, the fitted unconditional distributions are shown in Fig. 3.27. The description interval can be extracted from the distributions, and applied for the homoscedastic description.

The ARCH polynomial degree should be carefully selected to ensure accuracy. The BIC criterion is applied to measure fitness. Setting the searching area ranges from 1 to 4, the BIC values with the different ARCH polynomial degrees for wind speed and wind direction are shown in Fig. 3.28. From Fig. 3.28, it can be observed that the best order of the ARCH with the ARIMA for wind direction data is 4, the best order of the ARCH with the ARIMA for wind speed data is 2, and the best orders of the ARCH with the SARIMA for wind speed and direction data are all 1.

Given the selected model orders, the estimated formula of the ARCH for wind speed with the ARIMA is shown as follows:

$$\begin{cases} x_t = f + \varepsilon_t \\ \varepsilon_t = \sigma_t e_t \\ \sigma_t^2 = 0.1284 + 0.0940\sigma_{t-1}^2 + 0.2968\sigma_{t-2}^2 + 0.0817\sigma_{t-3}^2 + 0.0818\sigma_{t-4}^2 \end{cases} \tag{3.27}$$

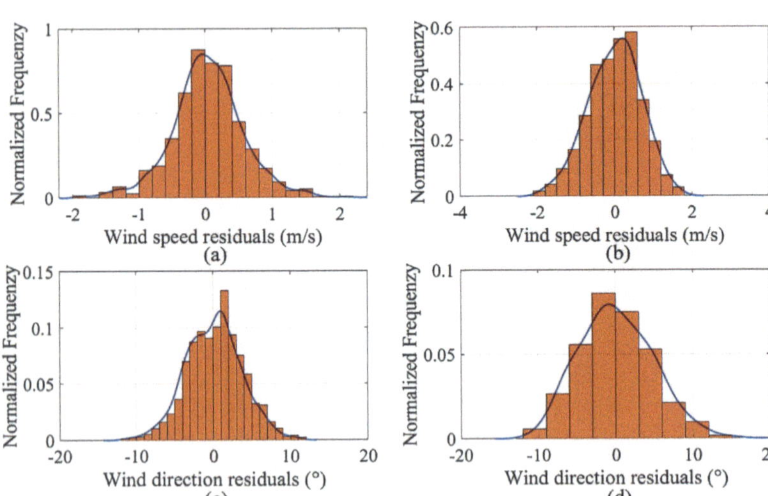

Figure 3.27 The unconditional distributions: (A) wind speed with ARIMA, (B) wind speed with SARIMA, (C) wind direction with ARIMA, and (D) wind direction with SARIMA.

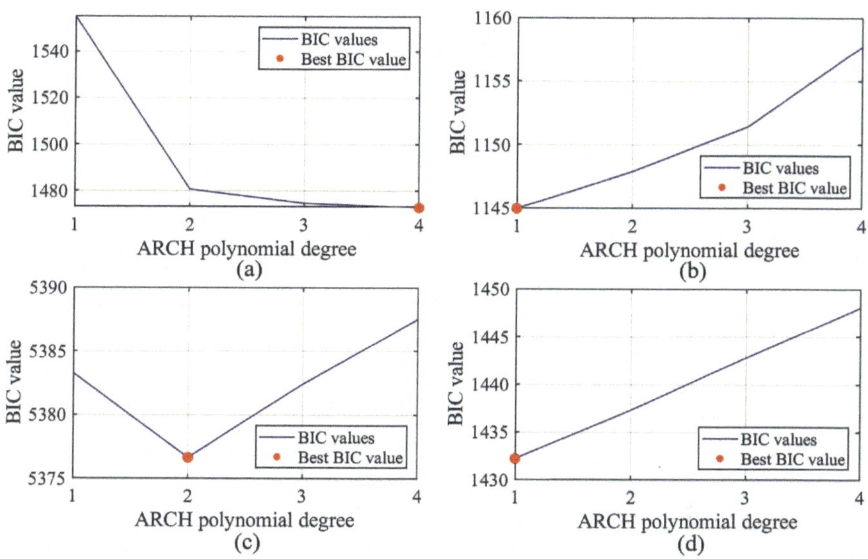

Figure 3.28 The BIC values with different ARCH polynomial degrees: (A) wind speed with ARIMA, (B) wind speed with SARIMA, (C) wind direction with ARIMA, and (D) wind direction with SARIMA.

The estimated formula of the ARCH for wind speed with the SARIMA is shown as follows:

$$\begin{cases} x_t = f + \varepsilon_t \\ \varepsilon_t = \sigma_t e_t \\ \sigma_t^2 = 0.3966 + 0.1371\sigma_{t-1}^2 \end{cases} \tag{3.28}$$

The estimated formula of the ARCH for wind direction with the ARIMA is shown as follows:

$$\begin{cases} x_t = f + \varepsilon_t \\ \varepsilon_t = \sigma_t e_t \\ \sigma_t^2 = 9.8619 + 0.1185\sigma_{t-1}^2 + 0.1302\sigma_{t-2}^2 \end{cases} \tag{3.29}$$

The estimated formula of the ARCH for wind direction with the SARIMA is shown as follows:

$$\begin{cases} x_t = f + \varepsilon_t \\ \varepsilon_t = \sigma_t e_t \\ \sigma_t^2 = 20.0222 + 0.0656\sigma_{t-1}^2 \end{cases} \tag{3.30}$$

3.5.3.3 Description results

3.5.3.3.1 Description results of wind speed ARCH model

Description results of the wind speed ARIMA-ARCH and SARIMA-ARCH model are all given in Figs. 3.29 and 3.30.

It can be seen from Fig. 3.29 that all actual wind speed data points are covered by the description interval. It proves that the ARIMA-ARCH model has satisfactory performance in probabilistic wind speed description.

Similarly, it can be seen from Fig. 3.30 that all actual wind speed data points are covered by the description interval of the SARIMA-ARCH model. It proves that the SARIMA-ARCH model also has satisfactory performance in probabilistic wind speed description. However, the main probabilistic description results of the SARIMA-ARCH model concentrate on the significance level $\alpha = 0.05$. As a contrast, the confidence interval of the SARIMA-ARCH model is wider than that of the ARIMA-ARCH model. It means that the ARIMA-ARCH model can outperform the SARIMA-ARCH model in probabilistic wind speed description, which is caused by the better performance of the ARIMA model in deterministic wind speed description.

3.5.3.3.2 Description results of wind direction ARCH model

Description results of the wind direction ARIMA-ARCH and SARIMA-ARCH model are given in Figs. 3.31 and 3.32.

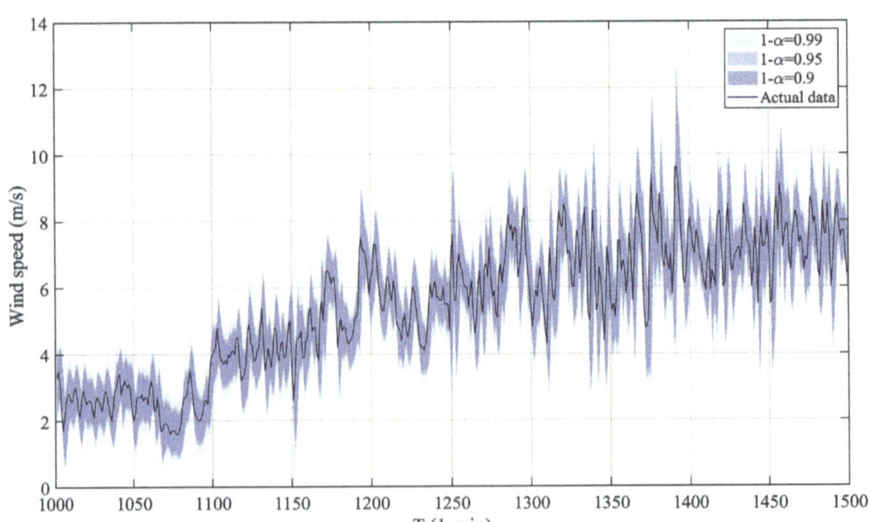

Figure 3.29 Description results of wind speed ARIMA-ARCH model.

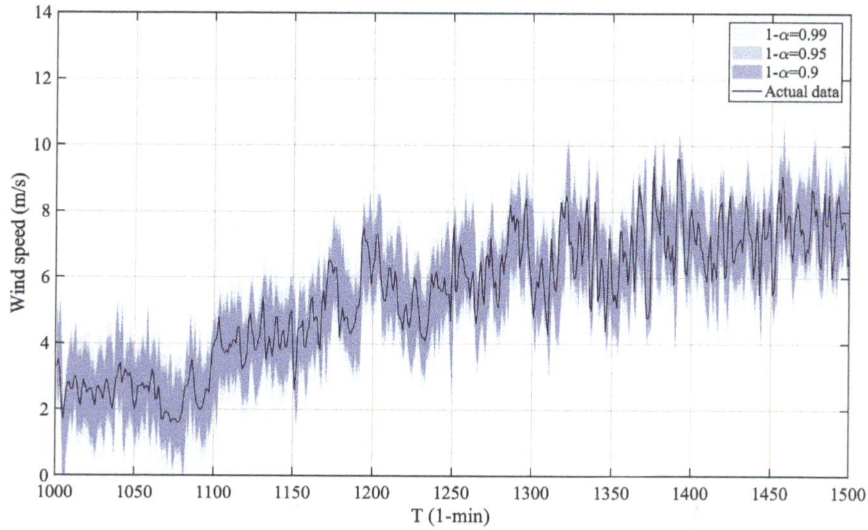

Figure 3.30 Description results of wind speed SARIMA-ARCH model.

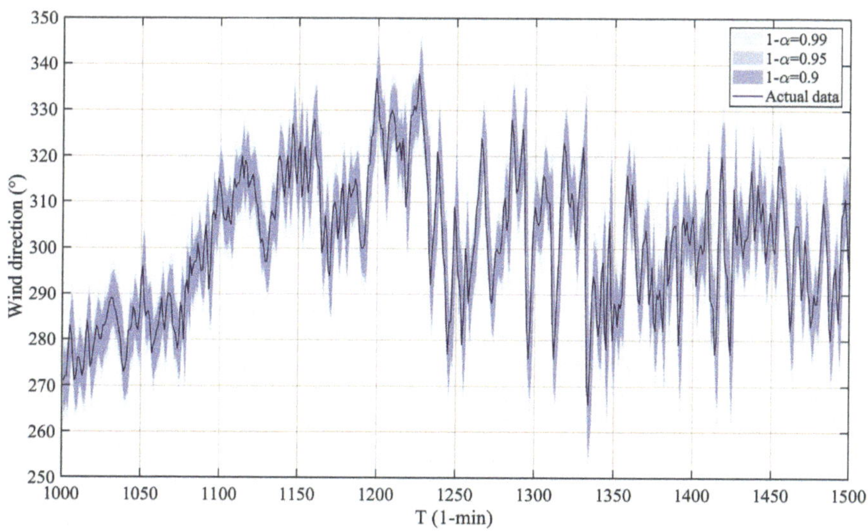

Figure 3.31 Description results of wind direction ARIMA-ARCH model.

As shown in Fig. 3.31, all wind direction data points are covered by the description interval of the ARIMA-ARCH model. Besides, the confidence interval is very narrow. It shows the good performance of the ARIMA-ARCH model in probabilistic wind direction description. Main probabilistic description results concentrate on the significance level $\alpha = 0.1$.

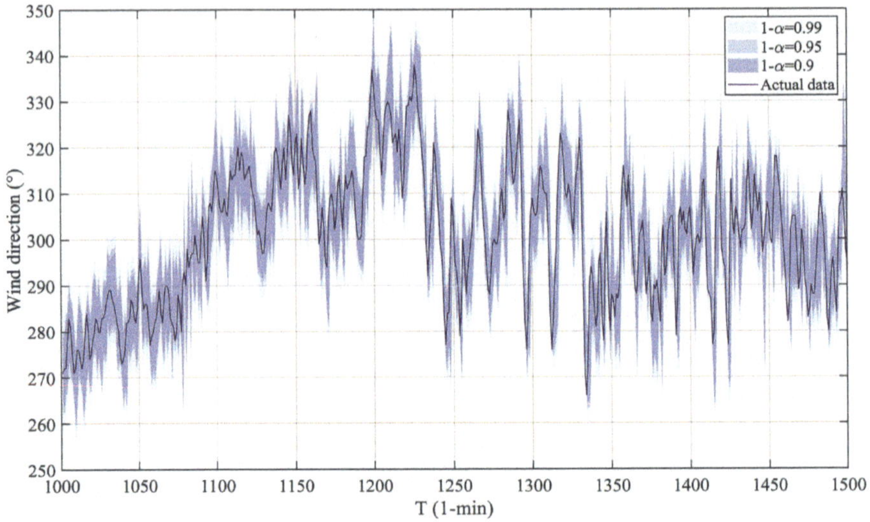

Figure 3.32 Description results of wind direction SARIMA-ARCH model.

It can be seen from Fig. 3.32 that the confidence interval of the SARIMA-ARCH model covers all actual wind direction data points. It shows the satisfactory performance of the SARIMA-ARCH model in probabilistic wind direction description. However, the main probabilistic description results of the SARIMA-ARCH model concentrate on the significance level $\alpha = 0.05$. The difference between the confidence intervals of ARIMA-ARCH and SARIMA-ARCH is not obvious from Figs. 3.31 and 3.32.

3.5.4 Generalized autoregressive conditionally heteroscedastic model

The ARCH model only describes the moving average characteristic of conditional variance and only applies to the heteroscedasticity process with short-term autoregressive characteristics. When the ARCH is used to fit the heteroscedasticity process with long-term autoregressive characteristics, the moving average coefficient will be very high, which will affect the model accuracy. The Generalized Autoregressive Conditionally Heteroscedastic (GARCH) model can effectively fit the heteroscedasticity function that has long-term memory.

The GARCH model describes the volatility and provides an effective analysis method for the time series. It is the most commonly used heteroscedasticity model for series fitting. But it is very restrictive on parameters. The unconditional variance must satisfy the requirement of nonnegative, and the conditional variance must satisfy the requirement of stationary.

As an effective heteroscedasticity model, the GRACH model has been widely used in many fields. Niu et al. proposed a model combining the ARMA, GRACH, ANN, and SVM to predict short-term power load [91]. Morana used the GRACH to predict the distribution of oil prices over a short period [92]. Fang et al. used a hybrid model combined with the GRACH to predict future gold market volatility [93].

To expand the application scope and improve the performance of the GARCH model, relevant scholars built several GARCH models derived models from different perspectives, such as the Integrated GARCH (IGARCH) model [94], the Fractionally Integrated GARCH (FIGARCH) model [95], and the Exponential GARCH (EGARCH) model [96].

3.5.4.1 Theoretical basis

Based on the ARCH, the GARCH model adds the description of the autoregressive relation of conditional variance. The GARCH model can be expressed as follows [97]:

$$\begin{cases} x_t = f(t, x_{t-1}, x_{t-2}, \cdots) + \varepsilon_t \\ \varepsilon_t = \sigma_t e_t \\ \sigma_t^2 = w + \sum_{i=1}^{p} \eta_j \sigma_{t-i}^2 + \sum_{j=1}^{q} \lambda_j \varepsilon_{t-j}^2 \end{cases} \tag{3.31}$$

where σ_t represents the conditional variance, e_t represents an independently distributed sequence of white noise, and ε_t is the predicted residual.

To ensure the accuracy and stability of the model, the parameters of the GRACH model are required. The requirements for the GARCH model parameters are shown below [97]:

$$\begin{cases} w > 0 \\ \eta_j \geq 0, \lambda_j \geq 0 \\ \sum_{i=1}^{p} \eta_j + \sum_{j=1}^{q} \lambda_j < 1 \end{cases} \tag{3.32}$$

3.5.4.2 Modeling steps

To find the optimal GARCH polynomial degree, the BIC criterion method is also applied to measure the fitness. Setting the searching area ranges from 1 to 4 for both ARCH degree and GARCH degree, the BIC values for wind speed and wind direction are shown in Fig. 3.33.

From Fig. 3.33, it can be observed that the wind speed ARIMA-GARCH model gets the lowest BIC value at the second ARCH degree and first GARCH degree. The wind speed SARIMA-GARCH model, wind direction ARIMA-GARCH model, and wind direction SARIMA-GARCH model all get the lowest BIC value at the first ARCH degree and first GARCH degree. Given the selected model orders, the estimated formula of the GARCH for wind speed with ARIMA is shown as follows:

$$\begin{cases} x_t = f + \varepsilon_t \\ \varepsilon_t = \sigma_t e_t \\ \sigma_t^2 = 0.0552 + 0.0734\sigma_{t-1}^2 + 0.2095\sigma_{t-2}^2 + 0.05229\varepsilon_{t-1}^2 \end{cases} \tag{3.33}$$

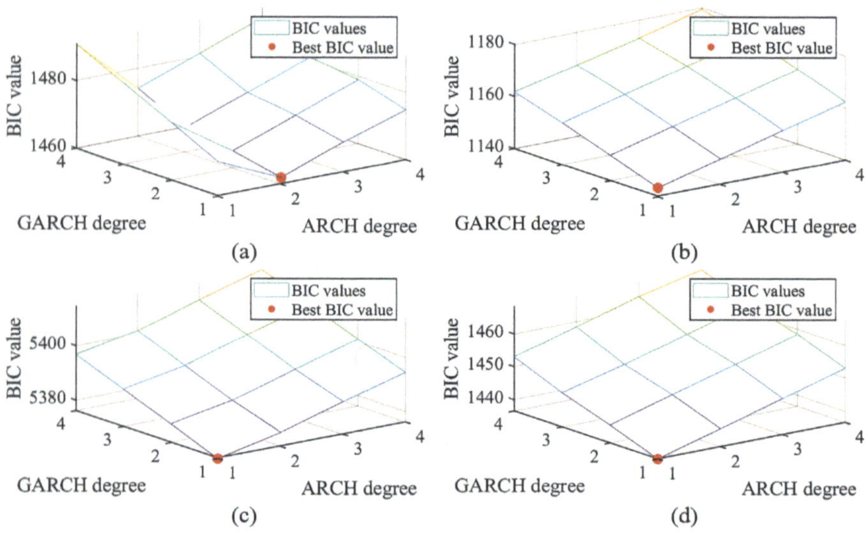

Figure 3.33 The BIC values with different GARCH polynomial degrees: (A) wind speed with ARIMA, (B) wind speed with SARIMA, (C) wind direction with ARIMA, and (D) wind direction with SARIMA.

The estimated formula of the GARCH for wind speed with SARIMA is shown as follows:

$$\begin{cases} x_t = f + \varepsilon_t \\ \varepsilon_t = \sigma_t e_t \\ \sigma_t^2 = 0.0513 + 0.0840\sigma_{t-1}^2 + 0.8025\varepsilon_{t-1}^2 \end{cases} \tag{3.34}$$

The estimated formula of the GARCH for wind direction with ARIMA is shown as follows:

$$\begin{cases} x_t = f + \varepsilon_t \\ \varepsilon_t = \sigma_t e_t \\ \sigma_t^2 = 3.5244 + 0.1434\sigma_{t-1}^2 + 0.5903\varepsilon_{t-1}^2 \end{cases} \tag{3.35}$$

The estimated formula of the GARCH for wind direction with SARIMA is shown as follows:

$$\begin{cases} x_t = f + \varepsilon_t \\ \varepsilon_t = \sigma_t e_t \\ \sigma_t^2 = 2.8576 + 0.0599\sigma_{t-1}^2 + 0.8100\varepsilon_{t-1}^2 \end{cases} \tag{3.36}$$

3.5.4.3 Description results

3.5.4.3.1 Description results of wind speed GARCH model

Description results of the wind speed ARIMA-GARCH and SARIMA-GARCH models are given in Figs. 3.34 and 3.35.

It can be seen from Fig. 3.34 that the wind speed ARIMA-GARCH has a description interval that can cover all wind speed data points. It proves that the ARIMA-GARCH model has satisfactory performance in probabilistic wind speed description. Main probabilistic description results concentrate on the significance level $\alpha = 0.1$.

It can be seen from Fig. 3.35 that all actual wind speed data points are covered by the description interval of the SARIMA-GARCH model. However, the confidence interval of the SARIMA-GARCH model is wider when compared with the ARIMA-GARCH model. It means that

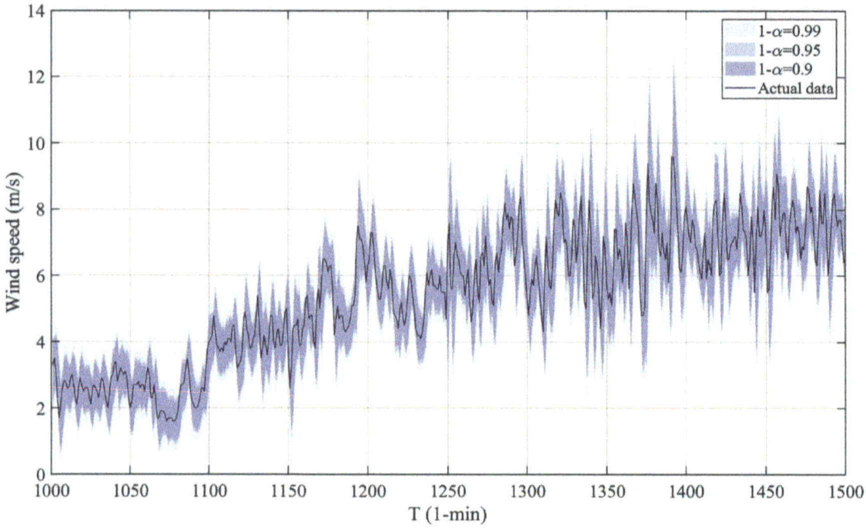

Figure 3.34 Description results of wind speed ARIMA-GARCH model.

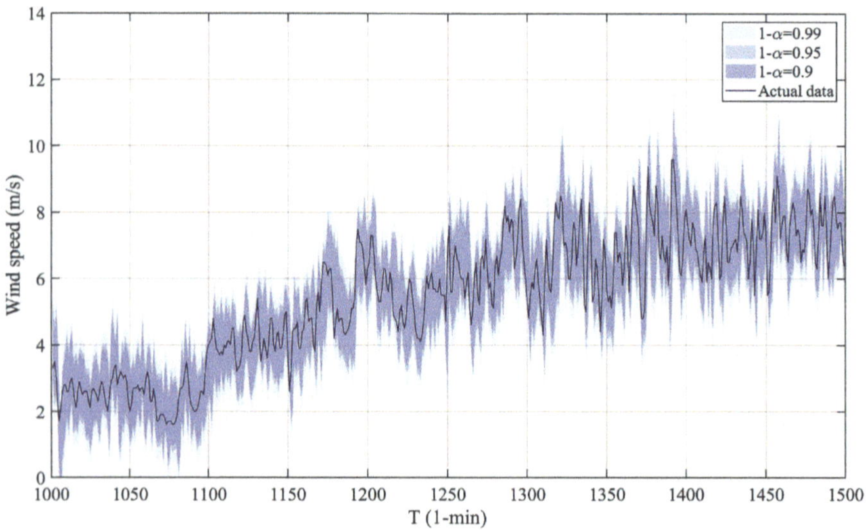

Figure 3.35 Description results of wind speed SARIMA-GARCH model.

the SARIMA-GARCH model performs worse than the ARIMA-GARCH model in probabilistic wind speed description, which may be explained by the better performance of the ARIMA model in deterministic wind speed description. Besides, the main probabilistic description results of the SARIMA-GARCH model concentrate on the significance level $\alpha = 0.05$.

3.5.4.3.2 Description results of wind direction GARCH model

Description results of the wind direction ARIMA-GARCH and SARIMA-GARCH models are all given in Figs. 3.36 and 3.37.

As shown in Figs. 3.36 and 3.37, both the ARIMA-GARCH model and the SARIMA-GARCH model have the confidence interval that covers all wind direction data points. It shows that both models have satisfactory performance in probabilistic wind direction description. The difference between the performance of the ARIMA-GARCH model and that of the SARIMA-GARCH model is not obvious. To analyze the performance more quantitatively, several probabilistic description accuracy evaluation indicators are proposed in Section 3.6.2.

3.6 Description accuracy evaluation indicators

3.6.1 Deterministic description accuracy evaluation indicators

Common deterministic description accuracy evaluation indicators include Mean Absolute Error (MAE), Mean Absolute Percentage Error (MAPE), and Root Mean Squared Error (RMSE).

The MAE is one of the most intuitive evaluation indicators, which measures the average bias between the description value and the actual

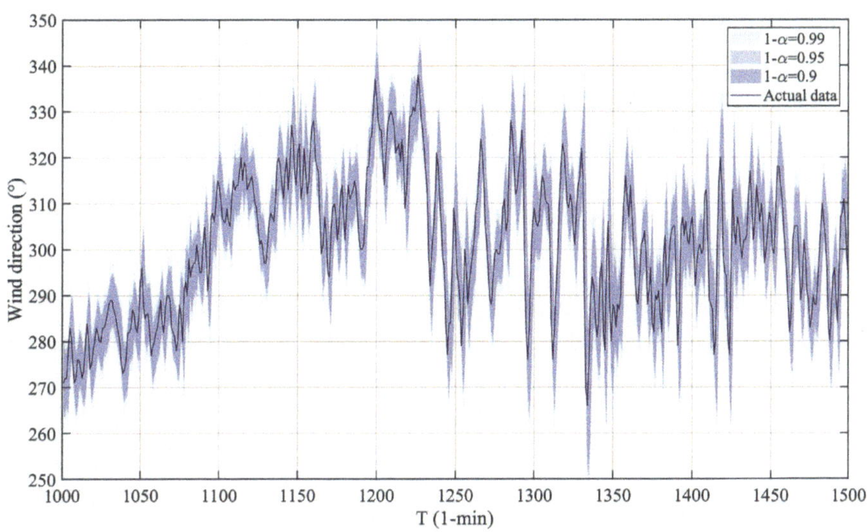

Figure 3.36 Description results of wind direction ARIMA-GARCH model.

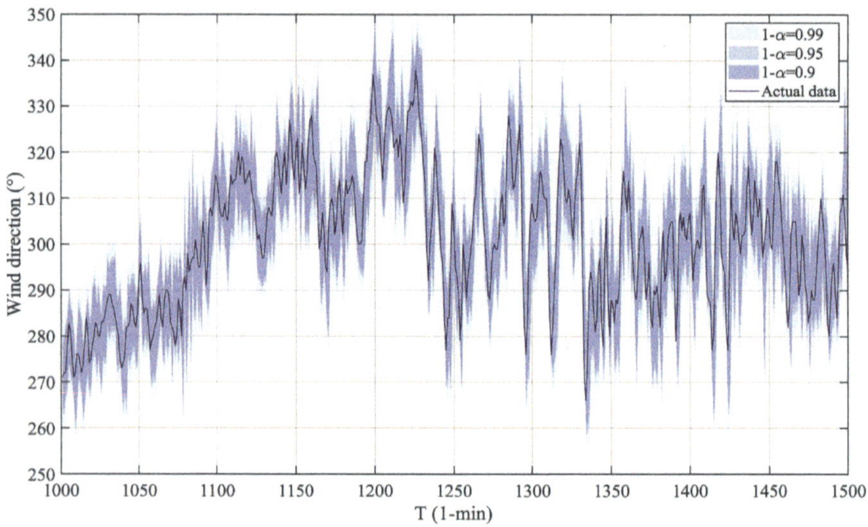

Figure 3.37 Description results of wind direction SARIMA-GARCH model.

value of each test point. The calculation formula of the MAE is as follows [98]:

$$\mathrm{MAE} = \frac{1}{n}\sum_{i=1}^{n}\left|x_i - \widehat{x}_i\right| \tag{3.37}$$

where x_i is the true value of the sample point of the test set, $\widehat{x}_i$ is the value of description output, and n is the number of sample points in the test set.

The MAPE reflects the average of the percentage of relative errors compared with the true value, and is a relative error evaluation indicator. The calculation formula of the MAPE is as follows [98]:

$$\mathrm{MAPE} = \frac{100\%}{n}\sum_{i=1}^{n}\left|\frac{x_i - \widehat{x}_i}{x_i}\right| \tag{3.38}$$

The RMSE is used to calculate the root mean square of the error. Compared with the MAE, it is more sensitive to large errors and small errors. The calculation formula is as follows [98]:

$$\mathrm{RMSE} = \sqrt{\frac{1}{n}\sum_{i=1}^{n}\left(x_i - \widehat{x}_i\right)^2} \tag{3.39}$$

3.6.1.1 Deterministic wind speed description results analysis

To quantitatively analyze the performance of the ARIMA model and SARIMA model in deterministic wind speed description, the deterministic description accuracy evaluation indicators are utilized to analyze the description results in Section 3.5. The deterministic description accuracy evaluation indicators of these models are given in Table 3.6 after calculation.

It can be seen from Table 3.6 that both the $ARIMA(1,1,0)$ and $SARIMA(2,1,0)\times(4,1,0)_{214}$ have satisfactory performance in deterministic wind speed description. All accuracy evaluation indicators are relatively low. It proves that both ARIMA and SARIMA methods are valid in deterministic wind speed description.

It can also be seen that the $ARIMA(1,1,0)$ can outperform the $SARIMA(2,1,0)\times(4,1,0)_{214}$ in deterministic wind speed description. However, changing percentages are not obvious. The phenomenon can be explained by that the seasonality of the wind speed data series is not obvious. Therefore, the ARIMA method has a better performance in wind speed description.

3.6.1.2 Deterministic wind direction description results analysis

To quantitatively analyze the performance of the ARIMA model and SARIMA model in deterministic wind direction description, the deterministic description accuracy evaluation indicators are utilized to analyze the description results in Section 3.5. The deterministic description accuracy evaluation indicators of these models are given in Table 3.7 after calculation.

Table 3.6 Deterministic wind speed description accuracy evaluation indicators.

	Evaluation indicators		
Models	**MAE (m/s)**	**MAPE (%)**	**RMSE (m/s)**
$ARIMA(1,1,0)$	9.357	0.506	0.682
$SARIMA(2,1,0)\times(4,1,0)_{214}$	10.328	0.529	0.698
Improving percentages			
SARIMA versus ARIMA	P_{MAE} (%) −10.378	P_{MAPE} (%) −4.534	P_{RMSE} (%) −2.315

Table 3.7 Deterministic wind direction description accuracy evaluation indicators.

	Evaluation indicators		
Models	**MAE (°)**	**MAPE (%)**	**RMSE (°)**
$ARIMA(1,1,0)$	1.621	4.863	6.274
$SARIMA(1,1,0)\times(2,1,0)_{375}$	1.583	4.745	6.120
Improving percentages			
SARIMA versus ARIMA	P_{MAE} (%) 2.313	P_{MAPE} (%) 2.414	P_{RMSE} (%) 2.454

Similarly, it can be seen from Table 3.7 that both the $ARIMA(1,1,0)$ and $SARIMA(1,1,0)\times(2,1,0)_{375}$ have satisfactory performance in deterministic wind direction description. All accuracy evaluation indicators are relatively low. It proves that both ARIMA and SARIMA methods are valid in deterministic wind direction description.

In the deterministic wind direction description, the $SARIMA(1,1,0)\times(2,1,0)_{375}$ has a better performance when compared with the $ARIMA(1,1,0)$ model. However, improving percentages are not obvious. The phenomenon can be explained by that the wind direction data series has stronger seasonality when compared with the wind speed data series. In this condition, the SARIMA model can outperform the ARIMA model to some extent.

3.6.2 Probabilistic description accuracy evaluation indicators

The probabilistic description accuracy evaluation indicators mainly include Prediction Interval Coverage Probability (PICP), Prediction Interval Normalized Average Width (PINAW), and Coverage Width-based Criterion (CWC).

The PICP calculates the proportion of actual value in the description interval. The calculation formula of the PICP is given as follows [99]:

$$\mathrm{PICP} = \frac{1}{n}\sum_{t=1}^{n} c_t \tag{3.40}$$

where if $t(x)\in[L_t, U_t]$, then $c_t = 1$, vice $c_t = 0$. If the PICP index is less than the corresponding confidence $1-\alpha$, then the model is invalid.

The PINAW calculates the width of the description interval. The variation range of the description value is considered and standardized during the calculation. The equation of the PINAW is given as follows [99]:

$$\text{PINAW} = \frac{1}{nD}\sum_{t=1}^{n}(U_t - L_t) \tag{3.41}$$

where D is the difference between the minimum and maximum description value, measuring the changing range of the description results.

The CWC combines the characteristics of the PICP and PINAW. The equation of CWC is given as follows [99]:

$$\text{CWC} = \text{PINAW}\left(1 + \xi(\text{PICP})e^{-\eta(\text{PICP}-\xi)}\right) \tag{3.42}$$

where if PICP $\geq \xi$, then $\xi(\text{PICP}) = 0$, vice $\xi(\text{PICP}) = 1$, ξ and η are constants.

3.6.2.1 Probabilistic wind speed description results analysis

To quantitatively analyze the performance of the ARCH model and GARCH model in probabilistic wind speed description, the probabilistic description accuracy evaluation indicators are utilized to analyze the description results in Section 3.5. The probabilistic description accuracy evaluation indicators of these models are given in Table 3.8 after calculation. Besides, the improving percentages between heteroscedastic models and homoscedastic models are given in Table 3.9.

As shown in Table 3.8, the PICP of ARIMA-ARCH and ARIMA-GARCH models are generally lower than the confidence interval, which shows that the performance of ARIMA-based heteroscedastic models cannot meet the description accuracy requirement. On the contrary, the PICP of the SARIMA-ARCH and the SARIMA-GARCH models are generally higher than the confidence interval, which verifies that the SARIMA-based heteroscedastic models have satisfactory performance. It also proves that the SARIMA model can provide probabilistic wind speed description results with satisfactory accuracy. The performance of ARCH and GARCH heteroscedastic models is changing with significant levels. The PICP increases with the increase of the confidence interval $1 - \alpha$, which means most heteroscedastic models perform better.

It can be seen from Table 3.9 that the heteroscedastic models can overall outperform the homoscedastic models in wind speed probabilistic description. The improving percentages of the CWC are significant between the ARIMA-based heteroscedastic models and homoscedastic

Table 3.8 Probabilistic wind speed description accuracy evaluation indicators.

	Evaluation indicators			
Models	**1 − α**	**PICP**	**PINAW (m/s)**	**CWC (m/s)**
ARIMA	0.9	0.826	0.218	9.029
	0.95	0.862	0.260	21.401
	0.99	0.942	0.341	4.101
SARIMA	0.9	0.908	0.279	0.279
	0.95	0.940	0.333	0.881
	0.99	0.990	0.437	0.437
ARIMA-ARCH	0.9	0.882	0.241	0.833
	0.95	0.948	0.287	0.604
	0.99	0.992	0.377	0.377
ARIMA-GARCH	0.9	0.876	0.243	1.051
	0.95	0.944	0.290	0.681
	0.99	0.988	0.381	0.802
SARIMA-ARCH	0.9	0.938	0.278	0.278
	0.95	0.990	0.331	0.331
	0.99	1.000	0.435	0.435
SARIMA-GARCH	0.9	0.920	0.275	0.275
	0.95	0.980	0.328	0.328
	0.99	0.998	0.431	0.431

Table 3.9 Improving percentages between heteroscedastic models and homoscedastic models in probabilistic wind speed description.

	Improving percentages			
Comparisons	**1 − α**	P_{PICP} (%)	P_{PINAW} (%)	P_{CWC} (%)
ARIMA-ARCH versus ARIMA	0.9	6.780	−10.582	90.770
	0.95	9.977	−10.582	97.177
	0.99	5.308	−10.582	90.803
ARIMA-GARCH versus ARIMA	0.9	6.053	−11.660	88.362
	0.95	9.513	−11.660	96.818
	0.99	4.883	−11.660	80.449
SARIMA-ARCH versus SARIMA	0.9	3.304	0.508	0.508
	0.95	5.319	0.508	62.438
	0.99	1.010	0.508	0.508
SARIMA-GARCH versus SARIMA	0.9	1.322	1.325	1.325
	0.95	4.255	1.325	62.746
	0.99	0.808	1.325	1.325

models. However, the PINAW shows a decrease in comparison groups ARIMA-ARCH versus ARIMA and ARIMA-GARCH versus ARIMA. Between the SARIMA-based heteroscedastic models and homoscedastic models, the improving percentages are overall positive. Besides, the improvement mainly concentrates on the confidence interval $1 - \alpha = 0.95$. The phenomenon can also be seen from the figures in Section 3.5. It proves that the SARIMA model can provide satisfactory performance for probabilistic description models.

3.6.2.2 Probabilistic wind direction description results analysis

The probabilistic description accuracy evaluation indicators of probabilistic wind direction description models are given in Table 3.10 after calculation. Besides, the improving percentages between heteroscedastic models and homoscedastic models are given in Table 3.11.

It can be seen from Table 3.10 that the PICP of all models is overall lower than the corresponding confidence interval. It shows that the performance of the proposed heteroscedastic models is not satisfactory in wind direction probabilistic description. Besides, the performance of ARCH and

Table 3.10 Probabilistic wind direction description accuracy evaluation indicators.

	Evaluation indicators			
Models	**$1 - \alpha$**	**PICP**	**PINAW (m/s)**	**CWC (m/s)**
ARIMA	0.9	0.648	0.165	48,875.112
	0.95	0.808	0.196	238.202
	0.99	0.892	0.258	34.916
SARIMA	0.9	0.772	0.212	127.533
	0.95	0.838	0.252	68.421
	0.99	0.916	0.331	13.731
ARIMA-ARCH	0.9	0.810	0.195	17.782
	0.95	0.906	0.233	2.334
	0.99	0.994	0.306	0.306
ARIMA-GARCH	0.9	0.858	0.208	1.908
	0.95	0.946	0.248	0.551
	0.99	1.000	0.326	0.326
SARIMA-ARCH	0.9	0.802	0.218	29.549
	0.95	0.870	0.260	14.470
	0.99	0.952	0.342	2.629
SARIMA-GARCH	0.9	0.850	0.245	3.228
	0.95	0.916	0.292	1.889
	0.99	0.976	0.383	1.156

Table 3.11 Improving percentages between heteroscedastic models and homoscedastic models in probabilistic wind direction description.

Improving percentages				
Comparisons	**$1-\alpha$**	**P_{PICP} (%)**	**P_{PINAW} (%)**	**P_{CWC} (%)**
ARIMA-ARCH versus ARIMA	0.9	25.000	−18.542	99.964
	0.95	12.129	−18.542	99.020
	0.99	11.435	−18.542	99.124
ARIMA-GARCH versus ARIMA	0.9	32.407	−26.327	99.996
	0.95	17.079	−26.327	99.769
	0.99	12.108	−26.327	99.066
SARIMA-ARCH versus SARIMA	0.9	3.886	−3.242	76.831
	0.95	3.819	−3.242	78.852
	0.99	3.930	−3.242	80.855
SARIMA-GARCH versus SARIMA	0.9	10.104	−15.759	97.469
	0.95	9.308	−15.759	97.239
	0.99	6.550	−15.759	91.583

GARCH heteroscedastic models is changing with significant levels. The PICP increases with the increase of the confidence interval $1-\alpha$, which means most heteroscedastic models perform better.

It can be seen from Table 3.11 that the improving percentages of PICP and CWC are all positive when comparing the heteroscedastic models and homoscedastic models in wind direction probabilistic description. Especially, the CWC shows an impressive improvement with the improving percentages all over 80%. However, the PINAW shows a decrease in all comparison groups between the heteroscedastic models and homoscedastic models. The phenomenon shows that the performance of these proposed heteroscedastic models cannot meet the requirements of wind direction probabilistic description.

3.7 Summary and outlook

In this chapter, the wind anemometer layout and railway safety are introduced. Then the time series seasonal analysis and heteroscedasticity analysis are described in detail. Based on the seasonal analysis, the ARIMA and SARIMA models are built for wind speed and wind direction deterministic description. And the ARCH and GARCH models are built for wind speed and wind direction probabilistic description based on the heteroscedasticity analysis.

In Section 3.2, the layout of wind anemometers along railways is introduced. Firstly, the development progress of high-speed railway safety in recent years is introduced. Besides, the numerical simulation methods and layout optimization algorithms are detailly introduced in the field of railway safety.

In Section 3.3, the single-point wind speed and wind direction data series are analyzed by the seasonal analysis methods. Firstly, the ADF method tests the stationarity of the wind data series. Secondly, the Hurst exponent can find out the seasonality hidden in the series. Thirdly, the FFT can find the seasonal cycle of the series in the frequency domain. Finally, the ACF, PACF, and BIC methods are utilized to find the optimal polynomial degrees of the ARIMA model and SARIMA model. The seasonality of wind speed data and wind direction data is proved and the seasonal cycle is found out in Section 3.3.

In Section 3.4, two main heteroscedastic testing methods are proposed for heteroscedasticity analysis: graphical test, and hypothesis tests, five hypothesis tests methods are introduced in detail, including the Goldfeld–Quandt test, Breusch–Pagan test, White test, Park test, and Glejser test. In the section, the wind speed and wind direction data series are proved to have heteroscedasticity. Five hypothesis tests methods are also analyzed as a contrast in Section 3.4.

In Section 3.5, the ARIMA and SARIMA models are built for wind speed and wind direction deterministic description and proved to be effective. The ARCH and GARCH models are built for wind speed and wind direction probabilistic description and proved to be effective. To analyze the performance of these models, Section 3.6 proposes the deterministic description accuracy evaluation indicators for the ARIMA and SARIMA models, and the probabilistic description accuracy evaluation indicators for the ARCH and GARCH models. The performance of these models is analyzed in detail in Section 3.6. These description accuracy evaluation indicators will also be used in the following chapters.

In the future, the steps of seasonal analysis can be modified by other novel methods. The efficiency of finding seasonality and the seasonal cycle can be improved. Besides, hybrid modeling methods can be added to the ARIMA and ARCH models to improve description performance. With the development of the big data analysis techniques, wind data series with a larger scale can be analyzed by seasonal methods and heteroscedastic methods.

References

[1] J. Zhang, J. Wang, X. Tan, et al., Detached eddy simulation of flow characteristics around railway embankments and the layout of anemometers, J. Wind Eng. Ind. Aerod. 193 (2019) 103968.

[2] Z. Yao, J. Xiao, F. Jiang, Characteristics of daily extreme-wind gusts along the lanxin railway in Xinjiang, China, Aeolian Res. 6 (2012) 31–40.

[3] Y. Jiang, Y. Gao, Z. Dong, et al., Simulations of wind erosion along the qinghai-tibet railway in north-central tibet, Aeolian Res. 32 (2018) 192–201.

[4] F. Ziel, C. Croonenbroeck, D. Ambach, Forecasting wind power—modeling periodic and non-linear effects under conditional heteroscedasticity, Appl. Energy 177 (2016) 285–297.

[5] J. Copley, The three-dimensional flow around railway trains, J. Wind Eng. Ind. Aerod. 26 (1987) 21–52.

[6] T. Chiu, A two-dimensional second-order vortex panel method for the flow in a cross-wind over a train and other two-dimensional bluff bodies, J. Wind Eng. Ind. Aerod. 37 (1991) 43–64.

[7] C. Baker, N. Humphreys, Assessment of the adequacy of various wind tunnel techniques to obtain aerodynamic data for ground vehicles in cross winds, J. Wind Eng. Ind. Aerod. 60 (1996) 49–68.

[8] C. Fauchier, E. Le Devehat, R. Gregoire, Numerical study of the turbulent flow around the reduced-scale model of an Inter-Regio, in: TRANSAERO—A European Initiative on Transient Aerodynamics for Railway System Optimisation, Springer, 2002, pp. 61–74.

[9] B. Diedrichs, S. Krajnović, M. Berg, On the aerodynamics of car body vibrations of high-speed trains cruising inside tunnels, Eng. Applicat. Comput. Fluid Mechanic. 2 (2008) 51–75.

[10] M. Suzuki, K. Tanemoto, T. Maeda, Aerodynamic characteristics of train/vehicles under cross winds, J. Wind Eng. Ind. Aerod. 91 (2003) 209–218.

[11] D. Flynn, H. Hemida, D. Soper, et al., Detached-eddy simulation of the slipstream of an operational freight train, J. Wind Eng. Ind. Aerod. 132 (2014) 1–12.

[12] J.A. Morden, H. Hemida, C. Baker, Comparison of RANS and detached eddy simulation results to wind-tunnel data for the surface pressures upon a class 43 high-speed train, J. Fluid Eng. 137 (2015) 041108.

[13] R.M. Colombo, M. Herty, M. Mercier, Control of the continuity equation with a non local flow, ESAIM Control, Optim. Calc. Var. 17 (2011) 353–379.

[14] A. Aziz, Hydrodynamic and thermal slip flow boundary layers over a flat plate with constant heat flux boundary condition, Commun. Nonlinear Sci. Numer. Simulat. 15 (2010) 573–580.

[15] P.J. Migliorini, A. Untaroiu, W.C. Witt, et al., Hybrid analysis of gas annular seals with energy equation, in: ASME Turbo Expo 2013: Turbine Technical Conference and Exposition, vol. 55263, 2013. V07AT26A003.

[16] Y. Pan, H. Zhang, Q. Zhou, Numerical prediction of submarine hydrodynamic coefficients using CFD simulation, J. Hydrodyn. 24 (2012) 840–847.

[17] M. Deligant, P. Podevin, G. Descombes, CFD model for turbocharger journal bearing performances, Appl. Therm. Eng. 31 (2011) 811–819.

[18] M.H. Zawawi, A. Saleha, A. Salwa, et al., A review: fundamentals of computational fluid dynamics (CFD), in: AIP Conference Proceedings, vol. 2030, 2018, p. 020252.

[19] W. Jeong, J. Seong, Comparison of effects on technical variances of computational fluid dynamics (CFD) software based on finite element and finite volume methods, Int. J. Mech. Sci. 78 (2014) 19–26.

[20] D. Walters, S. Bhushan, M. Alam, et al., Investigation of a dynamic hybrid RANS/LES modelling methodology for finite-volume CFD simulations, Flow, Turbul. Combust. 91 (2013) 643–667.
[21] M.S. Shadloo, G. Oger, D. Le Touzé, Smoothed particle hydrodynamics method for fluid flows, towards industrial applications: motivations, current state, and challenges, Comput. Fluid 136 (2016) 11–34.
[22] M.L. Hosain, R.B. Fdhila, Literature review of accelerated CFD simulation methods towards online application, Energy Proc. 75 (2015) 3307–3314.
[23] T. Tamai, S. Koshizuka, Least squares moving particle semi-implicit method, Comput. Particle Mechanic. 1 (2014) 277–305.
[24] E. Jahanbakhsh, C. Vessaz, A. Maertens, et al., Development of a finite volume particle method for 3-D fluid flow simulations, Comput. Methods Appl. Mech. Eng. 298 (2016) 80–107.
[25] L. Luo, M. Krafczyk, W. Shyy, Lattice Boltzmann method for computational fluid dynamics, Encyclopedia Aerospace Eng. 56 (2010) 651–659.
[26] O.G. Sutton, Atmospheric Turbulence, Routledge, 2020.
[27] Y. Ren, H. Huang, G. Xie, et al., Atmospheric turbulence effects on the performance of a free space optical link employing orbital angular momentum multiplexing, Optic Lett. 38 (2013) 4062–4065.
[28] C.J. Subich, K.G. Lamb, M. Stastna, Simulation of the Navier–Stokes equations in three dimensions with a spectral collocation method, Int. J. Numer. Methods Fluid. 73 (2013) 103–129.
[29] S.B. Poussou, S. Mazumdar, M.W. Plesniak, et al., Flow and contaminant transport in an airliner cabin induced by a moving body: model experiments and CFD predictions, Atmos. Environ. 44 (2010) 2830–2839.
[30] Y. Haroun, D. Legendre, L. Raynal, Direct numerical simulation of reactive absorption in gas–liquid flow on structured packing using interface capturing method, Chem. Eng. Sci. 65 (2010) 351–356.
[31] Z. Peng, E. Doroodchi, C. Luo, et al., Influence of void fraction calculation on fidelity of CFD-DEM simulation of gas-solid bubbling fluidized beds, AIChE J. 60 (2014) 2000–2018.
[32] Y. Tominaga, T. Stathopoulos, CFD modeling of pollution dispersion in a street canyon: comparison between LES and RANS, J. Wind Eng. Ind. Aerod. 99 (2011) 340–348.
[33] H. Aluie, Scale decomposition in compressible turbulence, Phys. Nonlinear Phenom. 247 (2013) 54–65.
[34] A. Leonard, Energy cascade in large-eddy simulations of turbulent fluid flows, in: Advances in Geophysics, Elsevier, 1975, pp. 237–248.
[35] M. Germano, U. Piomelli, P. Moin, et al., A dynamic subgrid-scale eddy viscosity model, Phys. Fluid. Fluid Dynam. 3 (1991) 1760–1765.
[36] J. Mo, A. Choudhry, M. Arjomandi, et al., Large eddy simulation of the wind turbine wake characteristics in the numerical wind tunnel model, J. Wind Eng. Ind. Aerod. 112 (2013) 11–24.
[37] H. Xiao, P. Cinnella, Quantification of model uncertainty in RANS simulations: a review, Prog. Aero. Sci. 108 (2019) 1–31.
[38] Z. Tian, M. Perlin, W. Choi, An eddy viscosity model for two-dimensional breaking waves and its validation with laboratory experiments, Phys. Fluids 24 (2012) 036601.
[39] J. Liu, M. Heidarinejad, G. Pitchurov, et al., An extensive comparison of modified zero-equation, standard k-ε, and LES models in predicting urban airflow, Sustain. Cities Soc. 40 (2018) 28–43.
[40] F.R. Menter, P.E. Smirnov, T. Liu, et al., A one-equation local correlation-based transition model, Flow, Turbul. Combust. 95 (2015) 583–619.

[41] X. Feng, J. Cheng, X. Li, et al., Numerical simulation of turbulent flow in a baffled stirred tank with an explicit algebraic stress model, Chem. Eng. Sci. 69 (2012) 30–44.
[42] P.R. Spalart, Comments on the feasibility of LES for wings, and on a hybrid RANS/LES approach, in: Proceedings of First AFOSR International Conference on DNS/LES, 1997, pp. 137–148.
[43] P.R. Spalart, S. Deck, M.L. Shur, et al., A new version of detached-eddy simulation, resistant to ambiguous grid densities, Theor. Comput. Fluid Dynam. 20 (2006) 181–195.
[44] M.L. Shur, P.R. Spalart, M.K. Strelets, et al., A hybrid RANS-LES approach with delayed-DES and wall-modelled LES capabilities, Int. J. Heat Fluid Flow 29 (2008) 1638–1649.
[45] J. Chen, P. Zhang, N. Zhou, et al., Application of detached-eddy simulation based on Spalart-Allmaras turbulence model, J. Beijing Univ. Aeronaut. Astronaut. 38 (2012) 905–909.
[46] F. Menter, M. Kuntz, R. Bender, A scale-adaptive simulation model for turbulent flow predictions, in: 41st Aerospace Sciences Meeting and Exhibit, vol. 767, 2003, pp. 1–11.
[47] Z. Li, H. Chen, Y. Zhang, Scale adaptive simulation based on a k-kL two-equation turbulence model, Eng. Mech. 33 (2016) 21–30.
[48] W. Zheng, C. Yan, Influence analysis on grid scale limiter of XY-SAS model, J. Beijing Univ. Aeronaut. Astronaut. 40 (2014) 1725–1729.
[49] B. Chaouat, Simulations of turbulent rotating flows using a subfilter scale stress model derived from the partially integrated transport modeling method, Phys. Fluids 24 (2012) 045108.
[50] R. Schiestel, A. Dejoan, Towards a new partially integrated transport model for coarse grid and unsteady turbulent flow simulations, Theor. Comput. Fluid Dynam. 18 (2005) 443–468.
[51] B. Chaouat, R. Schiestel, Progress in subgrid-scale transport modelling for continuous hybrid non-zonal RANS/LES simulations, Int. J. Heat Fluid Flow 30 (2009) 602–616.
[52] C.G. Speziale, Computing non-equilibrium turbulent flows with time-dependent RANS and VLES, in: Fifteenth International Conference on Numerical Methods in Fluid Dynamics, 1997, pp. 123–129.
[53] S.S. Girimaji, Partially-averaged Navier-Stokes model for turbulence: a Reynolds-averaged Navier-Stokes to direct numerical simulation bridging method, J. Appl. Mech. 73 (2006) 413–421.
[54] S.S. Girimaji, E. Jeong, R. Srinivasan, Partially averaged Navier-Stokes method for turbulence: fixed point analysis and comparison with unsteady partially averaged Navier-Stokes, J. Appl. Mech. 73 (2006) 422–429.
[55] B. Huang, G. Wang, Partially averaged Navier-Stokes method for time-dependent turbulent cavitating flows, J. Hydrodyn. 23 (2011) 26–33.
[56] J.A. Morden, H. Hemida, C.J. Baker, Comparison of RANS and detached eddy simulation results to wind-tunnel data for the surface pressures upon a class 43 high-speed train, J. Fluid Eng. 137 (2015) 041108.
[57] J. Liu, D. Wang, K. He, et al., Combining Wang–Landau sampling algorithm and heuristics for solving the unequal-area dynamic facility layout problem, Eur. J. Oper. Res. 262 (2017) 1052–1063.
[58] G.-J. Gao, J. Zhang, X.-H. Xiong, Location of anemometer along Lanzhou-Xinjiang railway, J. Cent. S. Univ. 21 (2014) 3698–3704.
[59] G. Mosetti, C. Poloni, B. Diviacco, Optimization of wind turbine positioning in large windfarms by means of a genetic algorithm, J. Wind Eng. Ind. Aerod. 51 (1994) 105–116.

[60] S. Grady, M. Hussaini, M.M. Abdullah, Placement of wind turbines using genetic algorithms, Renew. Energy 30 (2005) 259–270.
[61] A. Emami, P. Noghreh, New approach on optimization in placement of wind turbines within wind farm by genetic algorithms, Renew. Energy 35 (2010) 1559–1564.
[62] J.H. Lopez, The power of the ADF test, Econ. Lett. 57 (1997) 5–10.
[63] Y.-H. Dai, W.-X. Zhou, Temporal and spatial correlation patterns of air pollutants in Chinese cities, PloS One 12 (2017) e0182724.
[64] P. Chen, H. Yuan, X. Shu, Forecasting crime using the arima model, in: 2008 Fifth International Conference on Fuzzy Systems and Knowledge Discovery, vol. 5, 2008, pp. 627–630.
[65] J. Kuha, AIC and BIC, Comparisons of assumptions and performance, Socio. Methods Res. 33 (2004) 188–229.
[66] H. Liu, Z. Duan, C. Chen, et al., A novel two-stage deep learning wind speed forecasting method with adaptive multiple error corrections and bivariate Dirichlet process mixture model, Energy Convers. Manag. 199 (2019) 111975.
[67] S.S. Uyanto, Monte Carlo power comparison of seven most commonly used heteroscedasticity tests, Commun. Stat. Simulat. Comput. (2019) 1–18.
[68] A.G. Halunga, C.D. Orme, T. Yamagata, A heteroskedasticity robust Breusch–Pagan test for Contemporaneous correlation in dynamic panel data models, J. Econom. 198 (2017) 209–230.
[69] A. Klein, C. Gerhard-Lehn, R. Büchner, et al., The detection of heteroscedasticity in regression models for psychological data, Psychol. Test Assess. Model. 58 (2016) 567–592.
[70] V. Berenguer-Rico, I. Wilms, Heteroscedasticity testing after outlier removal, Econom. Rev. 1–35 (2020).
[71] M. Nwakuya, J. Nwabueze, Application of box-cox transformation as a corrective measure to heteroscedasticity using an economic data, Am. J. Math. Stat. 8 (2018) 8–12.
[72] H. Glejser, A new test for heteroskedasticity, J. Am. Stat. Assoc. 64 (1969) 316–323.
[73] G.E. Box, G.M. Jenkins, G.C. Reinsel, Time Series Analysis: Forecasting and Control, John Wiley & Sons, 2011.
[74] F.-M. Tseng, H.-C. Yu, G.-H. Tzeng, Combining neural network model with seasonal time series ARIMA model, Technol. Forecast. Soc. Change 69 (2002) 71–87.
[75] R.G. Kavasseri, K. Seetharaman, Day-ahead wind speed forecasting using f-ARIMA models, Renew. Energy 34 (2009) 1388–1393.
[76] E. Cadenas, W. Rivera, Wind speed forecasting in three different regions of Mexico, using a hybrid ARIMA–ANN model, Renew. Energy 35 (2010) 2732–2738.
[77] H. Liu, H.-Q. Tian, Y.-F. Li, Comparison of two new ARIMA-ANN and ARIMA-Kalman hybrid methods for wind speed prediction, Appl. Energy 98 (2012) 415–424.
[78] M.F. Akhter, D. Hassan, S. Abbas, Predictive ARIMA Model for coronal index solar cyclic data, Astronomy Comput. 32 (2020) 100403.
[79] M. Bouzerdoum, A. Mellit, A.M. Pavan, A hybrid model (SARIMA–SVM) for short-term power forecasting of a small-scale grid-connected photovoltaic plant, Sol. Energy 98 (2013) 226–235.
[80] Ö.Ö. Bozkurt, G. Biricik, Z.C. Taysi, Artificial neural network and SARIMA based models for power load forecasting in Turkish electricity market, PloS One 12 (2017) e0175915.
[81] D.B. Alencar, C.M. Affonso, R.C. Oliveira, et al., Hybrid approach combining SARIMA and neural networks for multi-step ahead wind speed forecasting in Brazil, IEEE Access 6 (2018) 55986–55994.
[82] Z. Guo, J. Zhao, W. Zhang, et al., A corrected hybrid approach for wind speed prediction in Hexi Corridor of China, Energy 36 (2011) 1668–1679.

[83] J. Wang, J. Hu, K. Ma, et al., A self-adaptive hybrid approach for wind speed forecasting, Renew. Energy 78 (2015) 374–385.
[84] Q. Mao, K. Zhang, W. Yan, et al., Forecasting the incidence of tuberculosis in China using the seasonal auto-regressive integrated moving average (SARIMA) model, J. Infect. Public Health 11 (2018) 707–712.
[85] T. Bollerslev, R.F. Engle, D.B. Nelson, ARCH models, Handb. Econom. 4 (1994) 2959–3038.
[86] A.K. Dhamija, V.K. Bhalla, Financial time series forecasting: comparison of neural networks and ARCH models, Int. Res. J. Finance Econom. 49 (2010) 185–202.
[87] S. Gao, Y. He, H. Chen, Wind speed forecast for wind farms based on ARMA-ARCH model, in: 2009 International Conference on Sustainable Power Generation and Supply, 2009, pp. 1–4.
[88] P. Lv, L. Yue, Short-term wind speed forecasting based on non-stationary time series analysis and ARCH model, in: 2011 International Conference on Multimedia Technology, 2011, pp. 2549–2553.
[89] M.-D. Wang, Q.-R. Qiu, B.-W. Cui, Short-term wind speed forecasting combined time series method and arch model, in: 2012 International Conference on Machine Learning and Cybernetics, vol. 3, 2012, pp. 924–927.
[90] M. Meitz, P. Saikkonen, Maximum likelihood estimation of a noninvertible ARMA model with autoregressive conditional heteroskedasticity, J. Multivariate Anal. 114 (2013) 227–255.
[91] D. Niu, Y. Wei, An Improved short-term power load combined forecasting with ARMA-GRACH-ANN-SVM based on FHNN similar-day clustering, J. Softw. 8 (2013) 716–723.
[92] C. Morana, A semiparametric approach to short-term oil price forecasting, Energy Econ. 23 (2001) 325–338.
[93] L. Fang, H. Yu, W. Xiao, Forecasting gold futures market volatility using macroeconomic variables in the United States, Econ. Modell. 72 (2018) 249–259.
[94] Z. Chen, S. Yang, L. Hou, Based IGARCH error correction of the PLS-SVR short-term load forecasting, in: Unifying Electrical Engineering and Electronics Engineering, Springer, 2014, pp. 199–207.
[95] S.H. Kang, S.-M. Kang, S.-M. Yoon, Forecasting volatility of crude oil markets, Energy Econ. 31 (2009) 119–125.
[96] J. Zhang, Z. Tan, Day-ahead electricity price forecasting using WT, CLSSVM and EGARCH model, Int. J. Electr. Power Energy Syst. 45 (2013) 362–368.
[97] W. Kristjanpoller, M.C. Minutolo, A hybrid volatility forecasting framework integrating GARCH, artificial neural network, technical analysis and principal components analysis, Expert Syst. Appl. 109 (2018) 1–11.
[98] L. Xiang, J. Li, A. Hu, et al., Deterministic and probabilistic multi-step forecasting for short-term wind speed based on secondary decomposition and a deep learning method, Energy Convers. Manag. 220 (2020) 113098.
[99] X. Peng, W. Zheng, D. Zhang, et al., A novel probabilistic wind speed forecasting based on combination of the adaptive ensemble of on-line sequential ORELM (Outlier Robust Extreme Learning Machine) and TVMCF (time-varying mixture copula function), Energy Convers. Manag. 138 (2017) 587–602.

CHAPTER 4

Single-point wind forecasting methods based on deep learning

Contents

Wind Forecasting in Railway Engineering
ISBN 978-0-12-823706-9
https://doi.org/10.1016/B978-0-12-823706-9.00004-1

4.1 Introduction

As a novel branch of machine learning, deep learning has had a rapid development since the year 2006. All deep learning algorithms are based on the basic framework of Artificial Neural Networks (ANNs). Multiple layers of neurons make up complex structures to model nonlinear high-dimensional data. Deep learning algorithms are proposed for the representational learning of data at first. With the development of data size and computing hardware, the structure of neural networks is getting more and more complex, which has attracted much attention in recent years [1].

Most deep learning algorithms use the Back-Propagation (BP) method to modify the internal parameters of the network structure by large datasets. Those parameters or weightings usually calculate the representation in each layer from the former layer [2]. Deep learning makes it possible for a machine to imitate human activities such as vision, hearing, and thinking. In this way, many complex tasks can be done by machine now. Nowadays, deep learning algorithms have already been utilized in various fields, including time series forecasting [3], network intrusion detection [4], speech recognition [5], etc.

Wind forecasting, including wind speed and wind direction forecasting, is an important and major application area of deep learning techniques. For example, many hybrid models using deep learning algorithms have been proposed for short-term wind speed forecasting. The Gated Recurrent Unit (GRU) deep learning neural network was enhanced by the Wavelet Soft Threshold Denoising (WSTD). Besides, the parameters of the hybrid model were optimized by the cross-validated grid-search. The results showed the effectiveness of the deep learning hybrid model [6]. Hong et al. utilized a series of deep learning neural networks to build a hybrid model. In this model, the Convolutional Neural Network (CNN) was selected to extract characteristics. Besides, a Radial Basis Function Neural Network (RBFNN) activated by the Double Gaussian Function (DGF) was utilized to deal with the uncertain characteristics of wind data. The performance of the hybrid model was verified to be effective [7].

In this chapter, three mainstream deep learning neural networks will be introduced and studied, including Long Short-Term Memory (LSTM) neural network, GRU neural network, and Seriesnet (SN) algorithm. The performance of those deep learning algorithms is investigated in wind forecasting. Then the influence of hybrid modeling methods on deep learning algorithms will be further studied.

4.2 Wind data description

The single-station wind prediction can help the management to issue operation commands, to avoid possible threats and improve safety. In this chapter, wind speed and direction data are both collected from the wind measuring stations along the strong wind railway line. The wind forecasting studies can improve the safety of rail transit operation. The real wind speed and direction data utilized in this chapter contain 2000 data samples, where the 1st–1800th are the training set and the 1801th–2000th are the testing set. These data are used to analyze the performance of deep learning algorithms for single-point wind forecasting. In different models, the training set and testing set are utilized to train the deep learning algorithms and verify the forecasting performance of the algorithm.

The wind speed data series have 5 min temporal interval, while the wind direction data series has 15 s temporal interval. The diagrams of wind speed and direction data are, respectively, given in Figs. 4.1 and 4.2. The statistical descriptions of the wind speed and direction data series are presented in Table 4.1. As can be seen from Figs. 4.2 and 4.3, the wind speed and direction have significant fluctuation.

In this chapter, three evaluation indices are utilized to verify and estimate the forecasting performance of the machine learning models, quantitatively. These indices are widely used in the single-point wind forecasting field, which are Mean Absolute Percentage Error (MAPE), Mean Absolute Error (MAE), and Root Mean Square Error (RMSE). The details about these indicators are presented in Section 3.6.

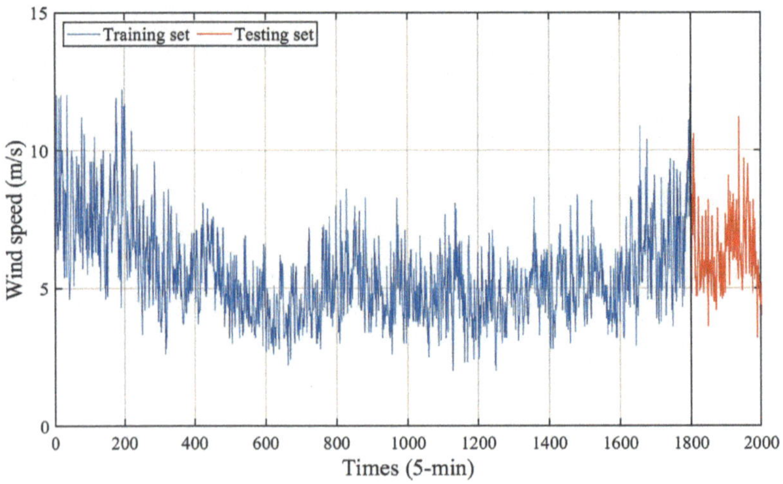

Figure 4.1 The wind speed data series.

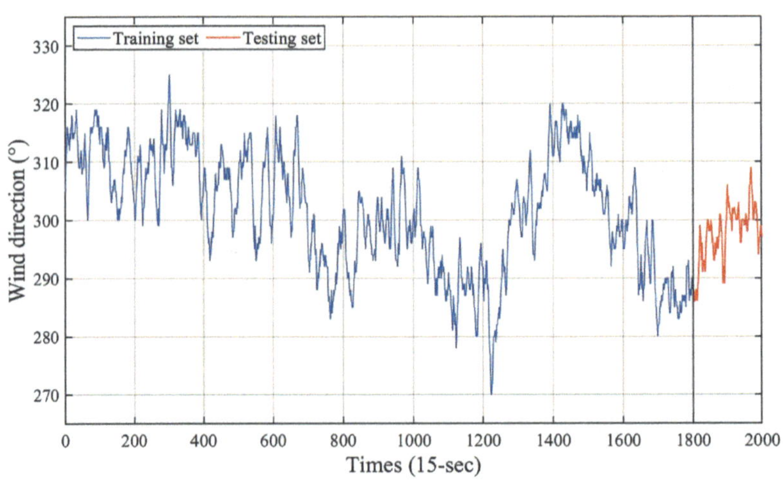

Figure 4.2 The wind direction data series.

Table 4.1 The statistical descriptions of the wind speed and direction data.

Data	Mean	Standard deviation	Minimum	Maximum	Skewness	Kurtosis
Wind speed	5.599 m/s	1.661 m/s	2.000 m/s	12.400 m/s	0.885	4.022
Wind direction	301.088 degrees	9.873 degrees	270.000 degrees	325.000 degrees	−0.144	2.318

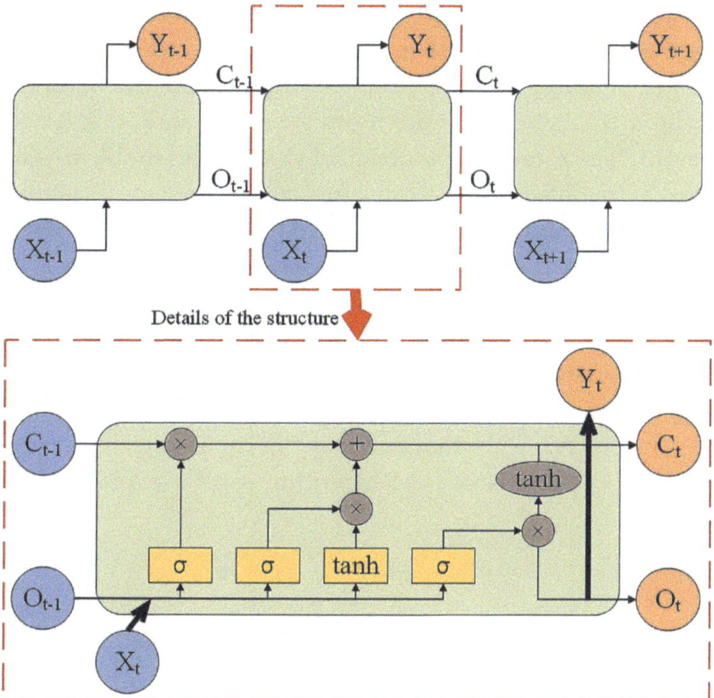

Figure 4.3 The structure of the LSTM wind speed forecasting model.

4.3 Single-point wind speed forecasting algorithm based on LSTM

4.3.1 Single LSTM wind speed forecasting model

4.3.1.1 Theoretical basis

LSTM neural network was firstly proposed by Hochreiter and Schmidhuber in 1997 [8]. LSTM can be regarded as a modified Recurrent Neural Network (RNN). The memory of past input is important in the sequence learning problem. LSTM was specially designed and modified to solve the vanishing gradient and explosion gradient problem in long-term training. Due to the memory cell, the LSTM can maintain the error values and continue the gradient flow. Therefore, the vanishing problem is eliminated and information can be learned from hundreds-time steps long sequences. LSTM has better performance than other RNN architectures because the vanishing gradient problem is alleviated.

The main difference between the LSTM and RNN is that a series of memory gates are added, including the forget gate, input gate, and output gate [9]. In the training process, some less important information is thrown away for the memory cell to make room for information that is newer and more relevant. The forget gate is designed to solve this problem. The forget gate deletes or maintains the information by multiplying the value in the memory cell by a number of 0 or 1. If a value needs to be preserved for many steps in the memory cell, the input gate or write gate is added to the LSTM structure. And the output gate is added to solve the problem that multiple memories are against each other. After knowing the memory gates, the LSTM neural network can be expressed.

The input data are set as x^t for the time t. The number of memory cells is set as N in LSTM. The dimension of the input data is set as M. The feedforward process of the LSTM algorithm can be explained as follows [10]:

$$i^t = g(\omega_i x^t + p_i y^{t-1} q_i \cdot c^{t-1} + b_i) \tag{4.1}$$

$$l^t = \sigma(\omega_l x^t + p_l y^{t-1} + q_l \cdot c^{t-1} + b_l) \tag{4.2}$$

$$f^t = \sigma\left(\omega_f x^t + p_f y^{t-1} + q_f \cdot c^{t-1} + b_f\right) \tag{4.3}$$

where i^t is the activation of the input gate, l^t is the activation of the output gate, and f^t is the activation of the forget gate. $\omega_i, \omega_l, \omega_f \in R^{N \times M}$ are the input weightings of the LSTM. $p_i, p_l, p_f \in R^{N \times M}$ are the output weightings of the LSTM. $q_i, q_l, q_f \in R^{N \times M}$ are the weightings of the memory cells, and $b_i, b_l, b_f \in R^{N \times M}$ are the biases of the corresponding activations. After that, the value of the output gate and the state of the memory cell can be calculated as follows [10]:

$$c^t = i^t \cdot l^t + c^{t-1} \cdot f^t \tag{4.4}$$

$$o^t = \sigma(\omega_o x^t + p_o y^{t-1} + q_o \cdot c^t + b_o) \tag{4.5}$$

where c^t is the state of the memory cell and o^t is the value of the output gate, and ω_o, p_o, q_o, b_o are the weightings and bias of the output gate. Finally, the output value of the cell can be calculated as follows [10]:

$$y^t = h(c^t) \bullet o^t \tag{4.6}$$

Among the above equations, $g(x), \sigma(x), h(x)$ are the activation functions of the input gate, forget gate, and output gate, respectively. In general, the sigmoid function and tanh function are set as the activation functions, which are given as follows [10]:

$$\sigma(x) = \frac{1}{1 + e^{-x}} \tag{4.7}$$

$$g(x) = h(x) = \tanh(x) \tag{4.8}$$

4.3.1.2 Model structure

The specific structure of the LSTM wind speed forecasting model is given in Fig. 4.3. From Fig. 4.3, it can be seen that the input information includes the input data X_t and information C_{t-1}, O_{t-1} from the last step. This information multiplies with the activation functions σ or tanh to generate the new information in this step. The state from the last step is also decided to be reserved or deleted in this way. Finally, the output Y_t of this step is obtained. And the state information C_t and output values can affect the next step.

4.3.1.3 Modeling steps

In the training process of the LSTM wind speed forecasting model, the BP algorithm is utilized. In wind speed forecasting, the LSTM processes the time series, thus the error from the whole series should be propagated back. Therefore, the error propagation in the LSTM can also be called Back Propagation Through Time (BPTT). It can be seen that the current cell is affected by the last cell when training the LSTM.

In the BPTT of the training process, C_{t-1} is determined by both C_t and O_{t-1}. And C_t contains influence from two aspects: C_{t+1} and O_t. When calculating the error of step t, both C_{t+1} and O_t are needed. In this way, the weightings can be updated according to the stochastic gradient descend.

In this study, the LSTM is built for wind speed forecasting model. In the training set, the number of the input is set as 15 for the LSTM model. To realize the multi-step ahead forecasting, the number of the output is set as 3. The Multiple-Input Multiple-Output (MIMO) strategy is applied in this section. After training, the LSTM model can forecast 3-steps ahead at the same time. The loss curve of the LSTM multi-step wind speed forecasting model is given in Fig. 4.4.

It can be seen from Fig. 4.4 that the training loss of the model shows a downward trend. It is because that the weighting of the model is changing according to the stochastic gradient descend. Moreover, the decreasing

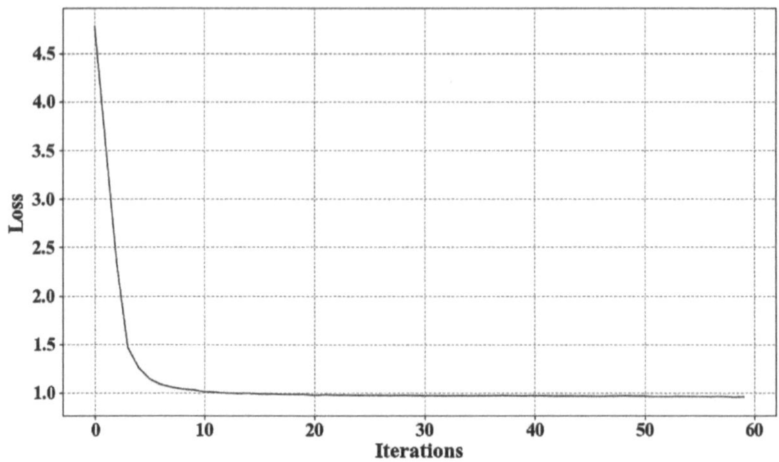

Figure 4.4 The loss curve of the LSTM multi-step wind speed forecasting model.

degree of the loss becomes smaller and smaller with the increase of the iteration times. Finally, the training loss gradually stabilizes around a certain level. The process conforms to the general trend of training loss of the deep networks.

After training, the weightings and different parameters of the LSTM wind speed forecasting model are determined. The model can forecast wind speed data now.

4.3.1.4 Result analysis

The LSTM wind speed forecasting model can forecast wind speed data in 3-steps ahead after training. All forecasting results are shown in Fig. 4.5 in different steps. The specific evaluation indices of the LSTM model are given in Table 4.2.

As shown in Fig. 4.5, the LSTM wind speed forecasting model can predict the trend of the actual series. Trends of the forecasting results and actual series are basically the same, which means the LSTM model can predict the basic variation of the wind speed series. However, when compared to the actual series, the model cannot predict the extreme points accurately. With the increase of the forecasting steps, the deviation between the forecasting values and actual series is also increasing. It shows that the performance of the proposed LSTM model can be affected by the forecasting steps to some extent.

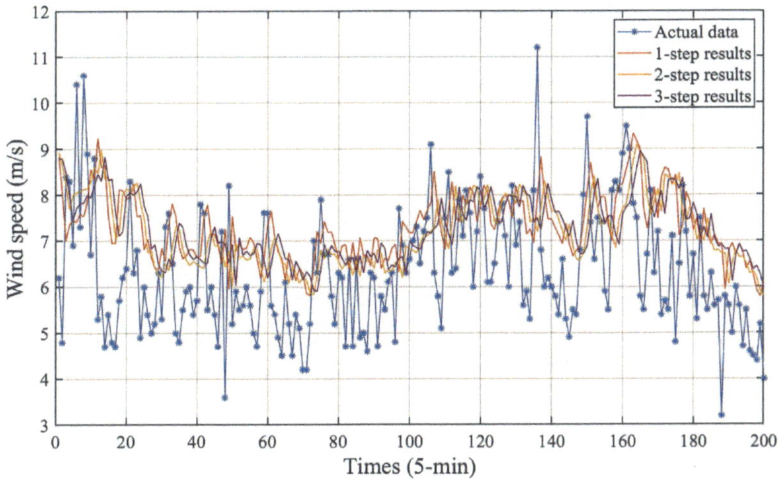

Figure 4.5 Forecasting results of the LSTM wind speed forecasting model.

Table 4.2 Evaluation indices of the LSTM wind speed forecasting model.

Steps	MAE (m/s)	MAPE (%)	RMSE (m/s)
1-Step	1.319	23.139	1.577
2-Step	1.357	24.014	1.611
3-Step	1.361	24.090	1.634

It can also be seen from Fig. 4.5 that the deviation between the forecasting values and the actual data increases greatly at the extreme points and slope values. This indicates that the generalization ability of the single LSTM model is limited. The performance of a single LSTM model cannot meet the accuracy requirement.

It can also be seen from Table 4.2 that the forecasting accuracy is decreasing with the increase of the forecasting steps. However, the differences between different forecasting steps are not significant. It is because the MIMO strategy is utilized in the LSTM wind speed multi-step forecasting model. The model is trained to generate results of different forecasting steps at the same time. The influence of the forecasting steps can be alleviated in this way. On the whole, the single LSTM neural network can predict the wind speed trend. However, its performance at extreme points is not satisfactory. To solve this problem, hybrid modeling is combined with the LSTM algorithm to improve the accuracy performance.

4.3.1.5 Conclusions

According to the above experiments, some conclusions about the LSTM wind speed forecasting model can be drawn as follows:

(a) The loss of the single LSTM wind speed forecasting model tends to be stable at around the iteration of 20. The weightings of different memory gates are settled in this way. It proves the learning ability of the LSTM neural network in the wind forecasting.

(b) The single LSTM model can forecast the basic changing trend of the original wind speed data series. However, the forecasting performance at extreme points is not satisfactory. It shows that the single LSTM model cannot meet the requirement of forecasting accuracy.

(c) The forecasting step affects the forecasting performance of the LSTM wind speed forecasting performance. The forecasting accuracy of the model is decreasing with the increase of the forecasting steps. However, different results show that the effect is not obvious. The MIMO strategy helps to alleviate the influence of forecasting steps to some extent.

4.3.2 Hybrid WPD-LSTM wind speed forecasting model

4.3.2.1 Theoretical basis

In the hybrid WPD-LSTM multi-step wind speed forecasting model, the LSTM predictor is the same as introduced in Section 4.3.1. In this section, the Wavelet Packet Decomposition (WPD) method is introduced. WPD is a variant of the wavelet decomposition. It divides the approximate and detailed components of each layer to obtain a complete decomposed binary tree. WPD is a more sophisticated method for signal analysis. Moreover, it introduces the concept of optimal basis selection based on the wavelet analysis theory. After the frequency band is divided into multiple levels, the best basis function is adaptively selected according to the characteristics of the analyzed signal to make it match with the signal. The analysis ability of the signal is improved in this way. Therefore, WPD has a wide range of application value. In this study, the WPD method is utilized to reduce the complexity of wind speed data and improve the LSTM algorithm.

Set the $\alpha(x)$ as scale function and $\beta(x)$ as the wavelet function [11]:

$$\begin{cases} \beta_0(x) = \alpha(x) \\ \beta_1(x) = \beta(x) \end{cases} \tag{4.9}$$

$$\begin{cases} \alpha_{2l}(x) = \sum_{k=-\infty}^{+\infty} h_k \alpha_l(2x - k) \\ \beta_{2l+1}(x) = \sum_{k=-\infty}^{+\infty} g_k \alpha_l(2x - k) \end{cases} \tag{4.10}$$

Then the wavelet packet function $\{\beta_n(x)\}$ of scale function is obtained. And the function family $\left\{2^{-\frac{j}{2}}\beta_n(2^j x - k)\right\}$ is the wavelet library of the scale function $\alpha(x)$.

4.3.2.2 Model structure

The structure of the hybrid WPD-LSTM wind speed multi-step forecasting model is given in Fig. 4.6 in detail.

As shown in Fig. 4.6, the hybrid model can be mainly divided into two parts: the decomposition component and predictors. The WPD method decomposes the original wind speed data into eight subseries. Then eight LSTM predictors are built for each subseries. After the training process,

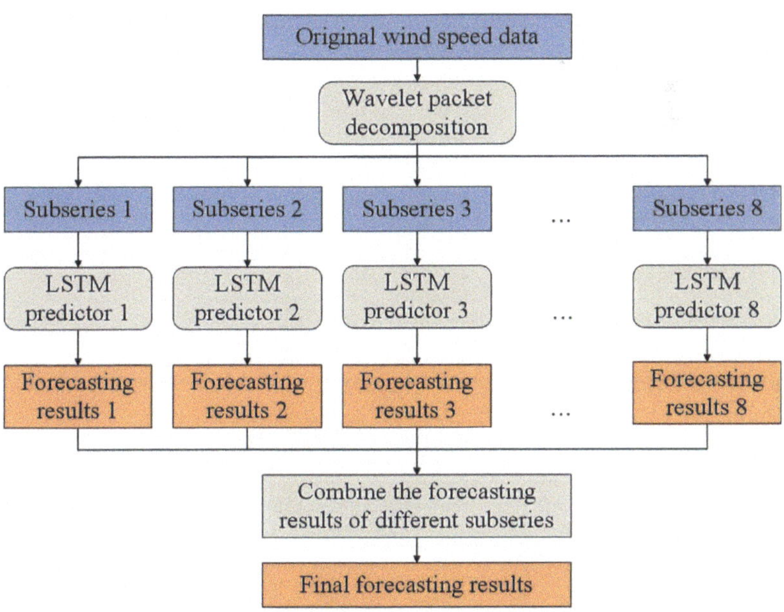

Figure 4.6 The structure of the hybrid WPD-LSTM wind speed forecasting model.

each LSTM can obtain the forecasting results corresponding to the subseries. Finally, the forecasting results of different subseries are combined to get the final forecasting results.

4.3.2.3 Modeling steps

Use the WPD method to decompose the wind speed original data, then eight subseries are obtained. The decomposed results are given in Fig. 4.7.

It can be seen from Fig. 4.7 that the S1 varies from 2 to 10 m/s. It concentrates most of the original wind speed data and can reflect the trend change of the original signal. The S2–S4 varies between −2 and 2 m/s. Subseries S2–S4 have the second amplitude range. Most of the values of S5–S8 are varying from −1 to 1 m/s. The decomposed subseries are arranged from low frequency to high frequency, and its amplitude decreases with the increase of frequency. This phenomenon indicates that the main energy of the wind speed data is concentrated in the low-frequency subseries.

After decomposition, a group of LSTM predictors are built and trained for each subseries. In total, eight LSTM predictors are built in the hybrid WPD-LSTM model. The decomposed results of WPD are used to train each LSTM predictor. In each LSTM predictor, the output is set as 3 to realize the multi-step ahead forecasting. After iterations, each LSTM predictor can obtain the minimum loss and errors. The LSTM predictors group is trained in this way.

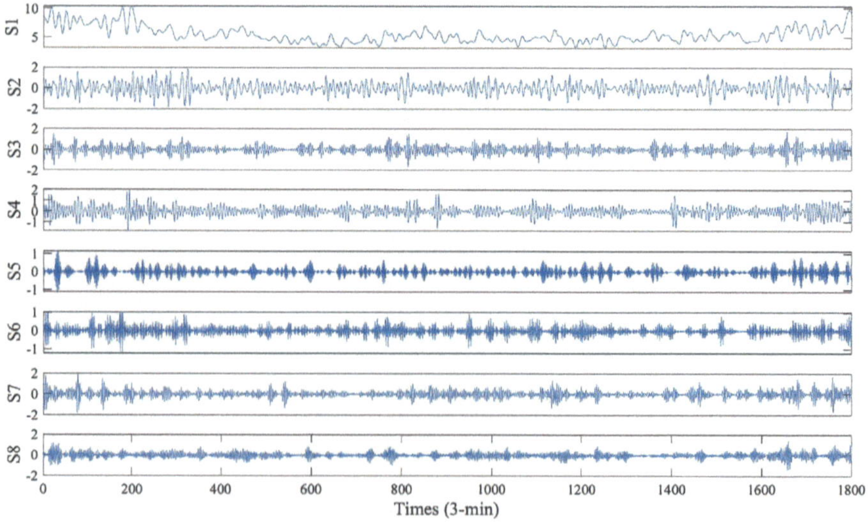

Figure 4.7 Decomposition results of wind speed data after WPD.

After the training process, the LSTM predictors can generate forecasting results of each decomposed subseries of the testing set. Those forecasting results are obtained and combined as the final forecasting results.

4.3.2.4 Result analysis

The hybrid WPD-LSTM wind speed forecasting model can forecast wind speed data in 3-steps ahead after training. All forecasting results are shown in Fig. 4.8 in different steps. The specific evaluation indices of the model are given in Table 4.3. Besides, Table 4.4 shows the improvement of the hybrid WPD-LSTM model when comparing to the single LSTM model.

It can be seen from Fig. 4.8 that the hybrid WPD-LSTM model can predict the wind speed data more accurately when comparing to the single LSTM model. The hybrid WPD-LSTM model can not only fit the trend of the actual data but also predict the extreme points well. However, the forecasting results at minimum points are not satisfactory when comparing to the actual data.

Besides, the deviation between the forecasting results and the actual data increases at the extreme points and slope values. This shows that the generalization ability of the LSTM model is still limited. However, the limitation is released by the hybrid model framework. The decomposition method makes that deviation smaller.

It can also be seen from Fig. 4.8 that the differences between the results of different forecasting steps are not large. The influence of the forecasting

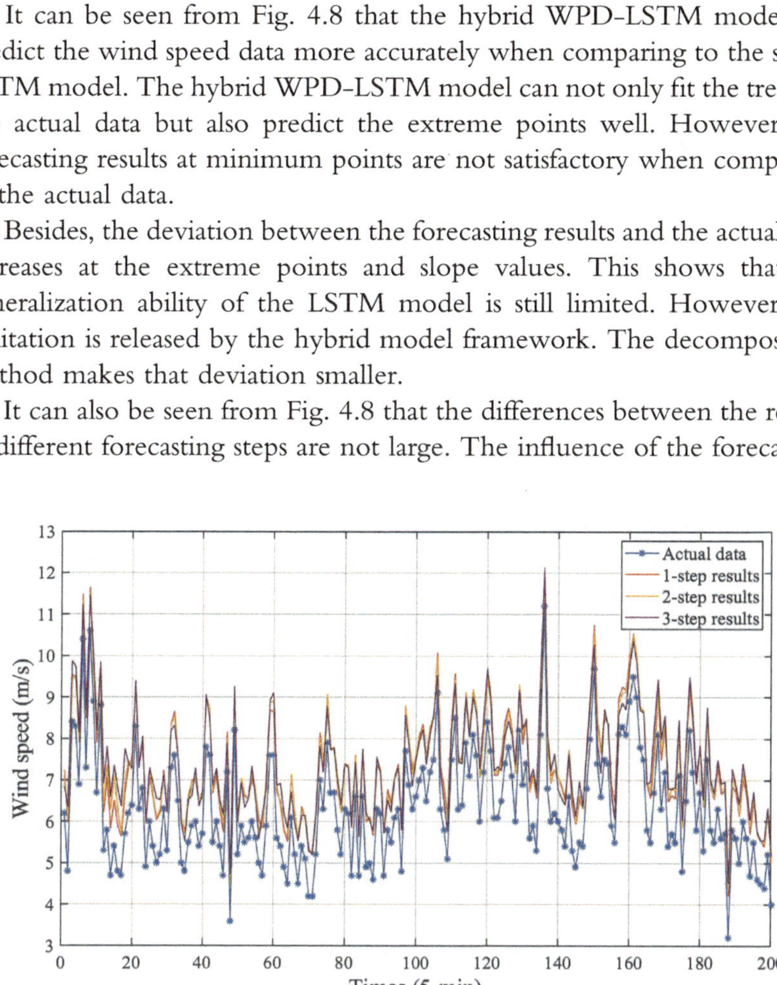

Figure 4.8 Forecasting results of the hybrid WPD-LSTM model.

Table 4.3 Evaluation indices of the hybrid WPD-LSTM model.

Steps	MAE (m/s)	MAPE (%)	RMSE (m/s)
1-Step	1.051	17.337	1.053
2-Step	1.078	17.858	1.089
3-Step	1.104	18.491	1.144

Table 4.4 Improving percentages of the hybrid WPD-LSTM model versus LSTM model.

Steps	P_{MAE} (%)	P_{MAPE} (%)	P_{RMSE} (%)
1-Step	20.330	25.074	33.210
2-Step	20.571	25.634	32.379
3-Step	18.884	23.240	30.017

steps is getting smaller when comparing to the single LSTM model. It is because that the WPD decomposition method reduces the complexity of the original wind data, which makes the LSTM predictors have better forecasting performance. Due to the MIMO strategy adopted in this study, the improvement of LSTM predictors is stronger than the performance decrease caused by increasing forecasting steps.

As shown in Table 4.3, the hybrid WPD-LSTM model has a satisfactory performance. The evaluation indices of all forecasting steps are acceptable. It can also be seen that the accuracy is still declining with the increase of forecasting steps. However, the decrease is small, which proves the effectiveness of hybrid modeling.

As shown in Table 4.4, the hybrid WPD-LSTM model can outperform the single LSTM model in all forecasting steps. The improving percentages of MAPE can be 23%—25% when comparing to the single LSTM model. It demonstrates the improvement and effectiveness of the WPD decomposition method in hybrid modeling. The hybrid WPD-LSTM wind speed forecasting model has a satisfactory multi-step ahead wind speed forecasting performance.

4.3.2.5 Conclusions

According to the above experiments, the following conclusions about the hybrid WPD-LSTM wind speed forecasting model can be drawn:

(a) The WPD data decomposition method can reduce the complexity of the wind speed data series. Different subseries contains a part of the vibration component of the wind speed data series. It is easier for the LSTM predictors to learn different subseries.

(b) The hybrid WPD-LSTM model can predict the short-term wind speed data series accurately. It has a better forecasting performance than the single LSTM model, which proves the effectiveness of the hybrid modeling method. The performance of the hybrid WPD-LSTM model is satisfactory.

(c) The forecasting step still has an effect on the forecasting performance of the wind speed forecasting performance. The forecasting accuracy of the hybrid WPD-LSTM model is decreasing when the forecasting steps increase. However, the MIMO strategy adopted by different LSTM base predictors has an improving effect on the decreasing performance.

4.4 Single-point wind speed forecasting algorithm based on GRU

4.4.1 Single GRU wind speed forecasting model

4.4.1.1 Theoretical basis

GRU is also a modified RNN algorithm which was proposed by Cho et al. in 2014 [12]. Just like the LSTM, GRU can also solve the vanishing gradient problem and explosion gradient problem of traditional RNN [13]. In Section 4.3, the LSTM algorithm is introduced and studied. However, the structure of LSTM is complex. The corresponding training time and forecasting time are relatively long. In general, the hardware cannot meet the requirement of the complicated computing caused by LSTM.

To solve that problem, GRU modified the structure of memory gates. GRU integrates the forget gate and input gate of the LSTM model into one gate: update gate. The number of memory gates are reduced from 3 to 2 in this way. The complexity of the model structure is decreased and the computing speed of the algorithm is increased. Therefore, GRU is preferred in many cases.

Although GRU is also a chain model formed by repeated combination of multiple neural cells, it is much more complex than the general RNN. The neural cell of general RNN is just a simple tanh or Rectified Linear Unit (ReLU) function, etc. But the neural cell of GRU is a relatively complex memory gate structure. The feedforward of GRU are shown as follows [14]:

$$z_t = \sigma(W_z \cdot [o_{t-1}, x_t]) \tag{4.11}$$

where z_t is the update gate, o_{t-1} is the output of the last neural cell, and x_t is the input of this cell. W_z are the weightings of the update gate. $\sigma(x)$ is the sigmoid function. Therefore, the update gate can be calculated by the output of the last cell and the input of this cell as follows [14]:

$$r_t = \sigma(W_r \cdot [o_{t-1}, x_t]) \tag{4.12}$$

where r_t is the reset gate and W_r are the weightings of the reset gate. It can be seen that the calculation process of reset gate is similar with that of the update gate. However, the difference is that the input information from the last cell is discarded when the reset gate value is 0. Only the input information of this cell is used in this situation [14].

$$\widehat{o}_t = \tanh(W_h \cdot [r_t * o_{t-1}, x_t]) \tag{4.13}$$

$$o_t = (1 - z_t)^* o_{t-1} + z_t{}^* \widehat{o}_t \tag{4.14}$$

$$y_t = \sigma(W_o \times o_t) \tag{4.15}$$

where $\widehat{o}_t$ is the output value to be determined in this cell, and o_t is the determined output value in the cell.

It can be seen from the above process that each neural cell in the GRU network is making decisions on the output information. There is a dependency relationship between each neural cell in GRU. In general, the reset gate is more active for short-distance learning and the update gate is more active for long-distance learning.

4.4.1.2 Model structure

The specific structure of the GRU wind speed forecasting model is given in Fig. 4.9. It can be seen from Fig. 4.9 that the structure of the GRU model is simpler than the LSTM model.

As shown in Fig. 4.9, each neural cell only has 2 input: the output information O_{t-1} from the last cell and the input data X_t. Besides, the output O_t and result Y_t are the same. The output is utilized to join the computing of the next cell. In each neural cell, there are only two memory gates: the update gate z_t and the reset gate r_t. Moreover, the number of sigmoid functions is also reduced from 3 to 2.

Therefore, the decrease of the input information, functions, and memory gates results in the decrease of the computation and running time. In this way, the GRU model has better training efficiency and faster computing speed without sacrificing the forecasting performance.

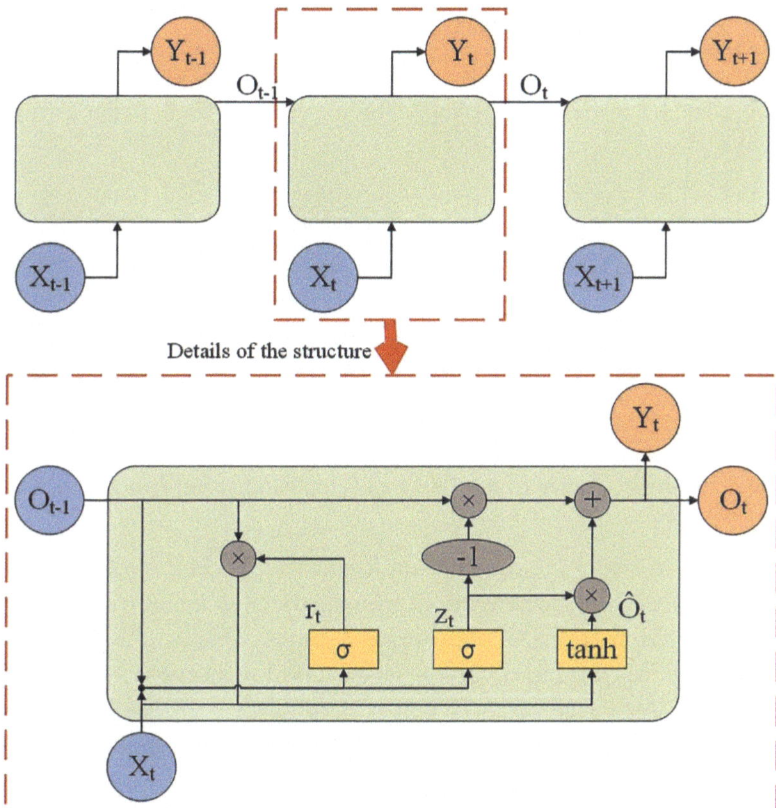

Figure 4.9 The structure of the GRU wind speed forecasting model.

In general, the difference between LSTM and GRU is not great. Both of them are obviously better than the traditional RNN and solve the gradient problems. The difference is that the number of memory gates is reduced from 3 in LSTM to 2 in GRU. This change can reduce part of matrix multiplication. In the case of large input training data, the training and forecasting speed of the model can be significantly improved in this way.

4.4.1.3 Modeling steps

Just like the LSTM model, there are two parts of the GRU training: feedforward and backpropagation. In the training process of the GRU wind speed forecasting model, the BP method is also utilized. In this study, the loss curve of the GRU multi-step ahead wind speed forecasting model is given in Fig. 4.10.

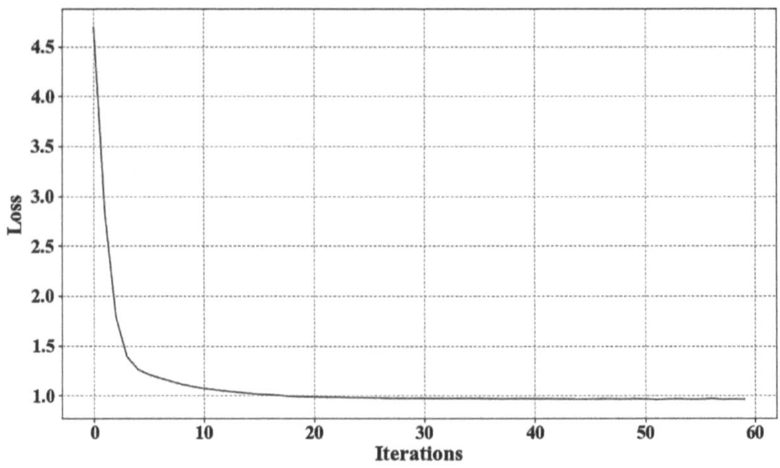

Figure 4.10 The loss curve of the GRU multi-step wind speed forecasting model.

As shown in Fig. 4.10, the training loss of the GRU shows a downward trend. Moreover, the decreasing degree of the loss becomes smaller and smaller with the increase of the iteration times. Finally, the training loss gradually stabilizes around a certain level. The loss curve of GRU is very similar to the loss curve of LSTM. It is because that the weighting of the GRU model is also changing according to the stochastic gradient descend. The process conforms to the general trend of training loss of the deep networks. It also proves that the GRU model can achieve considerable results but is much easier to train.

After training, the weightings and different parameters of the GRU wind speed forecasting model are determined. The model can forecast wind speed data now.

4.4.1.4 Result analysis

The GRU wind speed forecasting model can forecast wind speed data in 3-steps ahead after training. All forecasting results are shown in Fig. 4.11 in different steps. The specific evaluation indices of the GRU model are given in Table 4.5. Besides, Table 4.6 shows the improvement of the GRU model when comparing to the LSTM model.

As shown in Fig. 4.11, the GRU model has a similar forecasting performance to the LSTM model. Both of them can predict the trend of the actual series. Trends of the forecasting results and actual series are basically the same. It proves that the GRU model has the basic ability to predict

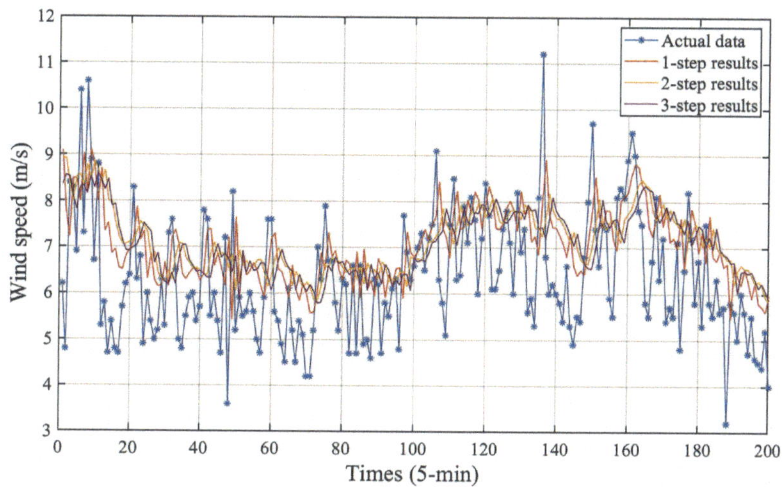

Figure 4.11 Forecasting results of the GRU model.

Table 4.5 Evaluation indices of the GRU model.

Steps	MAE (m/s)	MAPE (%)	RMSE (m/s)
1-Step	1.164	20.405	1.402
2-Step	1.240	21.876	1.488
3-Step	1.239	21.837	1.494

Table 4.6 Comparison results of the GRU and the LSTM model.

Steps	P_{MAE} (%)	P_{MAPE} (%)	P_{RMSE} (%)
1-Step	11.723	11.815	11.113
2-Step	8.621	8.904	7.654
3-Step	8.978	9.351	8.572

wind speed data. It also verifies the considerable performance of GRU when comparing to LSTM.

However, just like the LSTM model, the GRU model cannot predict the extreme points accurately. It can also be seen from Fig. 4.11 that the phenomenon of increasing deviation between the forecasting values and the actual data at the extreme points and slope values. This shows that the single GRU model also lacks generalization ability. The performance of a single GRU model cannot meet the accuracy requirement.

Another common phenomenon between the LSTM model and the GRU model is that the forecasting error is increasing with the increase of the forecasting steps. It shows that the performance of the proposed GRU model can be affected by the forecasting steps to some extent.

As shown in Table 4.5, the evaluation indices of the GRU wind speed forecasting model are acceptable. It can also be seen that the forecasting accuracy is decreasing with the increase of the forecasting steps. However, this decrease is not significant. The performance of 3-step is even better than 2-step. It is because the MIMO strategy is utilized in the GRU wind speed multi-step forecasting model. The influence of the forecasting steps is alleviated in this way.

As shown in Table 4.6, the GRU model can outperform the LSTM model in all forecasting steps. Although the improving percentages are not significant, which ranges from 7% to 11%. It proves that the performance of GRU is considerable when comparing to LSTM deep learning algorithm.

On the whole, the GRU model can predict the wind speed trend. However, its performance at extreme points is not satisfactory. Besides, the GRU model has considerable performance and accuracy when comparing to the LSTM model. And the GRU model has less training time and faster computing speed.

4.4.1.5 Conclusions

According to the above experiments, some conclusions about the GRU wind speed forecasting model can be drawn as follows:

(a) The loss of the single GRU wind speed forecasting model becomes stable before the iteration of 20. It is because the GRU neural network has a simpler structure than the LSTM neural network. The number of memory gate is reduced from 3 to 2. It also proves that the GRU model can achieve considerable performance but is much easier to train.

(b) Similarly, the single GRU model can only forecast the basic changing trend of the wind speed data. The forecasting performance at extreme points is not satisfactory. It shows that the single GRU model cannot meet the requirement of forecasting accuracy.

(c) Similarly, the single GRU model can also be influenced by the wind speed forecasting steps. The forecasting performance of the model sees a fall with the increase of the forecasting steps. However, different results show that the effect is not obvious. The MIMO strategy helps to alleviate the influence of forecasting steps to some extent.

4.4.2 Hybrid EMD-GRU wind speed forecasting model

4.4.2.1 Theoretical basis

In the hybrid EMD-GRU multi-step wind speed forecasting model, the GRU predictor is the same as introduced in Section 4.4.1. In this section, the Empirical Mode Decomposition (EMD) method is introduced [15]. The analysis of linear and stationary signals reflects the physical meaning of signals better than other time–frequency analysis methods.

EMD is a data decomposition method based on the time-scale characteristics without any preset basis function. EMD can decompose complex signals into several Intrinsic Mode Functions (IMFs). The IMFs contain the feature of different time scales. The EMD method identifies all the Intrinsic Oscillatory Mode (IOM) in the signals with the characteristic time scales. In this process, the characteristic time scale and IMFs are both empirical and approximate to some extent. Compared with other signal processing methods, the EMD method is intuitive, indirect, posterior, and adaptive. The mathematical processes are given as follows [11]:

$$m_1(x) = \frac{U_x + L_x}{2} \tag{4.16}$$

where U_x is the upper envelope of the maximum points and L_x is the lower envelope of the minimum points.

$$h_1(x) = S(x) - m_1(x) \tag{4.17}$$

where $S(x)$ is the original signal.

Introduce the IMF criteria: (1) The number of zeros and the number of extreme values in the signal are equal or differ by 1 at most; (2) The mean values of maximum and minimum envelopes are equal to 0.

It $h_1(x)$ cannot meet the IMF criteria, replace $S(x)$ with $h_1(x)$ and repeat processes to get $h_{1m}(x)$ [14]:

$$h_{1m}(x) = h_1(x) - m_{1m}(x) \tag{4.18}$$

Until $h_{1m}(x)$ meets the IMF criteria, mark it as $IMF_1(x)$ [11].

$$r_1(x) = S(x) - IMF_1(x) \tag{4.19}$$

where $r_1(x)$ is the rest signal data. Replace $S(x)$ with $r_1(x)$ and repeat the above processes to get the rest $IMF_i(x)$. In general, the EMD can be explained as follows [11]:

$$S(x) = \sum_{i=1}^{n} IMF_i(x) + r_n(x) \tag{4.20}$$

4.4.2.2 Model structure

The structure of the hybrid EMD-GRU wind speed multi-step forecasting model is given in Fig. 4.12 in detail.

As shown in Fig. 4.12, the hybrid model can be mainly divided into two parts: the decomposition component and predictors. The EMD decomposition method can decompose original wind speed data into eight subseries, including 7 IMFs and 1 residual. And eight GRU predictors are built for each subseries. After the training process, each GRU can obtain the forecasting results corresponding to the subseries. Finally, the forecasting results of different subseries are combined to get the final forecasting results.

4.4.2.3 Modeling steps

Firstly, the EMD method is utilized to decompose the original data. After decomposition, eight subseries are obtained, which are shown in Fig. 4.13.

As shown in Fig. 4.13, there are eight subseries after the EMD decomposition. Among them, S1–S7 are IMFs and S8 is the residual. It can be seen that the S1 varies from −4 to 8 m/s. The subseries S2–S7 all vary from −2 to 2 m/s. This phenomenon indicates that the main energy of the wind speed data is concentrated in the low-frequency subseries. Besides, the residual subseries S8 has the biggest amplitude from 4 to 8 m/s. Because the residual has the most of the original wind speed data and can reflect the trend change of the original signal.

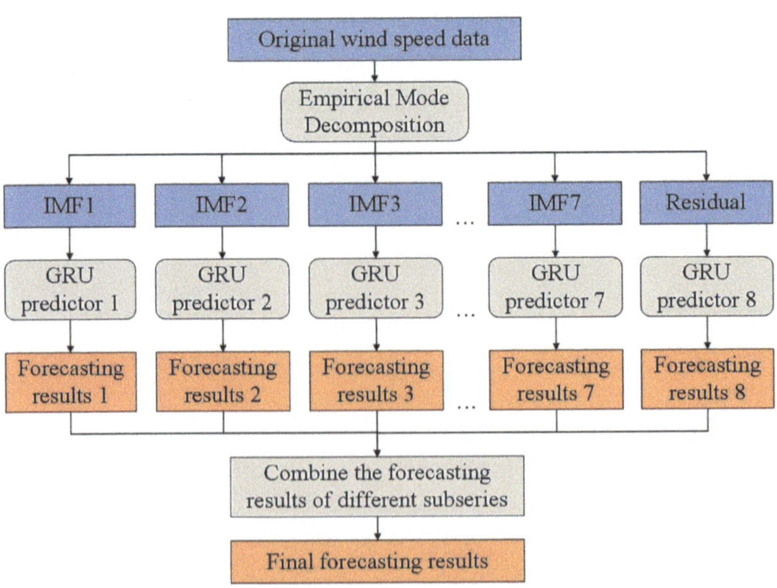

Figure 4.12 The structure of the hybrid EMD-GRU wind speed forecasting model.

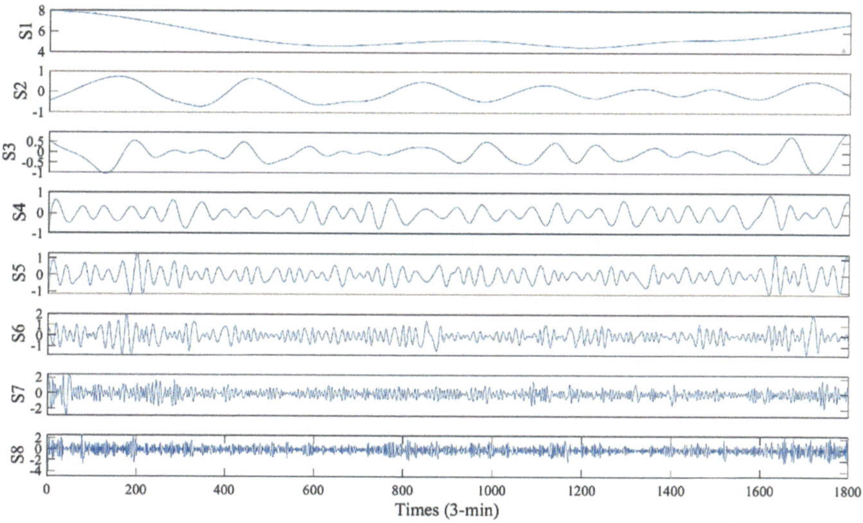

Figure 4.13 Decomposition results of wind speed data after EMD.

The EMD method can autonomously generate different basis functions according to different signals. It decomposes the original wind speed data into different subseries from high frequency to low frequency through the "screening" process. Each subseries contains different time-scale characteristics. In this way, the forecasting of each subseries can reduce the interference between the data with different scales and get more accurate results.

After decomposition, a group of GRU predictors are built and trained for each subseries. In total, eight GRU predictors are built in the hybrid EMD-GRU model. The decomposed results of EMD are used to train each GRU predictor. In each GRU predictor, the output is set as 3 to realize the multi-step ahead forecasting. After iterations, each GRU predictor can obtain the minimum loss and errors. After the training process, the GRU predictors can generate forecasting results of each decomposed subseries of the testing set. Those forecasting results are obtained and combined as the final forecasting results of the testing set.

4.4.2.4 Result analysis

The hybrid EMD-GRU wind speed forecasting model can forecast wind speed data in 3-steps ahead after training. All forecasting results are shown in Fig. 4.14 in different steps. The specific evaluation indices of the model are

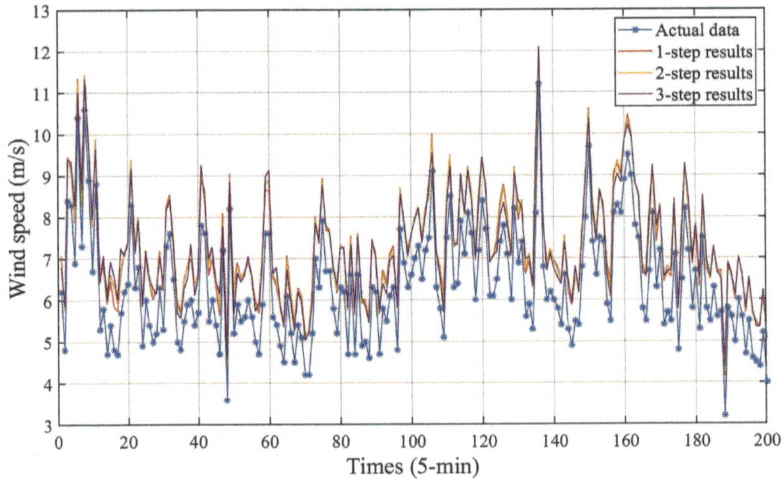

Figure 4.14 Forecasting results of the hybrid EMD-GRU model.

given in Table 4.7. Besides, Table 4.8 shows the improvement of the hybrid EMD-GRU model when comparing to the single GRU model.

It can be seen from Fig. 4.14 that the hybrid EMD-GRU model can forecast the wind speed data accurately. The hybrid model can fit the basic trend of the actual wind speed data. Besides, the hybrid model can also predict the extreme points accurately. Moreover, the deviation between the forecasting results and the actual data increases at the extreme points and slope values. It shows that the EMD decomposition increases the generalization ability of the GRU model. The decomposition method makes that deviation smaller. However, the deviation is still not satisfactory.

Table 4.7 Evaluation indices of the hybrid EMD-GRU model.

Steps	MAE (m/s)	MAPE (%)	RMSE (m/s)
1-Step	0.990	16.354	0.992
2-Step	1.015	16.788	1.023
3-Step	0.982	16.339	1.011

Table 4.8 Improving percentages of the hybrid EMD-GRU model versus GRU model.

Steps	P_{MAE} (%)	P_{MAPE} (%)	P_{RMSE} (%)
1-Step	14.927	19.852	29.241
2-Step	18.160	23.256	31.208
3-Step	20.692	25.177	32.372

Compared with wavelet-based decomposition methods, the EMD has the advantages of automatic generation of basis function, adaptive filtering characteristics, and adaptive multi-resolution. EMD can autonomously generate different basis functions according to different signals. The IMFs of the EMD method contain different characteristics of the original wind speed data. In this way, the prediction of each subseries can reduce the interference between the data with different scales and get more accurate results.

It can also be seen from Fig. 4.14 that the forecasting step still influences the performance and accuracy. However, differences between the results of different forecasting steps are not large. It is because that the decomposition method makes the performance of GRU predictors into full play. The EMD decomposition method reduces the complexity of the original wind data. Due to the MIMO strategy adopted in this study, the improvement of GRU predictors is stronger than the performance decrease caused by increasing forecasting steps.

As shown in Table 4.7, the hybrid EMD-GRU model has a satisfactory performance on wind speed forecasting. The evaluation indices of all forecasting steps are acceptable. The hybrid EMD-GRU model has a satisfactory performance. It can also be seen from Table 4.7 that the 3-step results have the best accuracy. It also proves that the influence of the forecasting step is alleviated by the data decomposition method.

As shown in Table 4.8, the hybrid EMD-GRU model can outperform the single GRU model in all forecasting steps. The improving percentages of MAPE can be 19%–25% when comparing to the single GRU model. It demonstrates the improvement and effectiveness of the EMD decomposition method in hybrid modeling.

4.4.2.5 Conclusions

According to the above experiments, the following conclusions about the hybrid EMD-GRU wind speed forecasting model can be drawn:

(a) The complexity of the wind speed data series can also be reduced by the EMD decomposition method. Different IMFs and the residual series include different time-scale characteristics of the original wind speed data. The GRU base predictors can learn the decomposed subseries more easily.

(b) The hybrid EMD-GRU model can predict the short-term wind speed data series accurately in multi-step. Compared with the single GRU model, the hybrid model has improving percentages of around 20%.

The hybrid modeling and EMD decomposition are verified to be effective with the GRU neural network. The performance of the hybrid EMD-GRU model is satisfactory.

(c) The forecasting step is still the major influence factor of the forecasting performance of the hybrid EMD-GRU model. The forecasting accuracy of the hybrid model is decreasing with the increase of the forecasting steps. However, the MIMO strategy adopted by different GRU base predictors has an improving effect on the decreasing performance.

4.5 Single-point wind speed direction algorithm based on Seriesnet

4.5.1 Single Seriesnet wind direction forecasting model

4.5.1.1 Theoretical basis

Aiming at time series forecasting, Seriesnet is designed based on the WaveNet architecture [16]. SN can also be regarded as a dilated casual convolutional neural network. It was found that the SN method can achieve comparable performance without data preprocessing and ensemble methods [16,17]. The SN network improves the convolutional neural networks' ability of time series forecasting in previous work.

SN can fully learn features and information of the time series in different interval lengths. The SN network was developed based on the WaveNet architecture and dilated casual convolution. The dilated casual convolutional neural network was presented to solve the loss of resolution and coverage problems generated by the down-sampling operation in image semantic segmentation [18]. The dilated convolutional neural network systematically aggregates multi-scale contextual information. The casual convolutional ensures the convolution operation in time series. The dilated casual convolutional neural network is shown in Fig. 4.15.

As shown in Fig. 4.15, the dilated causal convolution is stacked in the SN. The dilation is doubled from the input layer to output layer. As a result, the receptive field is increasing with the increase of the layers. In this study, there are 2 filter layers and 32 filters of each layer. The input data are transmitted to the dilated casual convolutional neural network.

4.5.1.2 Model structure

The structure of the SN model is given in Fig. 4.16 in detail.

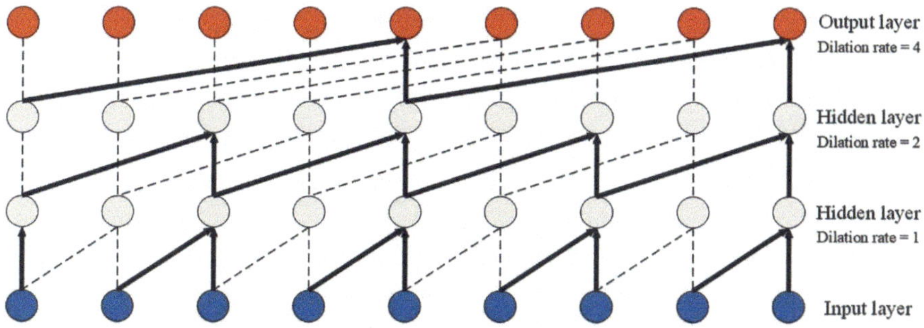

Figure 4.15 A stack of dilated casual convolution.

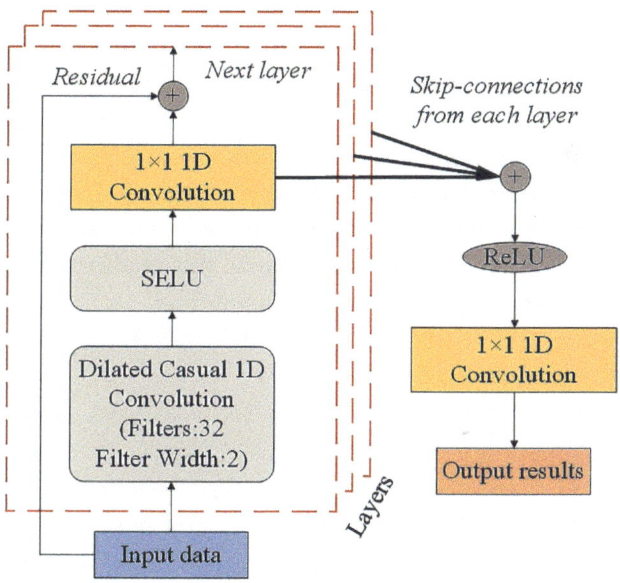

Figure 4.16 The structure of the Seriesnet wind direction forecasting model.

As shown in Fig. 4.16, the main body of each layer is a dilated casual convolutional network with 32 fitters. The filter width size of the network is 2 and the total receptive field of each layer is 128. It can also be seen from Fig. 4.16 that the residual block is added to each layer. Each residual block extracts features and information from the wind direction time series at different levels.

Besides, the Scaled Exponential Linear Unit (SELU) is adopted to remove the bias from the activation. When compared to the Leaky ReLU, the SELU was proved to have better stability, efficiency, and performance [16]. The SELU has the self-normalizing ability, therefore leads to better robustness.

The SN utilizes a long skip-connection operation to integrate the features from each dilated convolutional layer. Then the outputs of these layers are summed and passed to a ReLU activation. The ReLU function is shown as follows [16]:

$$\mathrm{ReLU}(x) = \begin{cases} x, & \text{if } x \geq 0 \\ 0, & \text{if } x < 0 \end{cases} \tag{4.21}$$

After the ReLU activation function, a 1×1 convolution is added to generate the final output results.

4.5.1.3 Modeling steps

In the training process, the Root Mean Square Propagation (RMSProp) algorithm is utilized to train the parameter settings of the SN neural network.

Set the input parameters and data as follows: input the data X, model parameter θ, learning rate η, small constant δ, decay rate ρ, and batch size m. Initiate the gradient accumulation variable $r = 0$ and sample a mini-batch of m examples from the input data. Then the gradient is calculated as follows [16]:

$$\begin{aligned} g &= \frac{1}{m}\nabla_\theta \sum \mathrm{loss} \\ \mathrm{loss}_{\min} &= \frac{1}{T}\sum_{t=1}^{T}\left(Y^{t+1} - x^{t+1}\right)^2 \end{aligned} \tag{4.22}$$

where Y is the forecasting results.

If the stopping criterion is not satisfied, the gradient is accumulated as follows [16]:

$$r = \rho r + (1 - \rho)g^2 \tag{4.23}$$

Then the new parameters are obtained and updated [16]:

$$\theta' = \theta - \frac{\eta}{\sqrt{\delta + r}}g \tag{4.24}$$

The iteration continues until the stopping criterion is satisfied.

The loss curve of the SN multi-step wind direction forecasting model is given in Fig. 4.17.

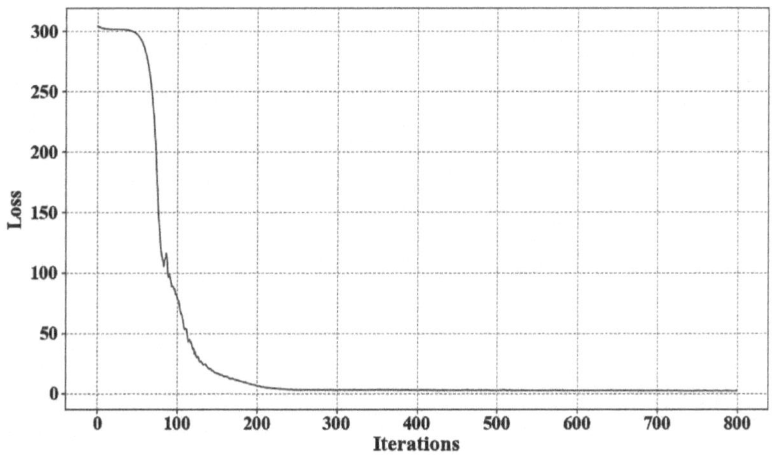

Figure 4.17 The loss curve of the SN multi-step wind direction forecasting model.

As shown in Fig. 4.17, the training loss of the SN model has a downward trend. Moreover, the decreasing degree of the loss becomes smaller and smaller with the increase of the iteration times. Finally, the training loss gradually stabilizes around a certain level. It can also be seen from Fig. 4.17 that the curve is not smoothly declined. The loss curve shows a rough fluctuation at about the 100th iteration. It is because that the weightings change of the SN model also adopts the stochastic gradient descent method.

After training, the weightings and different parameters of the SN wind direction forecasting model are determined. The model can forecast the wind direction data now.

4.5.1.4 Result analysis

The SN wind direction forecasting model can forecast wind speed data in 3-steps ahead after training. All forecasting results are shown in Fig. 4.18 in different steps. The specific evaluation indices of the SN model are given in Table 4.9.

As shown in Fig. 4.18, the SN model can forecast the wind direction accurately. The SN model not only can fit the basic trend of the wind direction data series but also can predict the extreme point well. It can be explained as the following reasons. On the one hand, the wind direction data are easier to fit and learn compared to the wind speed data.

On the other hand, the advantages of the SN network make the forecasting performance great. The SN network adopts dilated casual

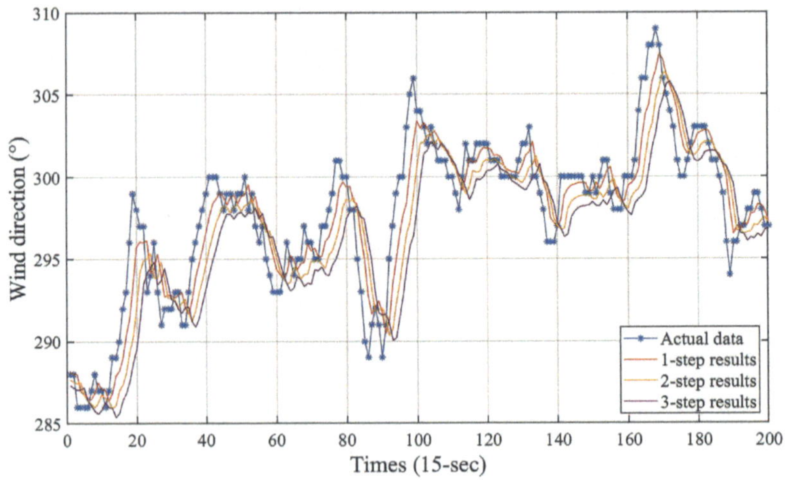

Figure 4.18 Forecasting results of the SN wind direction forecasting model.

Table 4.9 Evaluation indices of the SN wind direction forecasting model.

Steps	MAE (°)	MAPE (%)	RMSE (°)
1-Step	1.363	0.458	1.783
2-Step	2.033	0.682	2.703
3-Step	2.533	0.849	3.378

convolutional neural network and SELU to improve the robustness and stability. The structure of the SN model helps it to make full use of abundant information and hierarchical features in wind direction forecasting.

It can also be seen from Fig. 4.18 that the forecasting performance of the SN model is getting worse when the forecasting steps increase. It shows that the performance of the proposed SN model is affected by the forecasting steps in wind direction forecasting.

As shown in Table 4.9, the MAPE of the SN model can be significantly low. It is because of the stability of the wind direction data. Most wind direction data are stable at around 300 degrees. It is easy for the SN model to fit and learn. The SN model has a satisfactory performance in wind direction forecasting.

It can also be seen from Table 4.9 that the forecasting accuracy is decreasing when the forecasting steps increase. The forecasting step has a relatively significant influence on forecasting performance and accuracy.

On the whole, the single SN neural network model has a satisfactory performance in wind direction forecasting. Due to its advantages, even the single SN model can have better accuracy at extreme points. For further study, the influence of the data decomposition method on the SN model is investigated in the next section.

4.5.1.5 Conclusions

According to the above experiments, some conclusions about the SN wind speed forecasting model can be drawn as follows:

(a) Compared with the LSTM and GRU neural networks, the structure of the Seriesnet is much more complex. The loss of the SN wind speed forecasting model shows stability at the iteration of 300, which is significantly more than the LSTM and GRU models. As a result, the SN model requires more time for training and computing.

(b) The single SN model has a satisfactory performance in short-term wind direction forecasting. Because of the complex structure of the Seriesnet, the single SN model can predict the wind direction well even at extreme points. It proves the effectiveness and improvement of the Seriesnet compared with other earlier deep learning algorithms.

(c) Similarly, the major factor that affects the performance of the SN model is the forecasting step. The forecasting performance of the SN wind direction forecasting model sees a fall with the increase of the forecasting steps. The forecasting step has a relatively significant influence on forecasting performance and accuracy.

4.5.2 Hybrid WPD-SN wind direction forecasting model

4.5.2.1 Theoretical basis

In Section 4.5.1, the SN network and its performance are investigated in wind direction forecasting. In this section, the WPD decomposition will be combined with the SN network to further study its performance under the hybrid modeling framework.

As introduced in Section 4.3.2, WPD is a satisfactory data decomposition method. Therefore, the WPD is selected as the data decomposition method in a hybrid modeling framework. Since the theoretical basis of the WPD decomposition method has been explained in Section 4.3.2, this part will no longer repeat that.

To investigate the influence of the WPD method on the SN network, a hybrid WPD-SN wind direction forecasting model is proposed. The hybrid model is composed of two components: the WPD decomposition and the

SN predictor. SN can be a satisfactory wind direction forecasting model in Section 4.5.1. The theoretical basis of the SN network has been detailly explained in Section 4.5.1, this part will no longer repeat that.

4.5.2.2 Model structure

The structure of the hybrid WPD-SN wind speed multi-step forecasting model is given in Fig. 4.19 in detail.

As shown in Fig. 4.19, the hybrid WPD-SN wind direction model can be mainly divided into two parts: the decomposition component and predictors. The WPD method is used to decompose the original wind direction data into eight subseries. Then eight SN predictors are built for each subseries. After the training process of each SN network, each predictor can obtain the forecasting results corresponding to the subseries. Finally, the forecasting results of different subseries are combined to get the final forecasting results.

4.5.2.3 Modeling steps

Use the WPD method to decompose the wind direction original data, then eight subseries are obtained. The decomposed results are given in Fig. 4.20.

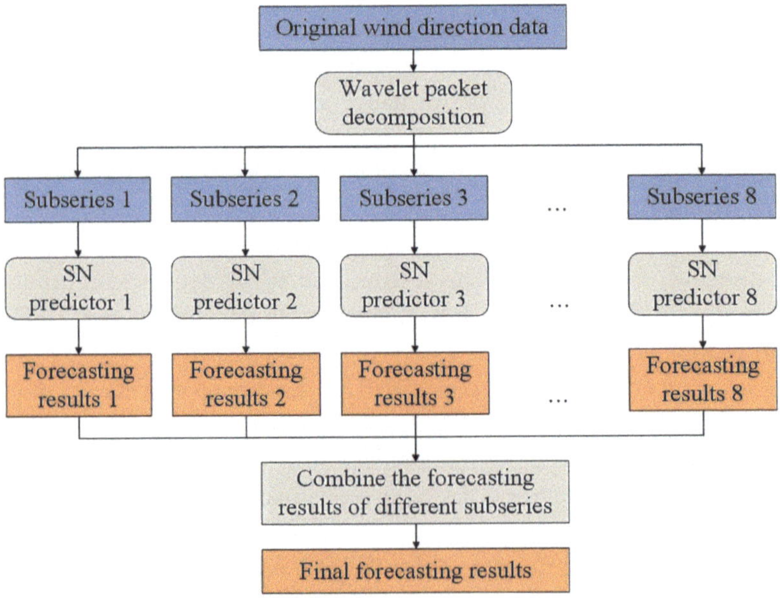

Figure 4.19 The structure of the hybrid WPD-SN wind direction forecasting model.

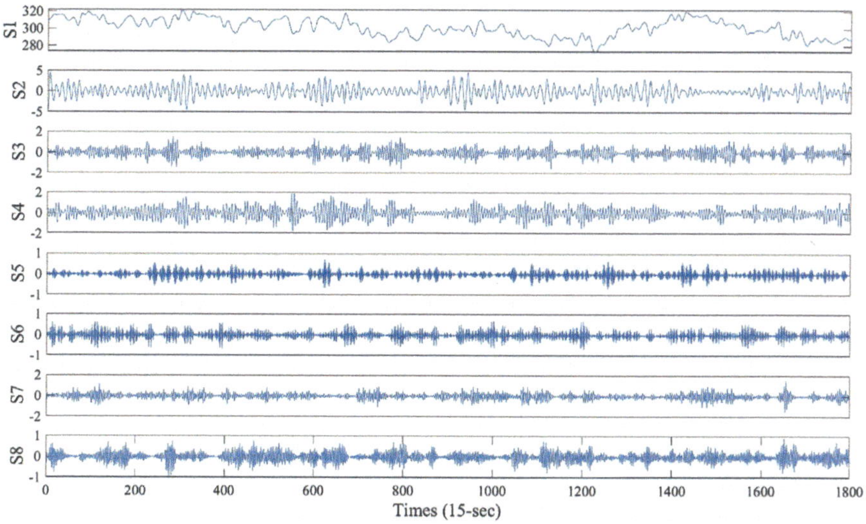

Figure 4.20 Decomposition results of wind direction data after WPD.

As shown in Fig. 4.20, the S1 subseries vary from 280 to 320 degrees. It shows that the S1 series reflects most of the original wind direction data and shows the basic changing trend of the signal. Besides, the rest S2−S8 subseries all vary in an interval of X-axis symmetry. The S2 series has the largest amplitude range from −5 to 5 m/s. The S3−S4 varies between −2 and 2 m/s, which have the second amplitude range. Most of the values of S5−S8 are varying from −1 to 1 m/s, which is the smallest amplitude range. It can be seen that the amplitude decreases with the increase of frequency. This phenomenon indicates that the main energy of the wind direction data is concentrated in the low-frequency subseries.

After the WPD decomposition, a group of SN predictors is built and trained for each subseries. In total, eight SN predictors are built in the hybrid model corresponding to eight subseries. The decomposed results of WPD are used to train each SN predictor. After iterations, each SN predictor can obtain the minimum loss and errors. The SN predictors group is trained in this way.

After the training process, the SN predictors can generate forecasting results of each decomposed subseries of the testing set. Those forecasting results are obtained and combined as the final forecasting results of the testing set.

4.5.2.4 Result analysis

The hybrid WPD-SN wind direction forecasting model can forecast wind direction data in 3-steps ahead after training. All forecasting results are shown in Fig. 4.21 in different steps. The specific evaluation indices of the model are given in Table 4.10. Besides, Table 4.11 shows the improvement of the hybrid WPD-SN model when comparing to the single SN model.

As shown in Fig. 4.21, the proposed hybrid WPD-SN model also has a satisfactory performance and accuracy in wind direction forecasting. However, it is hard to clearly see the difference between the single SN model and the hybrid WPD-SN model from Fig. 4.21. Especially, the hybrid WPD-SN model has an excellent performance in 1-step wind direction forecasting. The hybrid WPD-SN model not only can fit the trend of the actual data but also predict the extreme points well.

It can also be seen from Fig. 4.21 that the deviation between the forecasting values and the actual data increases at the extreme points and slope values is satisfactory. It proves the generalization ability of the hybrid WPD-SN model. Besides, the WPD decomposition method also makes the deviation smaller and improves forecasting performance and stability.

In addition, the forecasting performance of the hybrid WPD-SN model is getting worse with the increase of the forecasting steps. It shows that the performance of the proposed hybrid WPD-SN model can be influenced by the forecasting steps in wind direction forecasting. Even the WPD

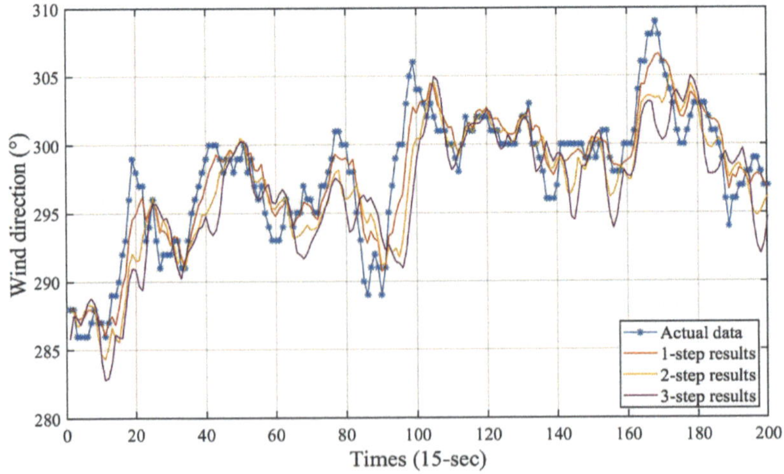

Figure 4.21 Forecasting results of the hybrid WPD-SN model.

Table 4.10 Evaluation indices of the hybrid WPD-SN model.

Steps	MAE (°)	MAPE (%)	RMSE (°)
1-Step	1.249	0.420	1.558
2-Step	2.048	0.687	2.642
3-Step	2.743	0.920	3.536

Table 4.11 Improving percentages of the hybrid WPD-SN model versus SN model.

Steps	P_{MAE} (%)	P_{MAPE} (%)	P_{RMSE} (%)
1-Step	8.394	8.373	12.627
2-Step	−0.744	−0.788	2.264
3-Step	−8.286	−8.438	−4.661

decomposition method can reduce the complexity of the original wind direction data, the forecasting step still has a significant influence on forecasting performance.

As shown in Table 4.10, the hybrid WPD-SN model has satisfactory evaluation indices in wind direction forecasting. The evaluation indices of all forecasting steps are acceptable. It can also be seen that the accuracy is decreasing with the increase of forecasting steps. The hybrid WPD-SN model has a satisfactory performance in wind direction forecasting.

It can also be seen from Table 4.10 that the MAPE of the forecasting results can be significantly low. It is because of the stability of the wind direction data. Most wind direction data are stable at around 300 degrees. It is easy for the hybrid WPD-SN model to fit and learn.

It can be seen from Table 4.11 that the hybrid WPD-SN model has a similar forecasting accuracy with the single SN model. When comparing the WPD-SN model with the SN model, the improving percentages are from −8% to 8%. In general, the influence of hybrid modeling and data decomposition on the SN network is not obvious. Because of the stability of the original wind direction data, the effect of the data decomposition method is not obvious. Besides, the single SN network has a satisfactory performance in wind direction forecasting. Due to the robust structure of the SN network, hybrid modeling has a smaller influence on it.

4.5.2.5 Conclusions

According to the above experiments, the following conclusions about the hybrid WPD-SN wind speed forecasting model can be drawn:

A. Even the wind direction data series is quite stable, the WPD decomposition method can also decompose the original wind direction data. The complexity is reduced. Besides, it is easier for different SN predictors to learn the decomposed subseries.

B. Compared with the single SN model, the improvement of the hybrid WPD-SN model is limited, even with some accuracy falling at the 3-step forecasting step. In general, the influence of hybrid modeling and data decomposition on the SN network is not obvious. Thanks to the robust structure and good stability of the Seriesnet, the single SN model is enough to provide satisfactory performance in short-term wind direction forecasting.

C. Similarly, the major factor that affects the performance of the hybrid WPD-SN model is still the forecasting step. The forecasting performance of the hybrid WPD-SN wind direction forecasting model sees a fall with the increase of the forecasting steps.

4.6 Summary and outlook

In this chapter, the deep learning algorithms in the field of wind forecasting are detailly introduced. Compared with the traditional machine learning algorithm, a deep learning algorithm has deeper network layers and a more complex structure. It has been widely used in image processing, speech recognition, and other fields. Besides, a deep learning algorithm can also be utilized and applied in wind forecasting. In this chapter, deep learning algorithms have been successfully proved to be effective and accurate in the field of both wind speed forecasting and wind direction forecasting. Moreover, the influence and improvement of hybrid modeling on deep learning networks are also studied.

In Section 4.3, the LSTM neural network is introduced and studied in wind speed forecasting. LSTM is a modified RNN that has an architecture of three memory gates, including input gate, forget gate, and output gate. The results show that the single LSTM model can have a satisfactory performance in wind speed forecasting. But the single LSTM model cannot predict the extreme points of wind speed data well. Besides, the effect of the WPD decomposition method is studied. A hybrid WPD-LSTM model is proposed to forecast wind speed in multi-step ahead. The effectiveness of the hybrid WPD-LSTM model has been verified. The WPD decomposition method can improve the LSTM neural network significantly. In general, the LSTM deep learning algorithm and its hybrid models can achieve satisfactory performance in wind forecasting.

In Section 4.4, the GRU neural network is introduced in detail. Its performance in wind speed forecasting has been studied. GRU can be seen as a simplified LSTM network. GRU is proposed to reduce the complexity of the LSTM structure to speed up the computation. The GRU network only has two memory gates: update gate and reset gate. GRU can achieve considerable performance when reducing the calculating time. The results show that the single GRU model can have a satisfactory performance in multi-step wind speed forecasting. Still, the single GRU model cannot predict the extreme points of wind speed data well. For further study, the effect of the EMD decomposition method is studied. A hybrid EMD-GRU model is proposed to forecast wind speed in multi-step ahead. The effectiveness of the hybrid EMD-GRU model has been verified. The EMD decomposition method can improve forecasting performance and accuracy, significantly. In general, the GRU deep learning algorithms can achieve a considerable forecasting performance in wind forecasting with shorter calculating and training time.

In Section 4.5, the SN neural network is introduced and studied in wind direction forecasting. SN has a more complex architecture when comparing to other deep networks. In the SN, the dilated casual convolutional neural network is used to learn different time intervals. The results show that the single SN model can have an accurate performance in multi-step wind direction forecasting. Even at extreme points, the single SN model has a satisfactory accuracy. However, it seems that the SN model can be more influenced by the forecasting step. Besides, the effect of the WPD decomposition method is studied. A hybrid WPD-SN model is proposed to forecast wind direction in multi-step. The results show that the improvement of the hybrid WPD-SN model is not significant. It proves that the SN method can achieve comparable performance without data decomposition methods. In general, the SN deep learning algorithm can achieve satisfactory performance in wind forecasting even without hybrid modeling.

In the future, other deep learning networks with more efficient architecture can be adopted in the field of wind forecasting. With the increase of the hardware basis, hybrid models with a more complex structure based on deep learning algorithms can be proposed. For example, other data preprocessing methods and ensemble methods can also improve the forecasting performance of deep neural networks. Optimization methods can optimize the parameter settings of deep algorithms.

Besides, with the development of big data analysis techniques, wind forecasting will also have a new promotion. Wind big data can further improve the performance of the hybrid wind forecasting models. More features and information can be learned by the neural networks from the wind big data. Moreover, different indices in wind forecasting, such as wind speed, wind direction, and wind pressure, can also be integrated to improve the forecasting performance.

References

[1] X. Hao, G. Zhang, S. Ma, Deep learning, Int. J. Semantic Comput. 10 (2016) 417–439.

[2] Y. Lecun, Y. Bengio, G. Hinton, Deep learning, Nature 521 (2015) 436.

[3] Y. Bai, Y. Li, B. Zeng, et al., Hourly PM2.5 concentration forecast using stacked autoencoder model with emphasis on seasonality, J. Clean. Prod. 224 (2019) 739–750.

[4] S. Gamage, J. Samarabandu, Deep learning methods in network intrusion detection: a survey and an objective comparison, J. Netw. Comput. Appl. 169 (2020) 102767.

[5] B. Liu, S. Nie, Y. Zhang, et al., Boosting noise robustness of acoustic model via deep adversarial training, in: 2018 IEEE International Conference on Acoustics, Speech and Signal Processing (ICASSP), 2018, pp. 5034–5038.

[6] Z. Peng, S. Peng, L. Fu, et al., A novel deep learning ensemble model with data denoising for short-term wind speed forecasting, Energy Convers. Manag. 207 (2020) 112524.

[7] Y.-Y. Hong, C.L.P.P. Rioflorido, A hybrid deep learning-based neural network for 24-h ahead wind power forecasting, Appl. Energy 250 (2019) 530–539.

[8] S. Hochreiter, J. Schmidhuber, Long short-term memory, Neural Comp. 9 (1997) 1735–1780.

[9] A. Graves, J. Schmidhuber, Framewise phoneme classification with bidirectional LSTM and other neural network architectures, Neural Network 18 (2005) 602–610.

[10] Y. Li, H. Wu, H. Liu, Multi-step wind speed forecasting using EWT decomposition, LSTM principal computing, RELM subordinate computing and IEWT reconstruction, Energy Convers. Manag. 167 (2018) 203–219.

[11] A. Subasi, S. Jukic, J. Kevric, Comparison of EMD, DWT and WPD for the localization of epileptogenic foci using Random Forest classifier, Measurement 146 (2019) 846–855.

[12] K. Cho, B. Van Merriënboer, C. Gulcehre, et al., Learning Phrase Representations Using RNN Encoder-Decoder for Statistical Machine Translation, 2014 arXiv preprint arXiv:14061078.

[13] L. Ding, W. Fang, H. Luo, et al., A deep hybrid learning model to detect unsafe behavior: integrating convolution neural networks and long short-term memory, Autom. ConStruct. 86 (2018) 118–124.

[14] B. Wang, J. Wang, Energy futures and spots prices forecasting by hybrid SW-GRU with EMD and error evaluation, Energy Econ. 90 (2020) 104827.

[15] N.E. Huang, Z. Shen, S.R. Long, et al., The empirical mode decomposition and the Hilbert spectrum for nonlinear and non-stationary time series analysis, Proc. Royal Soc. London Series A: Math. Phys. Eng. Sci. 454 (1998) 903–995.

[16] K. Papadopoulos, Seriesnet: a dilated causal convolutional neural network for forecasting, in: Proceedings of the International Conference on Pattern Recognition and Machine Intelligence, Union, NJ, USA, 2018, pp. 1–4.
[17] Z. Shen, Y. Zhang, J. Lu, et al., A novel time series forecasting model with deep learning, Neurocomputing 396 (2020) 302–313.
[18] F. Yu, V. Koltun, Multi-scale Context Aggregation by Dilated Convolutions, 2015 arXiv preprint arXiv:151107122.

CHAPTER 5

Single-point wind forecasting methods based on reinforcement learning

Contents

Wind Forecasting in Railway Engineering
ISBN 978-0-12-823706-9
https://doi.org/10.1016/B978-0-12-823706-9.00005-3

5.1 Introduction

Reinforcement Learning (RL) is a type of machine learning designed to solve optimal decision-making problems that cannot be solved by supervised learning and unsupervised learning. RL attempts to learn an optimal strategy through the constant interaction between the agent and the environment to maximize the final reward [1]. During the interaction with the environment, the agent selects an action to execute in the current state at each step according to the current strategy [2,3]. After the environment is affected by the action, it returns a new state and rewards the agent. The reward is used to instantly evaluate how well the agent took the action in the old state. Through continuous trial and strategy improvement, the agent can learn a good strategy that can maximize the final profit and meet the decision-making needs.

Q-learning algorithm and State-Action-Reward-State-Action (SARSA) algorithm are two classical RL algorithms, which are used to solve the RL problems defined in discrete state space and discrete action space. Deep Q-Network (DQN) and Deep Deterministic Policy Gradient (DDPG) are two Deep Reinforcement Learning (DRL) algorithms. Deep Neural Networks (DNNs) are introduced as function approximators to approximate strategy functions and value functions. The combination of the perception ability of Deep Learning (DL) and the decision-making ability of RL can solve the problem that the traditional RL algorithms cannot be applied in continuous space.

Wind speed forecasting is a supervised learning task. Therefore, RL algorithms are usually not directly used for forecasting. However, considering the excellent performance of RL in strategy search and decision-making, through appropriate problem transformation, RL algorithms can be used for optimization and decision-making in various stages of wind speed forecasting. The application of RL in wind speed forecasting mainly includes feature selection, model parameter optimization [4], optimization and selection of weight coefficients of ensemble model [5,6], model selection [7,8], forecasting model construction [9], which are summarized as shown in Fig. 5.1.

By rationally constructing the elements involved in RL: environment E, state space S, action space A, and reward function R, the above optimization and decision-making problems involved in wind speed forecasting can be transformed into RL problems. RL algorithms can then be used to solve the problem.

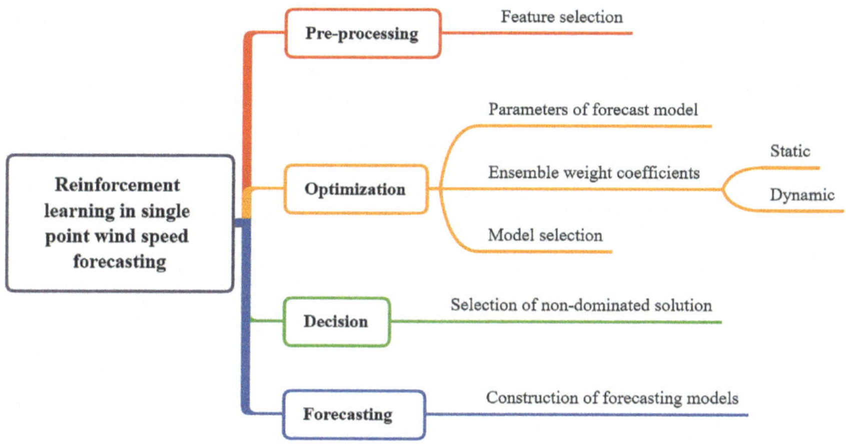

Figure 5.1 Applications of Reinforcement Learning in single-point wind speed forecasting.

For the optimization and decision-making problems involved in wind speed forecasting, since the four-tuple of the Markov decision process $E = \langle S, A, P, R \rangle$ corresponding to the task is usually difficult to determine, the environment cannot be accurately modeled, and it is impossible to simulate the same or similar situation as the environment [10]. Therefore, RL used to optimize wind speed forecasting is usually model-free RL.

5.2 Wind data description

The performance of wind speed forecasting based on RL is verified on the real wind speed dataset collected from a strong wind railway. The sampling interval of this dataset is 3 min, and the sampling unit is meters per second (m/s). In this chapter, 2000 consecutive samples are intercepted for experimentation. Among them, the data in the range of 1st−1500th are used to form the training set, the data from 1501th to 1800th are used to form the verification set, and the data from 1801th to 2000th are used to form the testing set. The wind speed time series and its division are shown in Fig. 5.2.

It is worth noting that due to the needs of specific experiments, in the subsequent sections, the division of each dataset has been reorganized. The specific explanation is given in the corresponding part of the subsequent sections.

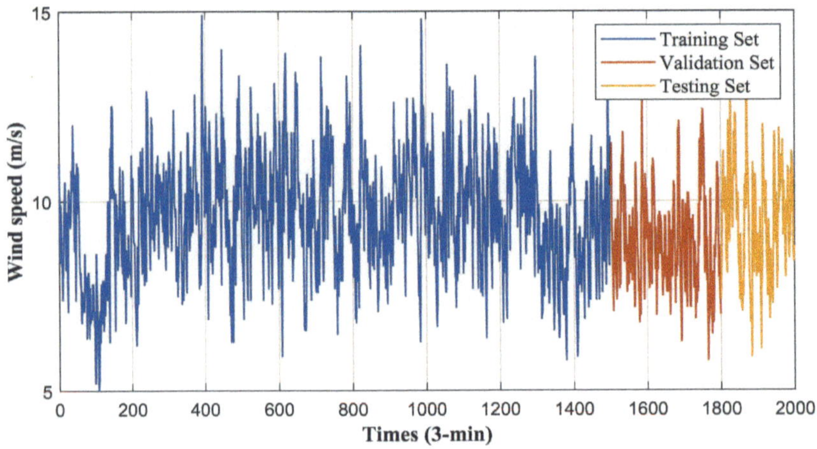

Figure 5.2 Wind speed time series and its division.

Several statistical parameters of the wind speed time series are shown in Table 5.1.

To evaluate the accuracy of the performance of wind speed forecasting, several evaluation metrics are adopted. These metrics include Root Mean Square Error (RMSE), Mean Absolute Percentage Error (MAPE), and Mean Absolute Error (MAE). The equations of these indicators are presented in Section 3.6.

5.3 Single-point wind speed forecasting algorithm based on Q-learning

This section will first introduce the Q-learning algorithm and analyze its typical applications in single-point wind speed forecasting. Since the Q-learning algorithm is designed for solving the RL problems defined in discrete state space and discrete action space, the subsequent state space and action space involved in this section are all obtained by discretizing the continuous space at a certain interval.

Table 5.1 Statistical characteristics of wind speed time series data.

Mean	Standard deviation	Minimum	Maximum	Skewness	Kurtosis
9.544 m/s	1.561 m/s	5.000 m/s	14.900 m/s	0.147	2.747

5.3.1 Q-learning algorithm

Q-learning algorithm belongs to temporal differential learning. Compared with the Monte Carlo RL algorithm, it can achieve more efficient model-free learning. Q-learning is an off-policy algorithm. In Q-learning, the experience learned by the agent is stored in the Q table, and the value in the table expresses the long-term reward value of taking specific action in a specific state. According to the Q table, the Q learning algorithm can tell the Q agent which action to choose in a specific situation to get the largest expected reward.

In a certain iteration, the agent of the Q-learning algorithm is in the state s_t at the time t and chooses an action a_t according to the strategy. It receives the reward r_t from the environment and enters the new state s_{t+1}, and then the Q table is updated according to the following equation [2]:

$$Q^{\text{new}}(s_t, a_t) = Q^{\text{old}}(s_t, a_t) + \alpha\left(r_t + \gamma \max_{a} Q^{\text{old}}(s_{t+1}, a) - Q^{\text{old}}(s_t, a_t)\right) \quad (5.1)$$

where α is the learning rate ($0 < \alpha \leq 1$), and γ is the discount factor ($0 < \gamma \leq 1$).

5.3.2 Single-point wind speed forecasting algorithm with ensemble weight coefficients optimized by Q-learning

The forecasting model based on ensemble learning has significantly superior performance compared to the single forecasting models [11]. There are two combination strategies for base forecasting models: assigning weight coefficients to the base models and using a new learning model to combine the output of the base models [12]. To optimize the combination strategy, the traditional method is to use the meta-heuristic optimization algorithm to search for the optimized weight coefficients or the key parameters of the secondary learning model. This section will provide a new combination optimization method from the perspective of RL. By reasonably transforming the weight coefficient search problem into the RL problem, it can be solved using RL algorithms.

In this section, the Q-learning algorithm is used to search and optimize the ensemble weight coefficients of N base forecasting models [5,6]. The optimized weight coefficients and the base forecasting models are then constituted to form a superior ensemble model for final forecasting. The framework of the ensemble model optimized by the Q-learning is demonstrated in Fig. 5.3.

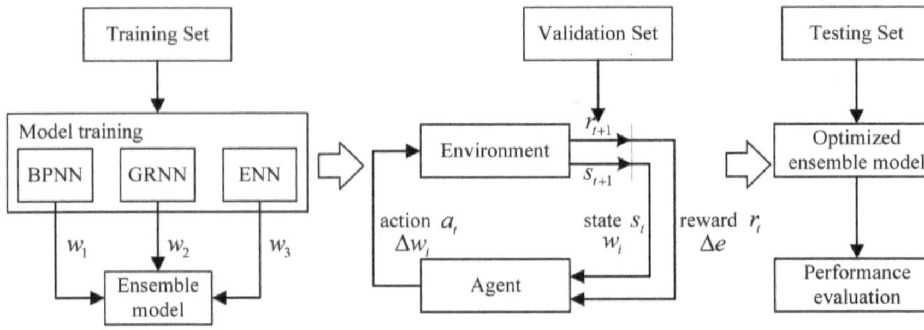

Figure 5.3 Static ensemble wind speed forecasting model with weight coefficients optimized by Q-learning algorithm.

5.3.2.1 Base forecasting models

5.3.2.1.1 Deep belief network

The Deep Belief Network (DBN) is a kind of Deep Neural Network, which is composed of stacked layers of Restricted Boltzmann Machines (RBMs). It is a generative model and was proposed by Geoffrey Hinton in 2006 [13]. DBN can be used to solve unsupervised learning tasks to reduce the dimensionality of features, and can also be used to solve supervised learning tasks to build classification models or regression models. To train a DBN, there are two steps, layer-by-layer training and fine-tuning. Layer-by-layer training refers to unsupervised training of each RBM, and fine-tuning refers to the use of error back-propagation algorithms to fine-tune the parameters of DBN after the unsupervised training is finished [14].

5.3.2.1.2 Long short-term memory

A detailed introduction and description of the long short-term memory (LSTM) network can be found in Chapter 4.

5.3.2.1.3 Gated recurrent units

A detailed introduction and description of the gated recurrent units (GRUs) network can also be found in Chapter 4.

5.3.2.2 Model abstraction

5.3.2.2.1 State *s*

The state s is composed of the ensemble weight coefficients of each base forecasting model, and the expression of the state s_t at the time t is presented as follows:

$$s_t = [w_1, w_2, w_3]$$

$$s.t. \begin{cases} \sum_{i=1}^{3} w_i = 1 \\ \exists x_i \ x_i \geq 0 \quad i = 1, 2, 3 \end{cases} \tag{5.2}$$

where w_1, w_2, and w_3 are the weight coefficients specified for three different single forecasting models.

5.3.2.2.2 Action *a*

The action a consists of the adjustments of each weighting coefficient, and the expression of the action a_t at the time t is shown as follows:

$$a_t = [\Delta w_1, \Delta w_2, \Delta w_3] \tag{5.3}$$

where Δw_1, Δw_2, and Δw_3 are the adjustments of the corresponding weight coefficients.

5.3.2.2.3 Reward *r*

The reward r is calculated based on the forecasting error e_t and e_{t+1}, which are calculated based on the weight coefficients corresponding to the state s_t and s_{t+1}. When e_t is less than e_{t+1}, the agent is punished, and when e_t is greater than or equal to e_{t+1}, the agent is rewarded. The expression of reward r_t is shown as follows:

$$r_t = \begin{cases} -1 + e_t - e_{t+1} & (e_t < e_{t+1}) \\ +1 + e_t - e_{t+1} & (e_t \geq e_{t+1}) \end{cases} \tag{5.4}$$

where the forecasting error e_t is calculated as the Mean Square Error (MSE), and can be presented as follows:

$$e_t = \text{MSE} = \frac{1}{N}\sum_{i=1}^{N} (y_i' - y_i)^2$$

5.3.2.2.4 Agent

Q agent is a value-based RL agent. It maintains a Q table as the critic to estimate the value function. The critic takes state s and action a as inputs and outputs the corresponding long-term reward expectations.

5.3.2.3 Experimental steps

5.3.2.3.1 Training of base forecasting models

Three models DBN, LSTM, and GRU are selected as the base forecasting models. Train the base forecasting models on the training set to obtain the trained models.

5.3.2.3.2 Training of agent

In this RL task, the training and deployment of the agent are completed at the same time. As the agent interacts with the environment, the agent uses strategies to achieve a compromise between exploration and exploitation while learning from the environment. In the end, the agent can maximize the reward and reach a relatively better final state (corresponding to the final optimized weight coefficients).

The training steps of the agent in the environment are given as follows [5,15]:

Step 1: Initialize discount factor γ, learning rate α, greedy parameter ε, Q table, initial state s_0, and initial strategy π.

Step 2: Execute action $a = \pi^{\varepsilon}(s)$ according to ε-greedy strategy π^{ε} of the strategy π. The expression of π^{ε} is expressed as follows:

$$\pi^{\varepsilon}(s) = \begin{cases} \text{action with the largest Q value} & \text{with probability } (1-\varepsilon) \\ \text{randomly select an action} & \text{with probability } \varepsilon \end{cases} \tag{5.5}$$

Step 3: Calculate the instant reward according to Eq. (5.4).

Step 4: Update the Q table according to Eq. (5.1).

Step 5: Set the current state to $s_t = s_{t+1}$.

Step 6: Repeat steps 2 to 5 until the termination condition of iteration is met.

Step 7: When the iteration ends, the weight coefficients corresponding to the current state are considered to be the optimal weight coefficients for combining the base forecasting models.

5.3.2.3.3 Testing of model performance

By combining the static optimal weight coefficients optimized by the Q-learning algorithm and the trained base models, the ensemble model can then be formed. The testing samples are input to the ensemble model for forecasting, and then the forecasting results and the effectiveness of the Q-learning algorithm are analyzed.

5.3.2.4 Result analysis

The forecasting results of each base forecasting model and ensemble model are shown in Fig. 5.4, and the evaluation metrics for forecasting errors are given in Table 5.2. Fig. 5.5 shows the scatter plots of each forecasting model.

It can be seen from Fig. 5.4 that each forecasting model can capture the trend of wind speed changes. However, the predicted value has a significant lag relative to the true value, and the forecasting performance is usually poor at extreme points. The scatter plot between the predicted and true wind speed shown in Fig. 5.5 also shows that the scattered points of the ensemble model are slightly more concentrated. As shown in Table 5.2, the forecasting accuracy of each forecasting model is relatively high. Compared with the various base forecasting models, the static ensemble model optimized by the Q-learning algorithm performs better and has achieved the best in each error evaluation metric.

5.3.3 Single-point wind speed forecasting algorithm with feature selection based on Q-learning algorithm

The feature selection algorithm can select a feature subset composed of relevant features useful for the current task from a high-dimensional original feature set, reducing the learning difficulty of machine learning algorithms. In wind speed forecasting, the feature is composed of historical wind speed, and since it is difficult to determine the correlation between future values

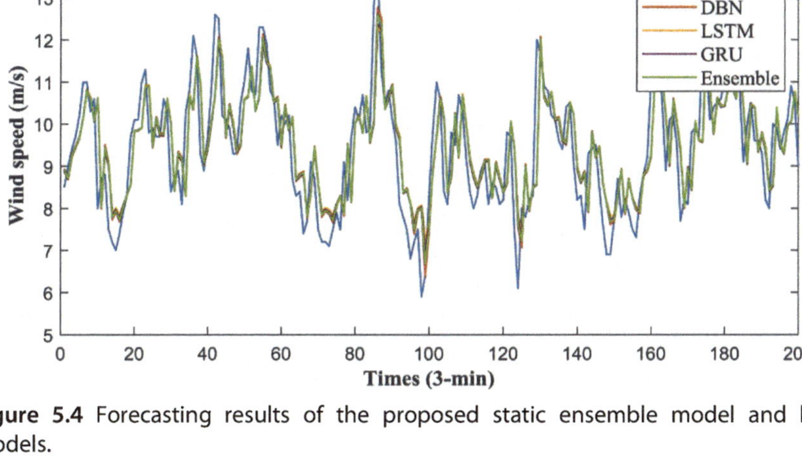

Figure 5.4 Forecasting results of the proposed static ensemble model and base models.

Table 5.2 Error metrics of the proposed static ensemble model and base models.

Forecasting models	MAE (m/s)	MAPE (%)	RMSE (m/s)
DBN	0.716	7.836	0.917
LSTM	0.720	7.895	0.919
GRU	0.716	7.847	0.916
Ensemble	**0.715**	**7.836**	**0.916**

Note: The values in bold represent the best performance.

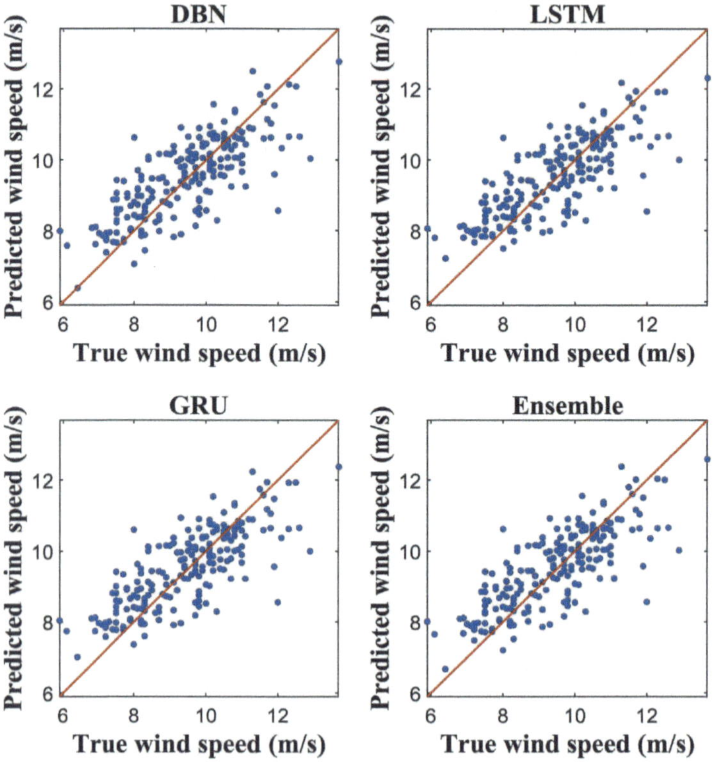

Figure 5.5 Scatter plots of the proposed static ensemble model and base models.

and historical values, it is necessary to conduct feature selection to select relevant features. According to the relationship between the feature evaluation method and learning algorithms, commonly used feature selection methods can be divided into filtering methods, wrapping methods, and embedded methods [16]. This section will combine the RL algorithm and the wrapping method to achieve feature selection. The RL algorithm mainly plays the role of searching for feature subset.

In this section, the original feature set consisting of m features is constructed first, and then the Q-learning algorithm is utilized to perform feature selection based on the wrapping method. After the feature selection stage, a feature subset composed of n features suitable for the current forecasting task is obtained.

According to the selected feature subset, train the wind speed forecasting model to obtain a trained model. The framework of the forecasting model combined with feature selection based on the Q-learning algorithm is shown in Fig. 5.6.

5.3.3.1 Forecasting model

The Elman Neural Network (ENN) is selected as the main forecasting model for this experiment. In the ENN model, the output status of the hidden layer at the current moment is stored and fed back as the input of the hidden layer at the next moment, which is different from the Jordan network where the feedback is taken from the output layer [17,18]. Both the ENN and the Jordan network belong to the Simple Loop Network (SRN). The structure that stores the current output of the hidden layer is called the context layer. Just like traditional feed-forward neural networks, the other three network layers in ENN are the input layer, hidden layer, and output layer. ENN can discover and capture time-varying patterns in

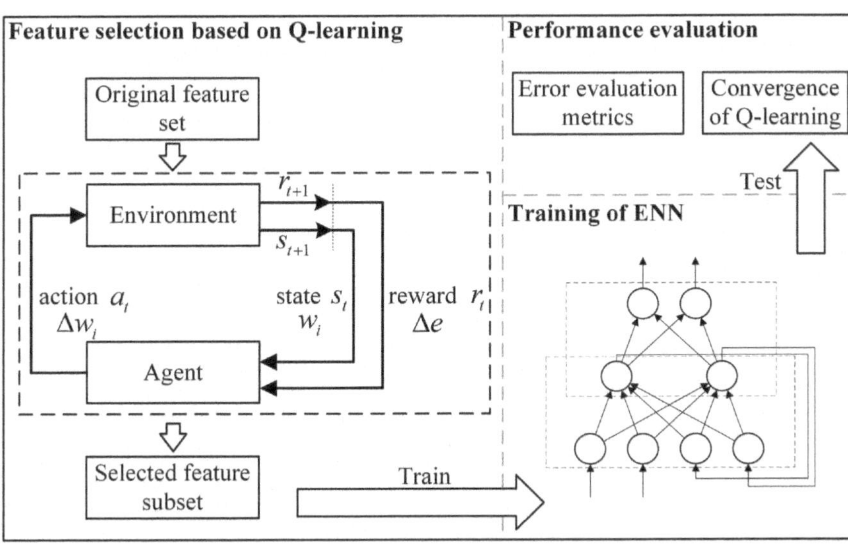

Figure 5.6 Wind speed forecasting model with feature selection based on Q-learning algorithm.

input data, which is also suitable for solving time series forecasting tasks. The equations of each network layer of ENN are given as follows:

$$\begin{aligned} h_t &= \sigma_h\left(W^{h,i}x_t + W^{h,c}h_{t-1} + b_h\right) \\ y_t &= \sigma_y\left(W^{o,h}h_t + b_y\right) \end{aligned} \tag{5.6}$$

where, x_t is the input vector. y_t is the output vector. h_t is the output vector of the hidden layer. $W^{h,i}$, $W^{h,c}$, and $W^{o,h}$ are the connection weights between the input layer and the hidden layer, the context layer and the hidden layer, and the hidden layer and the output layer, respectively. b_h and b_y are the biases of the hidden layer and output layer, respectively. σ_h and σ_y are the activation functions of the hidden layer and output layer, respectively.

5.3.3.2 Model abstraction

5.3.3.2.1 State *s*

The state s is composed of the selection status of each feature. The expression of the state s_t at time t is presented as follows:

$$s_t = [s_1, s_2, \cdots, s_m] \tag{5.7}$$

where s_i is the selection status of the ith feature f_i, and m is the total number of candidate features. The value of s_i can only be 1 or 0. When the ith feature is selected, the value of s_i is set to 1, otherwise, it is set to 0.

5.3.3.2.2 Action *a*

The action a consists of the operations (adding or removing) performed on each feature, and the expression of the action a_t at the time t is shown as follows:

$$a_t = [\Delta s_1, \Delta s_2, \cdots, \Delta s_m] \tag{5.8}$$

where Δs_i is the operation on the ith feature f_i. When the feature f_i has been selected, executing Δs_i will remove the feature from the feature subset, and when it is not selected, executing Δs_i will add the feature into the subset.

5.3.3.2.3 Reward *r*

According to the feature subset corresponding to the state s_t, the forecasting model can be trained and the predicted value can be obtained. Based on the forecasting error e_t, the reward r_t can then be calculated. The reward r_t at time t is calculated by the Pearson Correlation Coefficient (PCC) between the predicted and true values, as shown below:

$$r_t = \text{PCC} = \frac{\sum_{i=1}^{N}(y_i - \overline{y})(y_i' - \overline{y'})}{\sqrt{\sum_{i=1}^{N}(y_i - \overline{y})^2 \times \sum_{i=1}^{N}(y_i' - \overline{y'})^2}} \tag{5.9}$$

where N represents the total number of samples. y_i' and y_i represent the ith predicted value and true value, respectively. $\overline{y'}$ and $\overline{y}$ represent the average of the predicted values and true values, respectively.

5.3.3.3 *Experimental steps*

5.3.3.3.1 Initialization of candidate feature set

Initialize the candidate feature set F, from which the best feature subset is selected. The expression of the candidate feature set is shown as follows:

$$F = [f_1, f_2, f_3, \cdots, f_m] \tag{5.10}$$

5.3.3.3.2 Training of agent

The training steps of the agent in the environment are given as follows:

Step 1: Initialize discount factor γ, learning rate α, greedy parameter ε, Q table, initial state s_0, and initial strategy π.

Step 2: Execute action $a = \pi^{\varepsilon}(s)$ according to ε-greedy strategy π^{ε} of the strategy π. The expression of π^{ε} is expressed as follows:

$$\pi^{\varepsilon}(s) = \begin{cases} \text{action with the largest Q value} & \text{with probability } (1-\varepsilon) \\ \text{randomly select an action} & \text{with probability } \varepsilon \end{cases} \tag{5.11}$$

Step 3: According to the selected feature subset corresponding to the state, a new training set and a validation set can be constructed. Train the ENN on the new training set to obtain a trained neural network model. Then the instant reward r_t is calculated on the validation set according to Eq. (5.9).

Step 4: Update the Q table according to Eq. (5.1).

Step 5: Set the current state to $s_t = s_{t+1}$.

Step 6: Repeat steps 2 to 5 until the termination condition of iteration is met.

Step 7: When the iteration ends, the feature subset corresponding to the current state is considered the optimal feature selection result.

5.3.3.3.3 Testing of model performance

Firstly, use the feature subset optimized by the Q-learning algorithm to reconstruct the final training set and testing set. Secondly, train the forecasting model on the final training set to obtain a trained forecasting model. The testing samples are input to the ENN model for forecasting, and then the forecasting results and the effectiveness of the Q-learning algorithm are analyzed.

5.3.3.4 Result analysis

The forecasting results of the models with and without feature selection are shown in Fig. 5.7, and the evaluation metrics for forecasting errors are given in Table 5.3. Fig. 5.8 shows the scatter plots of each forecasting model.

It can be seen from Fig. 5.7 that each forecasting model can capture the trend of wind speed changes and meet the needs of forecasting. However, the predicted value has a significant lag relative to the true value, and the forecasting performance is usually poor at extreme points. Fig. 5.8 shows that after feature selection, the forecasting value of the ENN model has a better linear correlation with the true value. As shown in Table 5.3, the forecasting accuracy of each forecasting model is relatively high. Compared with the single ENN forecasting model, the ENN model combined with feature selection has better performance.

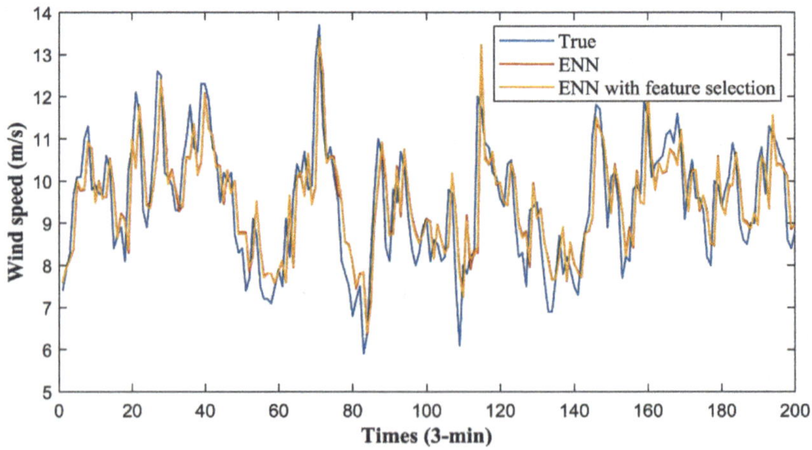

Figure 5.7 Forecasting results of the ENN model and ENN model with feature selection.

Table 5.3 Error metrics of the ENN model and ENN model with feature selection.

Forecasting models	MAE (m/s)	MAPE (%)	RMSE (m/s)
ENN	0.705	7.593	0.914
ENN with feature selection	**0.705**	**7.579**	**0.909**

Note: The values in bold represent the best performance.

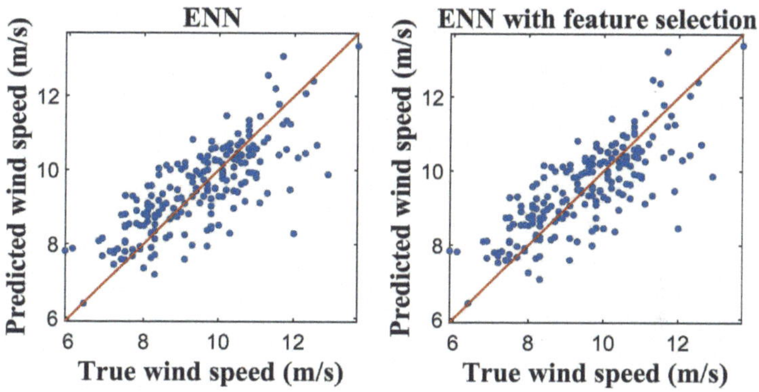

Figure 5.8 Scatter plots of the ENN model and ENN model with feature selection.

5.4 Single-point wind speed forecasting algorithm based on deep reinforcement learning

This section first describes the shortcomings of traditional RL algorithms, and then briefly introduces two DRL algorithms, DQN and DDPG, which are used to build the single-point wind speed forecasting model based on DRL. The DQN algorithm is used to select dynamic weight coefficients in the dynamic ensemble forecasting model, while the DDPG algorithm is directly used for the construction of the forecasting model.

5.4.1 Deep Reinforcement Learning algorithm

The traditional Q-learning algorithm realizes the representation of the RL strategy based on the Q table, which determines that the state space and action space involved are both discrete. This leads to the failure of the Q-learning algorithm when facing continuous states and continuous actions. To solve the above problems, DRL that combines Deep Learning and RL is proposed. With the feature extraction capabilities of Deep Learning and its ability to process continuous data, DRL can then be used to solve practical problems defined in continuous state space and continuous action space.

Typical algorithms in DRL include DQN [19] and DDPG [20]. The DQN algorithm introduces a neural network to realize the representation of the strategy and can solve the problem of RL in the continuous state space, but it is still only applicable to the discrete action space. Although the problem in continuous action space can be solved by discretizing continuous actions, the problem of the dimensional disaster still exists. The DDPG algorithm is proposed by combining the DQN algorithm and the Deterministic Policy Gradient (DPG) algorithm [21]. It can solve the RL problems both in continuous state space and continuous action space. There are four DNNs involved in DDPG, namely the current actor network, target actor network, current critic network, and target critic network. The actor network is used to output corresponding actions based on the input state, and the critic network is used to output the evaluation value of the critic based on two inputs of state and action. The target network will periodically obtain parameter updates from the current network according to the soft update strategy.

5.4.2 Single-point wind speed forecasting algorithm based on DQN

The static ensemble method uses the same weight coefficients for each step of the forecasting, and makes it hard to capture the time-varying characteristics of the fluctuating wind speed. Therefore, the forecasting accuracy of the static ensemble model can still be improved. Considering the different characteristics of wind speed time series in different periods, choosing different ensemble weight coefficients for each base forecasting model at different forecasting steps, will help improve the forecasting performance. In this way, the ensemble forecasting model is improved from a static ensemble forecasting model as shown in Eq. (5.12) to a dynamic ensemble forecasting model as shown in Eq. (5.13).

$$F_s(s) = \sum_{i=1}^{N} w_i f_i(s) \tag{5.12}$$

$$F_d(s) = \sum_{i=1}^{N} w_i(s) f_i(s) \tag{5.13}$$

where s is the feature composed of historical wind speed values and is also the state in RL. N is the number of base models. w_i is the static weight coefficient of the ith model in the static ensemble model, which is a fixed value. $w_i(s)$ is the dynamic weight coefficient of the ith model in the dynamic ensemble model, which is a function of feature s. $f_i(s)$ is the ith base model.

The dynamic model contains N base models. The alternative ensemble weight coefficients of base models are first calculated by the Non-dominated Sorting Genetic Algorithm II (NSGA-II). The multiobjective optimization algorithm NSGA-II is used to obtain the Pareto optimal solution set, from which the dynamic ensemble weight coefficients are selected. The framework of the dynamic ensemble model based on DQN is shown in Fig. 5.9.

5.4.2.1 Multiobjective optimization algorithm

NSGA-II is designed based on the robust Genetic Algorithm (GA), elitism principle, and diversity preservation mechanism. In the framework of wind speed forecasting, NSGA-II is often utilized to optimize the key parameters and ensemble weight coefficients of the forecasting models [12]. The flowchart of the NSGA-II algorithm is presented in Fig. 5.10 [22].

5.4.2.2 Model abstraction

5.4.2.2.1 State *s*

The state s is composed of continuous historical observations of wind speed, so that the characteristics of the wind speed time series over time can be

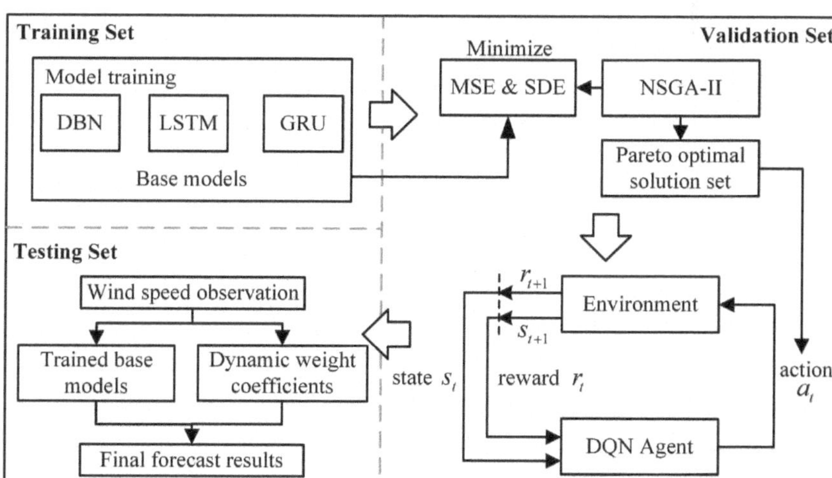

Figure 5.9 Dynamic ensemble wind speed forecasting model based on DQN.

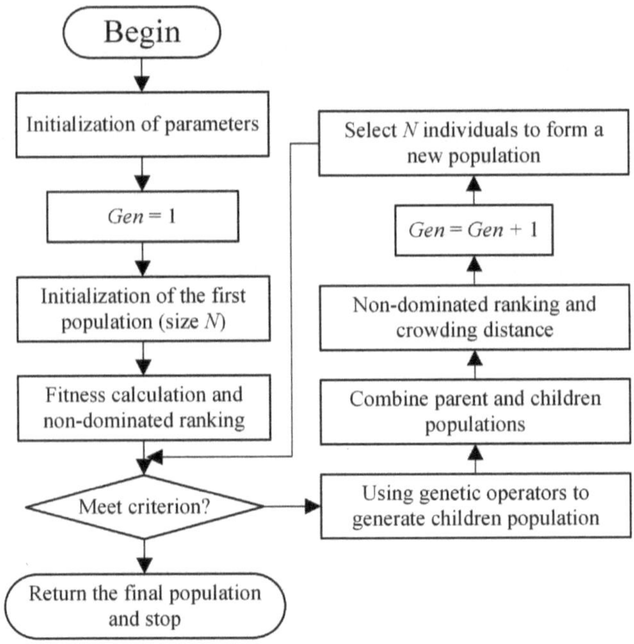

Figure 5.10 Flowchart of the NSGA-II.

described. All features are normalized to $[-1, 1]$ to eliminate the influence of amplitude. The expression of state s_t at the time t is shown as follows:

$$s_t = [x_{t-n}, \cdots, x_{t-3}, x_{t-2}, x_{t-1}] \tag{5.14}$$

where n is the number of historical values required for forecasting, which is equal to the dimension of the feature vector. x_i $(i = t-n, \ldots, t-2, t-1)$ is the true value of wind speed corresponding to time i. t is the time corresponding to the wind speed to be predicted.

5.4.2.2.2 Action *a*

The action a represents the selection of dynamic solutions, and the expression of action a_t at the time t is shown as follows:

$$a_t = [a_{t1}, a_{t2}, a_{t3}, \cdots a_{tp}] \tag{5.15}$$

where p is the total number of optional actions, which is also the number of solutions in the Pareto optimal solution set. If the ith solution is selected in one step, a_{ti} is set to 1, and the rest are set to 0.

5.4.2.2.3 Reward *r*

The reward r is used to evaluate whether the selected dynamic solution is better than the static solution, and is calculated according to the deviation from the forecasting error e_t to the reference forecasting error e_t^{ref}.

Firstly, according to the minimum $MAPE$ criterion, a static compromise solution $\mathbf{w}^* = [w_1^*, w_2^*, w_3^*, \ldots, w_N^*]$ can be selected in the Pareto optimal solution set. The reference error e_t^{ref} can then be calculated according to the forecasting results corresponding to $\mathbf{w}^*$. The expressions of the static solution and the reference error are shown as follows:

$$\mathbf{w}^* = \arg\min_{\mathbf{w}} MAPE\left(\sum_{i=1}^{N} w_i f_i(s)\right) \tag{5.16}$$

$$e_t^{ref} = \left|\sum_{i=1}^{N} w_i f_i(s_t) - y_t\right| \tag{5.17}$$

where N is the number of base forecasting models. w_i represents the ith component of $\mathbf{w}$, which is the static weight coefficient of the forecasting model f_i. y_t is the true value of wind speed at the time t.

Secondly, according to the action a_t selected at the time t and its corresponding weight $\mathbf{w}_t = [w_{t1}, w_{t2}, w_{t3}, \ldots, w_{tN}]$, the absolute value of the forecasting error e_t can be calculated, and the expression is shown as follows:

$$e_t = \left|\sum_{i=1}^{N} w_{ti} f_i(s_t) - y_t\right| \tag{5.18}$$

where w_{ti} is the weight coefficient of the forecasting model f_i at the time t.

By comparing the error e_t with the reference error e_t^{ref}, the reward can be calculated. When e_t is less than e_t^{ref}, the agent is rewarded with v, and when e_t is greater than or equal to e_t^{ref}, the agent is punished with a reward value of $-v$. The expression of reward r_t at the time t is shown as follows:

$$r_t = \begin{cases} v & e_t < e_t^{ref} \\ -v & e_t \geq e_t^{ref} \end{cases} \tag{5.19}$$

5.4.2.2.4 Agent

The structure of the critic network of the DQN agent is shown in Fig. 5.11. The critic network takes state and action as input and outputs the

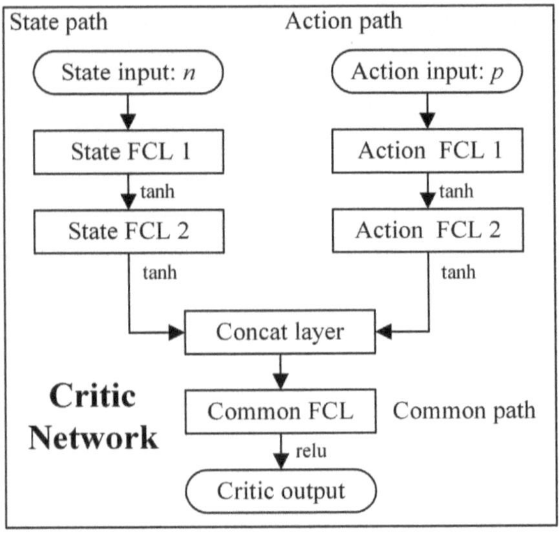

Figure 5.11 Deep network structures of the critic in DQN.

critic value. The two different types of inputs of the critic network are transmitted using independent network paths, and then connected by the concatenation layer.

5.4.2.3 Experimental steps

5.4.2.3.1 Training of base forecasting models

Three forecasting models DBN, LSTM, and GRU are selected as base models. Train these models on the training set to obtain the trained models. In this experiment, the validation and testing set were redivided equally, so that each has 250 samples.

5.4.2.3.2 Multiobjective optimization of ensemble weight coefficients

The NSGA-II is then utilized to search the ensemble weight coefficients of each base model. After the optimization stage, the Pareto optimal solution set from which the candidates are extracted is obtained. The optimization is based on the trained models and is completed in the validation set. Both the MSE and Standard Deviation of Error (SDE) are selected as objective functions to optimize the accuracy and stability of forecasting simultaneously [23]. The objective functions adopted are presented as follows:

$$\text{Minimize}\begin{cases} \text{MSE} = \dfrac{1}{N}\sum_{i=1}^{N}\left(y_i' - y_i\right)^2 \\ \text{SDE} = std\left(y_i' - y_i\right), \quad i = 1, 2, \cdots, N \end{cases} \tag{5.20}$$

where $std()$ means to calculate the standard deviation of its argument.

5.4.2.3.3 Training of agent

Construct an RL environment for agent training and define the agent, use the DQN algorithm to train the agent, and obtain the trained agent for optimal decision-making. The forecasting model optimized by the DQN algorithm is shown as Algorithm 5.1 [22].

Algorithm 5.1. Dynamic ensemble model based on DQN algorithm

Input:

M	Maximum number of episodes
T	Maximum number of steps in each episode
γ	Discount factor
τ	Soft update factor
R	Size of the experience replay pool
m	Number of samples for batch gradient descent
θ	Parameters of the current critic network $Q(s, a\|\theta)$
θ'	Parameters of the target critic network $Q'(s, a\|\theta')$
X	Pareto optimal solution set
$\mathbf{w}^*$	Selected static compromise solution
f_i	Trained forecasting model f_i $i = 1, 2, N$
D_v	Testing data set

Output:

θ	Parameters of the trained critic network
1:	Randomly initialize the current network $Q(s, a\|\theta)$ and target network $Q'(s, a\|\theta')$, where $\theta' = \theta$
2:	**for** episode = 1 to M **do**
3:	Reset the initial state s_1 as the first feature of the validation set
4:	**for** t = 1 to T **do**
5:	Select action a_t according to ε-greedy strategy
6:	Perform action a_t at state s_t, get reward r_t and move to next state s_{t+1}. The main steps include the following:
7:	Select the solution $\mathbf{w}_t$ according to Eq. (5.15)
8:	Calculate the errors e_t^{ref} and e_t according to Eqs. (5.17) and (5.18)
9:	Calculate the reward r_t according to Eq. (5.19)

(Continued)

Set the state as the next feature and get the next state s_{t+1}
Add the experience (s_t, a_t, r_t, s_{t+1}) to the experience playback pool R
Randomly select m samples from R to form a minibatch
Calculate the target value $y_i = r_i + \gamma \max Q'\left(s'_{i+1}, a' \middle| \theta'\right)$ of the value function $Q(s, a|\theta)$
Update current critic network by minimizing the following loss:
$L = \frac{1}{m}\sum_{i=1}^{m}\left(y_i - Q(s_i, a_i|\theta)\right)^2$
Update the target network parameters according to:
$\theta' = \tau\theta + (1 - \tau)\theta'$
end for
end for

5.4.2.3.4 Testing of model performance

The dynamic ensemble forecasting model is finally constructed by combining the trained agent and base forecasting models. At each forecasting time step, the actor network of the agent will be used to dynamically select the appropriate ensemble weight coefficients based on the input feature. The testing samples are sequentially input to the ensemble model for forecasting, and then the error of the forecasting and the effectiveness of the DQN algorithm are analyzed.

5.4.2.4 Result analysis

5.4.2.4.1 Training and deployment of the DQN agent

The rewards obtained by the DQN agent in each episode of training are shown in Fig. 5.12. The average reward is obtained by smoothing the episode reward with an averaging window length of five. The reward converges to the maximum at a faster speed and fluctuates slightly, indicating that it has a strong learning ability and can quickly learn a strategy for choosing superior actions based on the state. The instant reward of each step of the DQN agent in the training environment is shown in Fig. 5.13. It can be seen that the DQN agent can make better action choices in most situations and obtain positive rewards, which proves the effectiveness of the DQN algorithm. As shown in Fig. 5.14, the DQN algorithm selects variable actions (corresponding to variable weight coefficients) at different forecasting steps for each base model, which reflects the key difference between the dynamic ensemble model and static ensemble model. The red line in the figure represents the static solution that was selected for calculating the reference error.

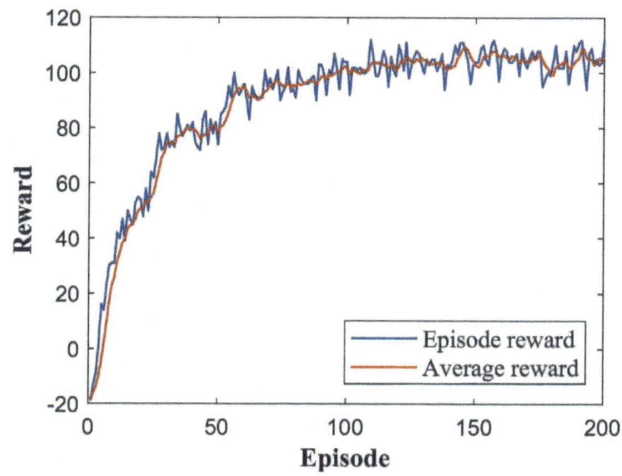

Figure 5.12 Episode reward of DQN agent during training.

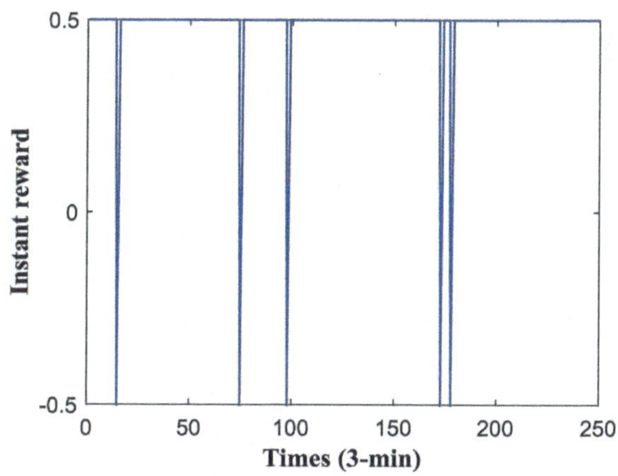

Figure 5.13 Reward for each step of the DQN agent in the training environment.

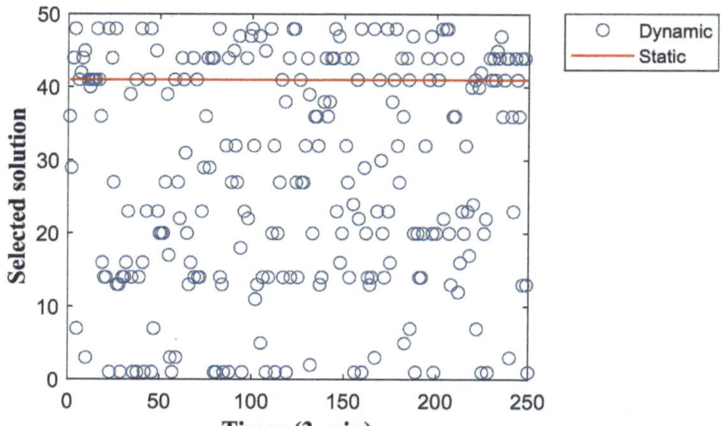

Figure 5.14 Selection results of the Pareto optimal solutions in the testing set.

5.4.2.4.2 Iteration conditions and optimization results of the NSGA-II algorithm

Fig. 5.15 shows the Pareto front calculated by the NSGA-II algorithm, which is convex toward the origin. The compromise solution selected according to the minimum *MAPE* criterion is also shown in the figure. This solution will be used as a static solution to calculate the reward. Fig. 5.16 shows the convergence curve of two objective functions. The value objective functions all reached convergence in a relatively short time.

5.4.2.4.3 Forecasting results and errors of the dynamic ensemble model

The forecasting results of the various forecasting models are demonstrated in Fig. 5.17, and the scatter plots of each forecasting result are shown in Fig. 5.18. The evaluation metrics are presented in Table 5.4.

Fig. 5.17 shows that each forecasting model can capture the trend of wind speed changes. However, the predicted value has a significant lag relative to the true value, and the forecasting performance is usually poor at extreme points. The scatter plot between the predicted and true wind speed shown in Fig. 5.18 also shows that the scattered points of the dynamic ensemble model are slightly more concentrated. As shown in Table 5.4, each forecasting model has relatively high accuracy. Compared with the various base forecasting models, the proposed dynamic ensemble model performs better and has achieved the best in each evaluation metric.

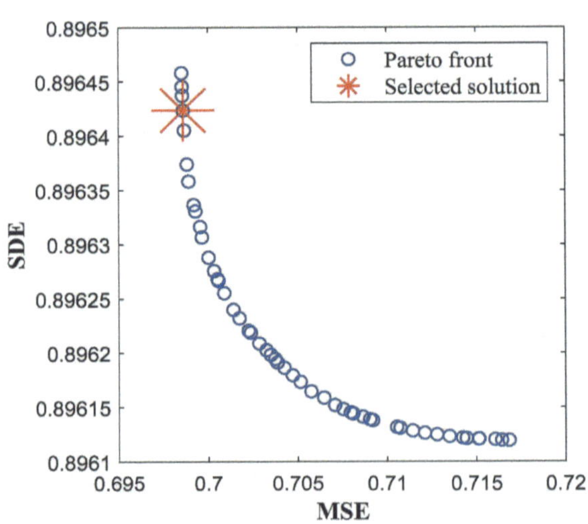

Figure 5.15 Pareto front of NSGA-II and the selected static solution.

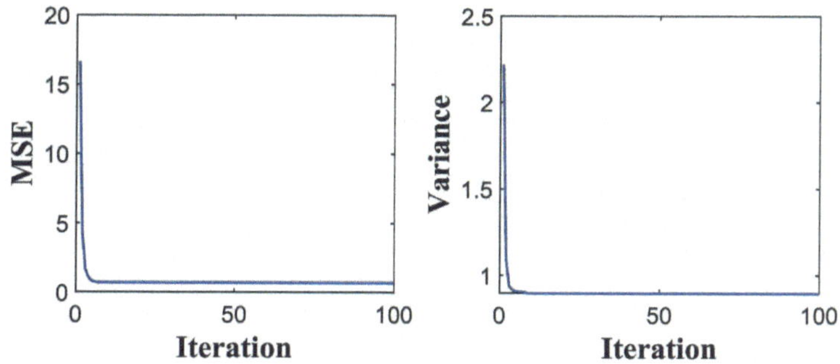

Figure 5.16 Convergence of the average objective function values of each generation during 100 iterations.

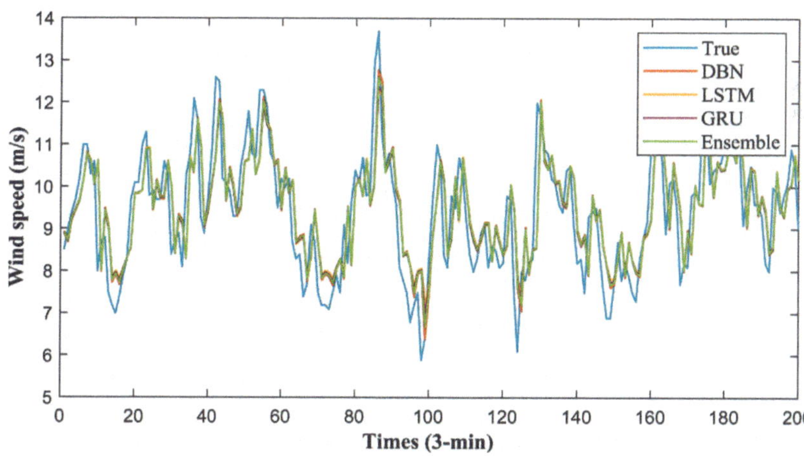

Figure 5.17 Forecasting results of the proposed dynamic ensemble model and base models.

5.4.3 Single-point wind speed forecasting algorithm based on DDPG

In addition to supervised learning methods, DRL algorithms based on actor-critic can also be directly used to build wind speed forecasting models [9]. Commonly used DRL algorithms include Asynchronous Advantage Actor-Critic (A3C), DDPG, and Recurrent Deterministic Policy Gradient (RDPG). Taking the wind speed feature as the state in the RL problem and the forecasting value of wind speed as the action, the time series forecasting problem, which belongs to the supervised learning, can then be solved through the RL method. Fig. 5.19 shows the wind forecasting framework, which includes four key parts: data preprocessing, forecasting model, data

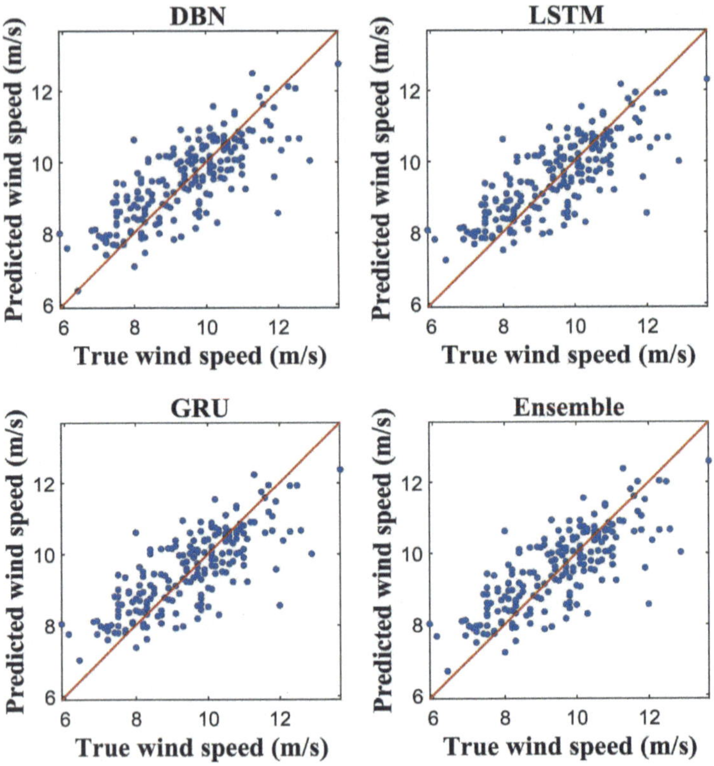

Figure 5.18 Scatter plots of the proposed dynamic ensemble model and base models.

Table 5.4 Error metrics of the proposed dynamic ensemble model and base models.

Forecasting models	MAE (m/s)	MAPE (%)	RMSE (m/s)
DBN	0.716	7.836	0.917
LSTM	0.720	7.895	0.919
GRU	0.716	7.847	0.916
Dynamic ensemble	**0.715**	**7.836**	**0.916**

Note: The values in bold represent the best performance.

postprocessing, and optimization algorithm [12,24,25]. In addition to the supervised learning models, the forecasting models in the framework have been supplemented with DRL-based models.

The wind forecasting based on DRL is done by the actor's function approximator (usually a neural network) in the algorithm, which can be constructed using a variety of superior models. The forecasting model

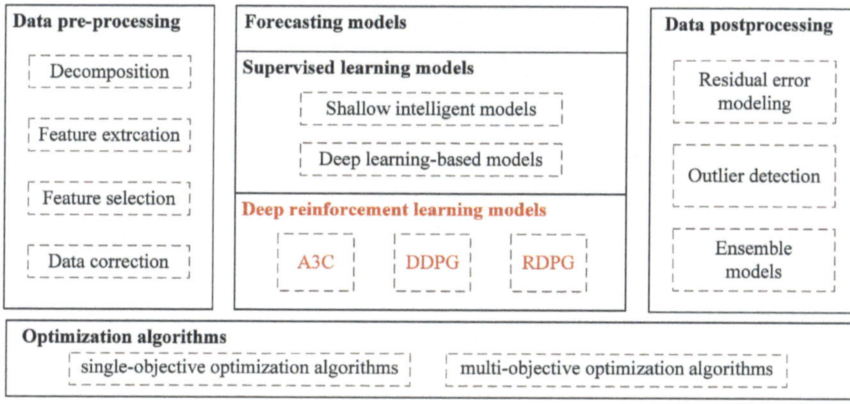

Figure 5.19 Wind speed forecasting framework supplemented with DRL-based forecasting models.

discussed in this section is constructed by the DDPG algorithm, in which the actor network is constructed by a Multi-Layer Perceptron (MLP). Fig. 5.20 shows a schematic diagram of the DDPG-based forecasting model.

5.4.3.1 Model abstraction

5.4.3.1.1 State *s*

The state s is composed of continuous historical observations of wind speed. All features are normalized to $[-1, 1]$ to eliminate the influence of amplitude. The expression of state s_t at the time t is shown as follows:

$$s_t = [x_{t-n}, \cdots, x_{t-3}, x_{t-2}, x_{t-1}] \tag{5.21}$$

where n is the number of historical values required for forecasting, which is equal to the dimension of the feature vector. x_i $(i = t-n, \ldots, t-2, t-1)$ represents the true value of wind speed corresponding to time i. t represents the time corresponding to the wind speed to be predicted.

5.4.3.1.2 Action *a*

The action a represents the predicted value of the DDPG-based wind speed forecasting model, and its dimension is the number of forecasting steps. The expression of action a_t at the time t is shown as follows:

$$a_t = [a_{t1}] \tag{5.22}$$

where the subscript 1 of a_{t1} represents one-step advance forecasting.

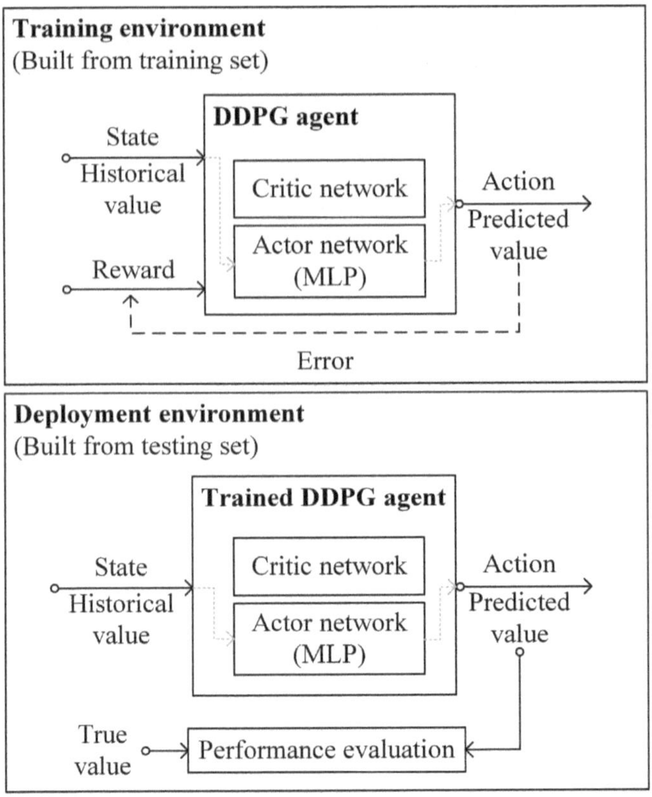

Figure 5.20 Schematic diagram of the DDPG-based forecasting model.

5.4.3.1.3 Reward *r*

The reward r evaluates the action a (predicted value), and is calculated based on the forecasting error. The expression of reward r_t at the time t is shown as follows:

$$r_t = -abs(y_t - a_t) \tag{5.23}$$

where $abs()$ means to take the absolute value of its argument, y_t and a_t represent the true and predicted value at the time t, respectively.

5.4.3.1.4 Agent

The DDPG agent contains critic and actor networks, and the network structures are shown in Fig. 5.21. The critic network has the same structure as the critic network in the DQN agent. However, the dimension of the action input of the critic network is different and defined as one, which is

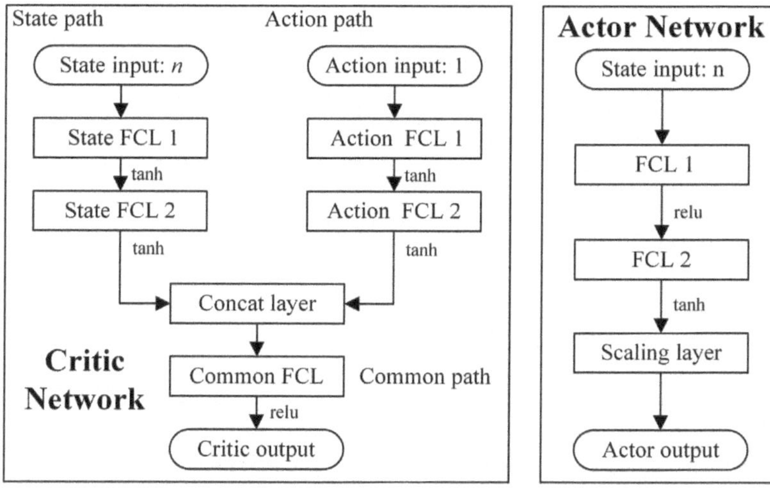

Figure 5.21 Deep network structures of the actor and critic in DDPG.

the number of advance steps. Compared with DQN, the DDPG agent also includes an actor network for receiving state input and outputting actions that should be performed in that state.

5.4.3.2 Experimental steps

5.4.3.2.1 The training process of the DDPG agent

Construct an RL environment for training and define the agent, use the DDPG algorithm to train the agent, and obtain the trained agent for performing forecasting. The DDPG-based forecasting model constructed in this experiment does not use the validation set. Therefore, in this experiment, the training set and validation set shown in Fig. 5.2 are merged as a new training set. The original testing set is still reserved for performance testing. The training process of the DDPG-based model is shown as Algorithm 5.2 [20].

Algorithm 5.2. Wind speed forecasting model based on DDPG algorithm

Input:

M	Maximum number of episodes
T	Maximum number of steps to run in each episode
γ	Discount factor
τ	Soft update factor
R	Size of the experience replay pool

(Continued)

m	Number of samples for batch gradient descent
ϕ	Parameters of the current actor network $\mu(s\|\phi)$
ϕ'	Parameters of the target actor network $\mu'(s\|\phi')$
θ	Parameters of the current critic network $Q(s, a\|\theta)$
θ'	Parameters of the target critic network $Q'(s, a\|\theta')$
P	The noise of the current actor network
f_i	Trained forecasting model f_i $i = 1, 2, N$
D_v	Testing data set

Output:

θ	Parameters of the trained actor network
w	Parameters of the trained critic network

```
Randomly initialize the actor network and critic network with parameters φ and θ, φ' = φ,θ' = θ
for episode = 1 to M do
Reset the initial state s_1 as the first feature of the validation set
for t = 1 to T do
Select action a_t according to the current actor network μ(s_t|φ) and noise P_t
Perform action a_t at states_t, get reward r_t and move to next state s_{t+1}. The main steps include the following:
Get the predicted value according to Eq. (5.22)
Calculate the reward r_t according to Eq. (5.23)
Set the state as the next feature vector and get the next state s_{t+1}
Add the experience (s_t, a_t, r_t, s_{t+1}) to the experience playback pool R
Randomly select m samples from R to form a minibatch
Calculate the target value y_i = r_i + γQ'(s_{i+1}, μ'(s_{i+1}|φ')|θ') of the value function Q(s, a|θ)
Update current critic network by minimizing loss:
    L = 1/m Σ_{i=1}^{m} (y_i − Q(s_i, a_i|θ))^2
Update the current actor network based on gradient loss:
    ∇_θ μ|_{s_i} ≈ 1/m Σ_{i=1}^{m} ∇_a Q(s, a|θ)|_{s=s_i, a=μ(s_i)} ∇_φ μ(s|φ)
Update the target network parameters according to:
    θ' = τθ + (1 − τ)θ'
    φ' = τφ + (1 − τ)φ'
end for
end for
```

5.4.3.2.2 Model performance verification

The trained agent is simulated in the RL environment constructed by the testing set for deployment. The action output of the agent in each step is then obtained, and the predicted value is obtained after denormalization. The analysis of the forecasting results and the effectiveness of the DDPG algorithm are then carried out.

5.4.3.3 Result analysis

5.4.3.3.1 Convergence and reward of the DDPG algorithm

Fig. 5.22 shows the cumulative rewards obtained by the DDPG algorithm in each episode. It can be seen from the figure that as the iteration progresses, the accumulated rewards of the agent in each episode gradually rise and tend to stabilize. Although there are still relatively large fluctuations in the later stages of the iteration, it has converged, which can also be seen from the average reward. The average reward is obtained by smoothing the episode reward with an average window length of 10. Figs. 5.23 and 5.24 show the instant rewards obtained by the trained DDPG agent at each step in the training and deployment environment, respectively. The instant reward represents the magnitude of the forecasting error at each step of the DDPG-based wind speed forecasting model. The rewards obtained by the trained agent at each step are mainly concentrated in the higher value,

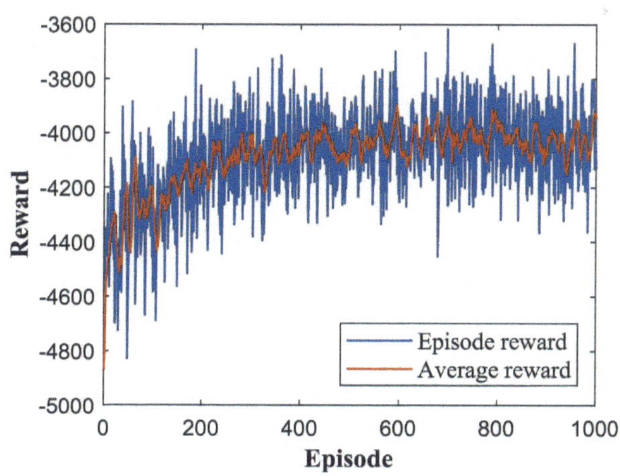

Figure 5.22 Episode reward of DDPG agent during training.

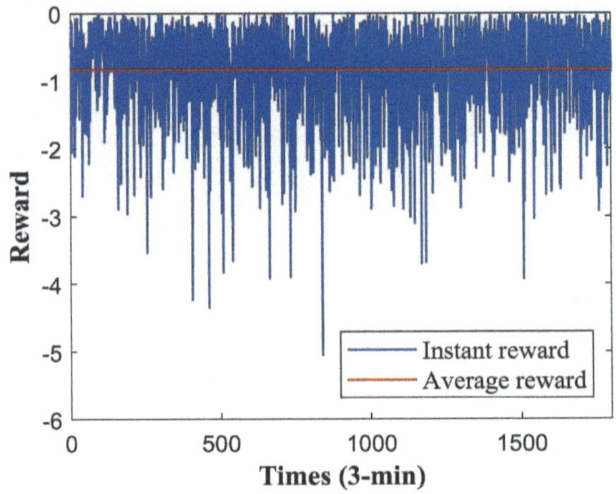

Figure 5.23 Instant reward for each step of the DDPG agent in the training environment.

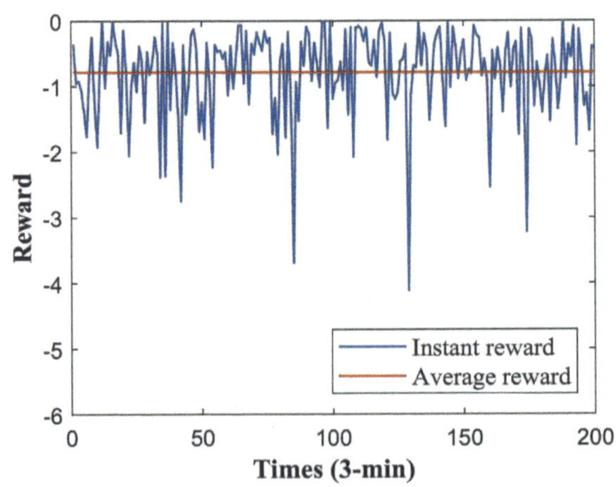

Figure 5.24 Instant reward for each step of the DDPG agent in the deployment environment.

which is close to zero. This shows that the actor network of the agent can produce relatively small forecasting errors at each step, and has high forecasting accuracy, which proves the feasibility of the DDPG algorithm.

5.4.3.3.2 Forecasting results and errors of the DDPG-based model

Fig. 5.25 demonstrates the forecasting results of the MLP model and DDPG-based model. The values of different performance metrics of the forecasting results are shown in Table 5.5. The MLP has the same network

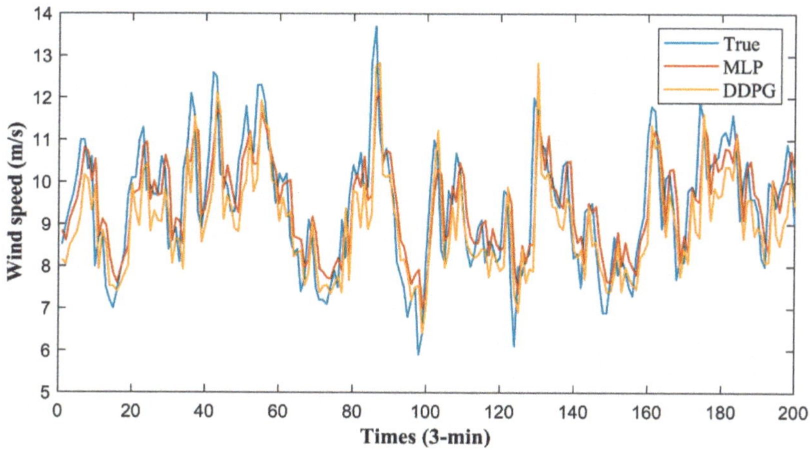

Figure 5.25 Forecasting results of the MLP model and proposed DDPG-based model.

Table 5.5 Error metrics of the MLP model and proposed DDPG-based model.

Forecasting model	MAE (m/s)	MAPE (%)	RMSE (m/s)
MLP	**0.718**	**7.822**	**0.934**
DDPG-based model	0.788	8.141	1.042

Note: The values in bold represent the best performance.

structure as the actor network adopted in the DDPG algorithm, and is trained in the traditional supervised learning method.

On the basis that the forecasting models have the same structure, the forecasting performance of the actor network in the DDPG agent is similar to that of MLP, but slightly worse, and the distribution of scattered points of MLP shown in Fig. 5.26 is relatively concentrated. Besides, the scatter plot also shows that the predicted value of the DDPG-based algorithm has a lower degree of linear correlation with the true value. All metrics listed in Table 5.5 have achieved relatively good values, but the DDPG-based forecasting model does not get the best in each metric, and the performance of the MLP is better.

5.5 Summary and outlook

The application scale of RL in wind speed forecasting is relatively small, and its powerful self-learning ability and decision-making ability still need to be further developed. Since wind speed time series forecasting needs to

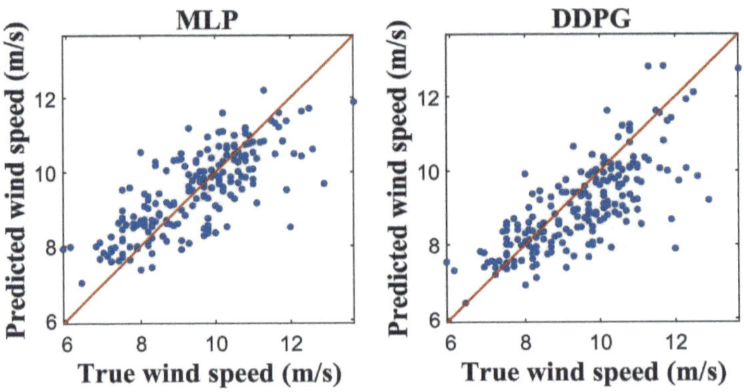

Figure 5.26 Scatter plots of the MLP model and proposed DDPG-based model.

be transformed into a supervised learning problem to be solved, the application of RL in this field mainly focuses on parameter optimization and selection, feature selection, model selection, etc. These applications all involve making optimal actions based on state.

For this reason, abstracting relevant problems reasonably and efficiently and transforming them into RL problems is the primary task of applying RL to solve single-point wind speed forecasting. By properly constructing the RL environment and defining the appropriate agent, state space, action space, reward function, and transfer function, it is possible to form a task suitable for solving with RL algorithms.

After the model is abstracted, an appropriate RL algorithm needs to be selected according to the nature of the abstracted problem. Traditional RL algorithms are only suitable for discrete state and discrete action spaces. However, due to the introduction of DNN, DRL can solve problems defined in continuous space.

Based on the introduction of the basic theory of RL and its possible application in single-point wind speed forecasting, several typical forecasting algorithms based on RL are then given. This chapter mainly gives the discussion and analysis of model abstraction, model framework, and algorithm flow. For some of the applications, relevant experiments are given to verify the performance and feasibility of the RL algorithm.

The RL algorithms used in this chapter mainly include the traditional Q-learning algorithm and the DRL algorithms: DQN and DDPG. In Section 5.3, the Q-learning algorithm is briefly introduced and explained, and the application examples of the algorithm in ensemble weight coefficients optimization and feature selection are given. The DQN algorithm is briefly introduced in Section 5.4, and then its application in the selection of dynamic ensemble weight coefficients is explained in detail. The detail and application of DDPG in forecasting are also presented in Section 5.4. The experiments carried out in this chapter have verified the feasibility of the application of RL in single-point wind speed forecasting.

In the future research work of RL-based forecasting methods, the following aspects can be further studied:

(a) Combination with commonly used auxiliary methods.

To illustrate the role of RL algorithms in singe-point wind speed forecasting and simplify the discussion, the forecasting models constructed in this chapter only contain the most important parts. However, as shown in Fig. 5.19, the conventional forecasting framework of wind speed also includes auxiliary methods such as data preprocessing, optimization, and data postprocessing, which are omitted in this chapter. In subsequent work, RL-based wind speed forecasting models should be considered to be combined with these auxiliary methods.

(b) Innovatively propose new model abstractions of RL in the wind speed forecasting problem.

When RL is used to solve problems such as optimization and decision-making in wind speed forecasting, it involves problem abstraction and transformation, which requires consideration and design of states, actions, rewards, and transitions. The design of these elements has an impact on the convergence ability and learning ability of the RL algorithm, so the sensitivity should be studied and discussed.

(c) Compare the performance of different RL algorithms on the same problem and discuss the reasons.

RL involves various specific algorithms based on strategy, value, and actor-critic. Their performance is different on different problems. Since the current research on RL in wind forecasting is still in its initial stage, it is necessary to fully discuss and research these RL algorithms. Based on the performance of different RL algorithms, some references can be provided for algorithm selection.

(d) Consider using RL algorithms to build wind speed forecasting models directly.

Traditional RL methods cannot directly solve supervised learning problems. The applications of RL in wind speed forecasting mentioned in this chapter are mainly used for optimization and decision-making. However, considering that some DRL algorithms, including A3C, DDPG, RDPG, can process continuous state space and continuous action space, if the feature vector composed of historical wind speed is taken as the state in RL tasks and the future wind speed is regarded as the action, then the RL algorithm can be used to directly construct the wind forecasting model [9]. Under this model abstraction, the forecasting work is done by the actor network, which can be implemented using any efficient Deep Neural Network structure. The excellent performance of these forecasting models can still be retained.

At the same time, due to the existence of the critic network, it can continuously and automatically evaluate the actor's forecasting results during the training process, which also leads to further improvement of the forecasting. Considering the powerful self-learning ability of the DRL algorithm, the DRL-based forecasting models should have a better performance in obtaining more accurate forecasting results. However, as demonstrated in Section 5.4.3, the forecasting performance of the DDPG-based model is even worse than that of the traditional supervised learning algorithm MLP. In future research, the DDPG-based model should be further adjusted and improved.

Due to the need to train two networks and maintain the RL process in the DRL algorithm, the training of the DRL-based forecasting model requires more computing time than the single forecasting model that has the same structure as the actor network. Besides, many hyperparameters involved in the algorithm, the definition of reward in RL problems, etc., all have a great influence on the convergence. Reasonable adjustments to these key influencing factors are very important and time-consuming.

(e) Dynamic ensemble model based on DDPG

In this chapter, a dynamic ensemble wind forecasting framework is proposed based on the DQN algorithm, and different ensemble weight coefficients are selected for each step according to input features. However, the candidate ensemble weights are extracted from the Pareto optimal solution set obtained by the NSGA-II. Since the number of candidates is still fixed, its performance can still be improved. In fact, this is because that DQN cannot output continuous actions.

However, if the wind speed feature is taken as the state, and the ensemble weight of each base forecasting model is regarded as multiple continuous actions, an ensemble model based on DDPG can be constructed. The agent will select different combinations of weight coefficients for each base forecasting model according to different wind speed features. This abstraction of RL problems may provide a new solution for the dynamic ensemble model based on RL.

References

[1] E. Mocanu, P.H. Nguyen, M. Gibescu, Chapter 7 - deep learning for power system data analysis, in: Big Data Application in Power Systems, Elsevier, 2018, pp. 125−158.

[2] J.R. Vázquez-Canteli, Z. Nagy, Reinforcement learning for demand response: a review of algorithms and modeling techniques, Appl. Energy 235 (2019) 1072−1089.

[3] M. Han, R. May, X. Zhang, et al., A review of reinforcement learning methodologies for controlling occupant comfort in buildings, Sust. Cities Soc. 51 (2019) 101748.

[4] H. Liu, C. Yu, C. Yu, et al., A novel axle temperature forecasting method based on decomposition, reinforcement learning optimization and neural network, Adv. Eng. Inf. 44 (2020) 101089.

[5] H. Liu, C. Yu, H. Wu, et al., A new hybrid ensemble deep reinforcement learning model for wind speed short term forecasting, Energy 202 (2020) 117794.

[6] Y. Li, Z. Liu, H. Liu, A novel ensemble reinforcement learning gated unit model for daily PM2.5 forecasting, Air Qual. Atmos. Health 1−11 (2020).

[7] C. Feng, J. Zhang, Reinforcement learning based dynamic model selection for short-term load forecasting, in: 2019 IEEE Power & Energy Society Innovative Smart Grid Technologies Conference, ISGT), 2019, pp. 1−5.

[8] C. Feng, M. Sun, J. Zhang, Reinforced deterministic and probabilistic load forecasting via Q-learning dynamic model selection, IEEE Trans. Smart Grid 11 (2020) 1377−1386.

[9] T. Liu, Z. Tan, C. Xu, et al., Study on deep reinforcement learning techniques for building energy consumption forecasting, Energy Build. 208 (2020) 109675.

[10] S. Zhifei, J. Er Meng, A review of inverse reinforcement learning theory and recent advances, in: 2012 IEEE Congress on Evolutionary Computation, 2012, pp. 1−8.

[11] A. Tascikaraoglu, M. Uzunoglu, A review of combined approaches for prediction of short-term wind speed and power, Renew. Sustain. Energy Rev. 34 (2014) 243−254.

[12] H. Liu, Y. Li, Z. Duan, et al., A review on multi-objective optimization framework in wind energy forecasting techniques and applications, Energy Convers. Manag. 224 (2020) 113324.

[13] G.E. Hinton, S. Osindero, Y.-W. Teh, A fast learning algorithm for deep Belief nets, Neural Comput. 18 (2006) 1527−1554.

[14] Z. Chen, W. Li, Multisensor feature fusion for bearing fault diagnosis using sparse autoencoder and deep Belief network, IEEE Trans. Instrum. Measure. 66 (2017) 1693−1702.

[15] H.a.A. Al-Rawi, M.A. Ng, K.-L.A. Yau, Application of reinforcement learning to routing in distributed wireless networks: a review, Artif. Intell. Rev. 43 (2015) 381−416.

[16] A. Jović, K. Brkić, N. Bogunović, A review of feature selection methods with applications, in: 2015 38th International Convention on Information and Communication Technology, Electronics and Microelectronics, MIPRO, 2015, pp. 1200−1205.

[17] J.L. Elman, Finding structure in time, Cognit. Sci. 14 (1990) 179−211.

[18] G. Ren, Y. Cao, S. Wen, et al., A modified Elman neural network with a new learning rate scheme, Neurocomputing 286 (2018) 11−18.

[19] V. Mnih, K. Kavukcuoglu, D. Silver, et al., Playing Atari with Deep Reinforcement Learning, 2013 arXiv preprint arXiv:13125602.

[20] T.P. Lillicrap, J.J. Hunt, A. Pritzel, et al., Continuous Control with Deep Reinforcement Learning, 2015 arXiv preprint arXiv:150902971.

[21] C. Qiu, Y. Hu, Y. Chen, et al., Deep deterministic policy gradient (DDPG)-Based energy harvesting wireless communications, IEEE Inter. Things J. 6 (2019) 8577−8588.

[22] C. Zhang, H. Wei, L. Xie, et al., Direct interval forecasting of wind speed using radial basis function neural networks in a multi-objective optimization framework, Neurocomputing 205 (2016) 53–63.
[23] C. Tian, Y. Hao, J. Hu, A novel wind speed forecasting system based on hybrid data preprocessing and multi-objective optimization, Appl. Energy 231 (2018) 301–319.
[24] H. Liu, C. Chen, Data processing strategies in wind energy forecasting models and applications: a comprehensive review, Appl. Energy 249 (2019) 392–408.
[25] H. Liu, C. Chen, X. Lv, et al., Deterministic wind energy forecasting: a review of intelligent predictors and auxiliary methods, Energy Convers. Manag. 195 (2019) 328–345.

CHAPTER 6

Single-point wind forecasting methods based on ensemble modeling

Contents

Wind Forecasting in Railway Engineering
ISBN 978-0-12-823706-9
https://doi.org/10.1016/B978-0-12-823706-9.00006-5

6.1 Introduction

Ensemble learning is a method of constructing and combining multiple individual learners to complete the target task [1]. Compared with a single learner, it can obtain better prediction results. Individual learners are obtained by training an existing algorithm using original data. Multiple learners can be the homogeneous ensemble of the same type of algorithms, or heterogeneous ensemble of different types of algorithms. Combining strategies mainly include the averaging method, voting method, and learning method.

Common ensemble models include Bagging, Boosting, Stacking, etc. They can handle classification and regression problems [2]. Because of their higher accuracy and better generalization performance, they have been deeply developed in the fields of transportation, finance, and medical care.

With the development of rail transit, people are paying more and more attention to the safety of rail transit. The strong wind along the railway seriously affected the safe operation of trains. If the wind speed exceeds the safety threshold, the train will face the risk of overturning, and even threaten the lives of the people on the train. Therefore, it is very important to accurately monitor and predict the wind speed and wind direction along the railway, which helps to make reasonable arrangements and adjustments to the train's running time and operating interval in advance to ensure the safe operation of trains. However, wind is random and volatile, and wind forecasting is still a subject worthy of further study.

Many scholars have applied ensemble models to wind forecasting. For example, Li et al. combined wavelet packet decomposition, Elman Neural Network (ENN), Boosting algorithms, and wavelet packet filter for large-scale multi-step wind speed prediction [3]. Qu et al. used a new hybrid bat-search flower-pollination algorithm for multi-objective ensemble wind speed forecasting [4]. Liu et al. proposed a novel multi-objective data ensemble wind speed prediction model, which obtains the combined weights of the base learners through a multi-objective multi-universe optimization algorithm, and can achieve high prediction accuracy [5].

This chapter will mainly introduce and study the three ensemble learning algorithms: multi-objective ensemble, Stacking, and Boosting, and apply them to the field of wind forecasting to explore their performance deeply.

6.2 Wind data description

In this chapter, real wind data from a strong wind railway will be used to analyze single-point wind forecasting methods based on ensemble modeling. The data sequences used in this chapter include wind speed sequence and wind direction sequence, with a time interval of 3 s, and each sequence contains 2000 data samples.

The following Figs. 6.1 and 6.2, respectively, show the visual display images of wind speed data and wind direction data.

Table 6.1 below lists the statistical descriptions of the wind speed series and wind direction series. It can be seen from the graphs and tables that both wind speed data and wind direction data have the characteristics of nonlinearity, unevenness, and large volatility, making it difficult to predict.

To quantitatively evaluate the stability and accuracy of the proposed ensemble model for wind forecasting, this chapter selects three major error evaluation indicators of prediction. Mean Absolute Error (MAE), Mean Absolute Percentage Error (MAPE), and Root Mean Squared Error (RMSE) are used to measure the prediction error. The smaller their value, the higher the accuracy of the model. The equations of these indicators are presented in Chapter 3.

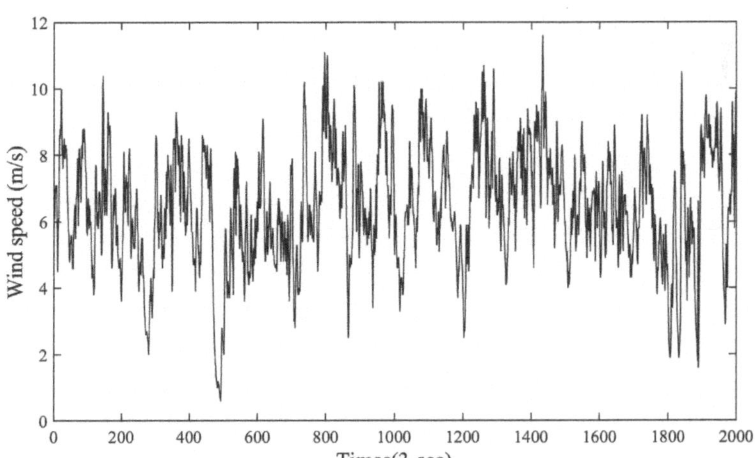

Figure 6.1 The wind speed data series.

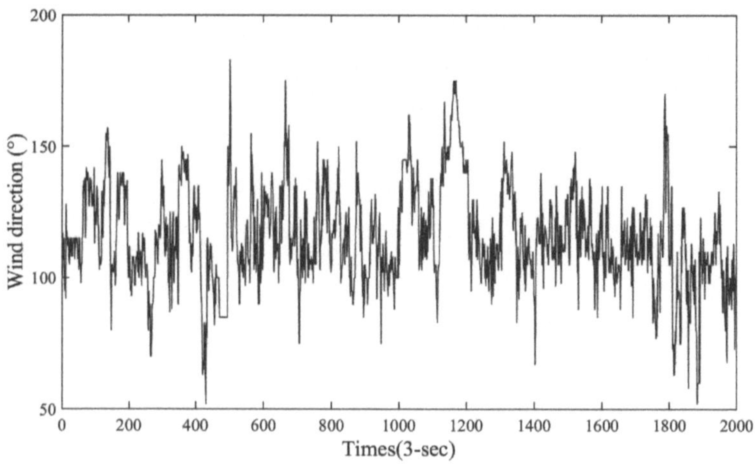

Figure 6.2 The wind direction data series.

Table 6.1 The statistical descriptions of the wind speed and direction data.

Data	Maximum	Minimum	Mean	Standard deviation	Skewness	Kurtosis
Wind speed	11.600 m/s	0.600 m/s	6.454 m/s	1.796 m/s	−0.243	3.078
Wind direction	183.000 degrees	52.000 degrees	115.409 degrees	18.933 degrees	0.232	3.528

6.3 Single-point wind speed forecasting algorithm based on multi-objective ensemble

The multi-objective ensemble algorithm is to construct multiple individual learners of the same or different types, and use a multi-objective optimization algorithm to optimize multiple objectives, find the best combination of multiple individual learners, and obtain better task results.

Railway wind speed data have strong nonlinearity and volatility, and it is difficult to predict. Therefore, it is necessary to select a suitable optimization algorithm to optimize and improve the wind speed forecasting model, and to reduce the adverse effects of wind speed sequence characteristics on the prediction process.

Optimization algorithms are usually divided into single-objective optimization algorithms and multi-objective optimization algorithms. The single-objective optimization algorithm needs to optimize only one objective function, and the optimal value is obtained according to the relevant constraints. The multi-objective optimization algorithm needs to optimize multiple objective functions, and these objective functions are not independent of each other, and there is no particularly strong correlation. There will be more or less conflicts between them, making it impossible for each objective function to reach the optimal at the same time [6].

Therefore, in the search of multi-objective optimization algorithms, the concept of Pareto dominance is generally used to obtain the optimal solution set. When solving a multi-objective optimization problem, it is necessary to find a set of solutions that are as close to the Pareto optimal front as possible, and find a set of solutions that are as different as possible [7].

In the wind speed prediction, to achieve better prediction results, it is not enough to consider only one optimization criterion. Using multi-objective optimization algorithms to consider multiple criteria such as accuracy and stability can establish a more effective prediction model. Commonly used multi-objective optimization algorithms include Multi-Objective Particle Swarm Optimization (MOPSO) algorithm, multi-objective bat algorithm, Multi-Objective Grey Wolf Optimization (MOGWO) algorithm, multi-objective dragonfly algorithm, multi-objective ant lion optimization algorithm, and Multi-Objective Grasshopper Optimization Algorithm (MOGOA).

The single-objective optimization algorithms used in this section mainly include Particle Swarm Optimization (PSO) algorithm, Bat Algorithm (BA) and Grey Wolf Optimization (GWO) algorithm, and multi-objective optimization algorithms include MOPSO, MOGWO, and MOGOA.

In wind speed prediction, the hybrid model can generally obtain higher prediction accuracy than other models [8]. Generally, the ensemble method or the decomposition method is used to obtain the hybrid model. This section combines the two methods to obtain the required hybrid model. The original wind speed sequence has strong volatility and randomness. The decomposition method can separate the wind speed sequence into different frequency bands, which makes the prediction model only need to pay attention to single-frequency components. Common decomposition methods include empirical mode decomposition, Wavelet Decomposition (WD), and maximum overlap discrete wavelet transform. This section uses WD to decompose the wind speed sequence into multiple subsequences, and determine their specific positions in the time domain [9]. Using WD, the original unsteady wind speed sequence is decomposed into a plurality of relatively stable wind speed subsequences, and corresponding prediction models can be established according to the characteristics of each subsequence, which can effectively improve the effect of wind speed prediction.

In the selection part of the wind speed forecasting model, the intelligent model has been studied deeply. This section uses Multi-Layer Perceptron (MLP) as the forecasting model. As a typical example of Artificial Neural Networks (ANNs), the MLP has an excellent performance in fitting nonlinear time series.

This section uses the wind speed data shown in Section 6.2 as the original sequence and divides it into two parts, namely D_1 and D_2. D_1 contains the 1st–500th data as training data, and D_2 contains the 1501th–2000th data as test data. Then this section combines the data decomposition algorithm with the intelligent model, and uses different single-objective optimization algorithms and multi-objective optimization algorithms for ensemble prediction. Finally, we use the error evaluation indicators mentioned in Section 6.2 to compare and analyze the prediction performance of different ensemble models that are established based on different optimization algorithms.

6.3.1 Model framework

The wind speed sequence along the railway is very random and unstable, which is difficult to predict. Therefore, this section first uses WD to decompose the original data into multiple subsequences, then trains the intelligent model MLP on each subsequence separately, and finally uses different optimization algorithms for ensemble prediction. The single-objective optimization algorithms used in this section include PSO, GWO, and BA, and multi-objective optimization algorithms include MOPSO, MOGWO, and MOGOA. For single-objective optimization algorithms, the minimum MAE of the predicted results is used as the optimization objective, and for multi-objective optimization algorithms, both MAE and RMSE of the predicted results are minimized as the optimization objectives. The overall model framework of the multi-objective ensemble is shown in Fig. 6.3.

6.3.2 Theoretical basis

The specific theoretical knowledge of related models and algorithms used in this section is introduced as follows.

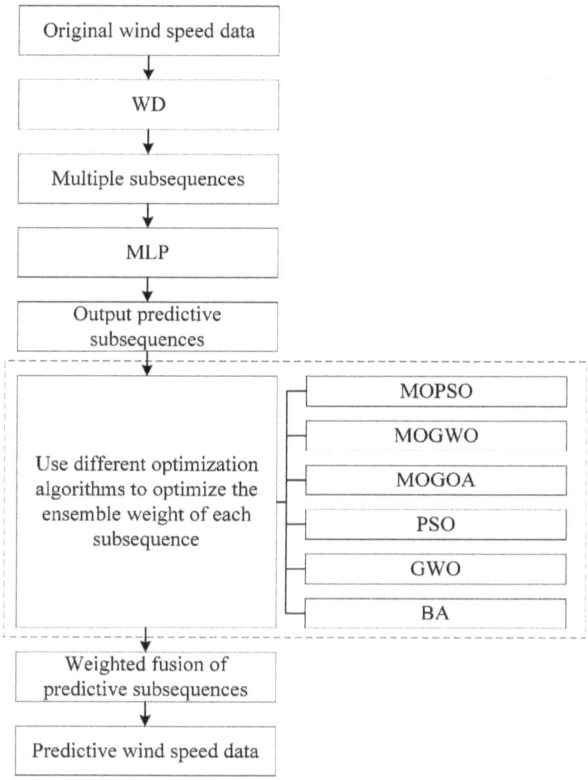

Figure 6.3 The model framework of multi-objective ensemble.

6.3.2.1 Wavelet decomposition

WD is a kind of data decomposition technology, the simplest form of which is wavelet transform. Wavelet transform is a theoretical method of time-frequency analysis proposed by Morlet in 1974. In 1989, Mallat proposed the Mallat algorithm, a fast wavelet transform algorithm [10], which further promoted the development of wavelet theory.

Wavelet transform improves the traditional Fourier transform, uses wavelet function to replace the traditional trigonometric function as the basis function, realizes the decomposition and reconstruction of the function, and can reflect the changes of signal frequency components over time. Wavelet transform can perform a multiscale detailed analysis of functions or signals through operations such as scaling and translation [9]. Therefore, the wavelet transform has been widely used in pattern recognition, signal processing, image processing, and so on.

Wavelet transform includes discrete wavelet transform and continuous wavelet transform. The wavelet basis function expansion factor and translation factor in the discrete wavelet transform are discretized data, which are often used for signal denoizing and compression. The continuous wavelet transform is continuously changing data and is often used for time—frequency analysis.

6.3.2.2 Multi-layer perceptron

The MLP is a very common ANN algorithm, and it can be used to solve classification and regression problems by simulating and simplifying biological neurons [11]. MLP is a feedforward ANN model. It can map multiple input datasets to output datasets, and can learn nonlinear functions.

6.3.2.3 Single-objective optimization algorithm

6.3.2.3.1 Grey wolf optimization algorithm

GWO is proposed by Mirjalili et al. in 2014, and it is a new type of optimization algorithm [12]. It draws on the characteristics of predation behavior of gray wolves, and achieves the purpose of optimization search through the processes of wolves tracking, encircling, hunting, and attacking prey. It has fewer parameters, strong convergence performance, and is easier to implement. Therefore, GWO has received in-depth research by scholars in recent years and has been widely used in image classification, function optimization, workshop schedule, and other fields. Wolves are very intelligent animals. When they prey on food, they often do not act alone, but are composed of several wolves. The GWO algorithm is proposed based on this predation behavior of wolves.

6.3.2.3.2 Particle swarm optimization algorithm

PSO is a single-objective optimization algorithm, proposed by Kennedy et al. in 1995 [13]. It originated from the study of bird predation behavior, simulating the flight foraging behavior of bird swarms, and using the sharing of information by individuals in the group to obtain the optimal solution [14]. PSO is relatively simple, and has a small number of parameters to be adjusted, which makes it easy to implement. Therefore, the PSO algorithm has great advantages and has been widely used in neural network training, function optimization, image processing, power system design, fuzzy system control, and other respects.

6.3.2.3.3 Bat algorithm

BA comes from the echolocation behavior of bats [15]. In the BA, to simulate the random search process of bats hunting prey and avoiding obstacles, the following idealized assumptions are made [16]:

(1) All bats in the population use the echolocation method to perceive distance.

(2) The flying speed v_i of bat at position x_i is random, and the bats have different frequencies $f_i \in [f_{\min}, f_{max}]$, impulse loudness A_i, and impulse emission rate r_i.

(3) When a bat searches and captures prey, it will change its parameters and search for the optimal solution until the target stops or the termination condition is met.

6.3.2.4 Multi-objective optimization algorithm

6.3.2.4.1 Multi-objective grey wolf optimization algorithm

MOGWO is proposed based on GWO in 2016 [17]. To form the MOGWO algorithm, two additional mechanisms are added to the GWO, namely the archive mechanism and the leader selection mechanism. Both of the mechanisms help the MOGWO to find the global optimal solution.

- Archive mechanism

  The first is the archive mechanism, and it is used to store the non-dominated Pareto optimal solution. The archiving process includes the following four different situations [6]:

  (1) When a new member is dominated by at least one member already stored in the archive, the new member is not allowed to enter the archive.

  (2) When a new member can dominate one or more archived members, the new member will become one of the members in the archive, and the dominated solution will be deleted from the archive.

  (3) When the new member and the members in the archive do not dominate each other, the new solution can be stored in the archive.

  (4) When the archive is full and there are new members to be deposited, the grid mechanism will run to find the most crowded section and delete members from it. New solutions can then be saved in the archive and increase the diversity of the final approximate Pareto optimal solution set.

- Leader selection mechanism

The second is the leader selection mechanism. According to the GWO, the optimal solution is divided into wolf α, wolf β, and wolf δ. These three leading wolves can guide other wolves to search for potential solutions in the search space. However, it is not easy to evaluate the optimal solution in multi-objective optimization.

To solve this problem, the leader selection mechanism selects wolf α, wolf β, and wolf δ in a specific solution section, and realizes the selection of a specific solution section through the roulette method. The less crowded the section, the greater the probability of being selected. The probability of a particular solution segment being selected is calculated as follows [17]:

$$P_i = \frac{c}{N_i} \tag{6.1}$$

where c is a constant and $c > 1$, and N_i is the ordinal number of the solution in the section.

Except for some special cases, a less crowded section has a greater probability of being selected.

Through the above two mechanisms, the MOGWO can avoid the repeated selection of wolf α, wolf β, and wolf δ, and it can search the unsearched area in the search space.

The specific implementation flow chart of the MOGWO algorithm is shown in Fig. 6.4.

6.3.2.4.2 Multi-objective particle swarm optimization algorithm

MOPSO is a multi-objective optimization algorithm based on PSO, which is proposed in 2002 [18]. PSO can only solve single-objective problems, while MOPSO can solve multi-objective problems.

Similar to MOGWO, MOPSO also uses an archive mechanism and a leader mechanism. These two mechanisms help MOPSO to search for the global optimal solution. Compared with PSO, when MOPSO chooses the *pBest*, it randomly selects one of them as the historical best when it is impossible to strictly compare which is better. For the selection of the *gBest*, MOPSO selects a leader in the optimal set according to the degree of congestion. MOPSO applies an adaptive grid method when selecting a leader and updating the archive [19].

The specific implementation flow chart of the MOPSO is shown in Fig. 6.5.

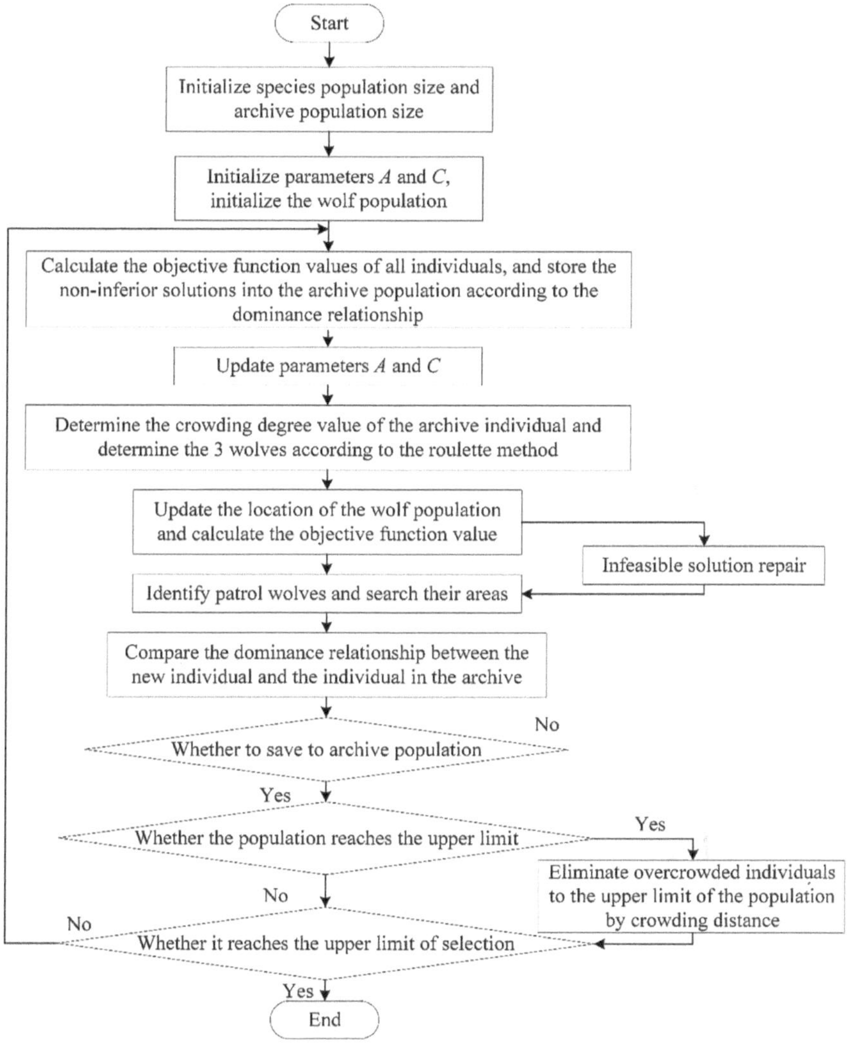

Figure 6.4 The flow chart of the MOGWO.

6.3.2.4.3 Multi-objective grasshopper optimization algorithm

MOGOA is proposed by Mirjalili et al. based on the Grasshopper Optimization Algorithm (GOA) in 2018 and is also a multi-objective optimization algorithm [20].

The GOA is an intelligent optimization algorithm with good global search capabilities. The special adaptive mechanism in the algorithm balances the individual search capabilities well, making the algorithm have a

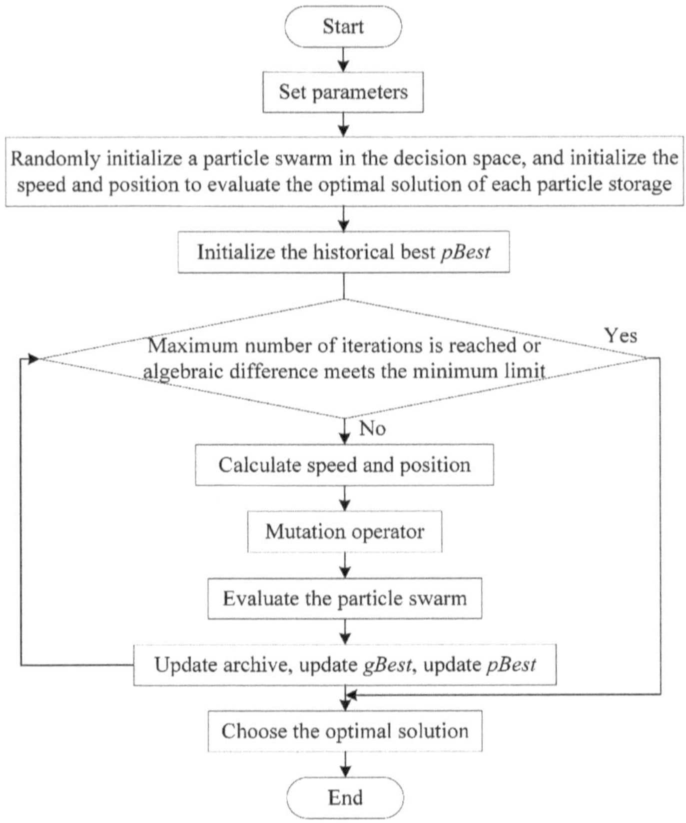

Figure 6.5 The flow chart of the MOPSO.

better convergence speed [21]. The algorithm parameter setting of GOA is simple and the GOA is convenient to use. The MOGOA improves based on GOA, not only has the advantages of GOA but also can solve the multi-objective optimization problem. Therefore, MOGOA simulates the population migration and foraging behavior of grasshoppers in nature, and has good global search capability. It is a novel and excellent multi-objective optimization algorithm.

The specific flow chart of the MOGOA algorithm is shown in Fig. 6.6.

6.3.3 Result analysis

The wind speed multi-step prediction results of the multi-objective ensemble and single-objective ensemble models are shown in Figs. 6.7–6.9. To display the model prediction results more clearly, a part of the interval is

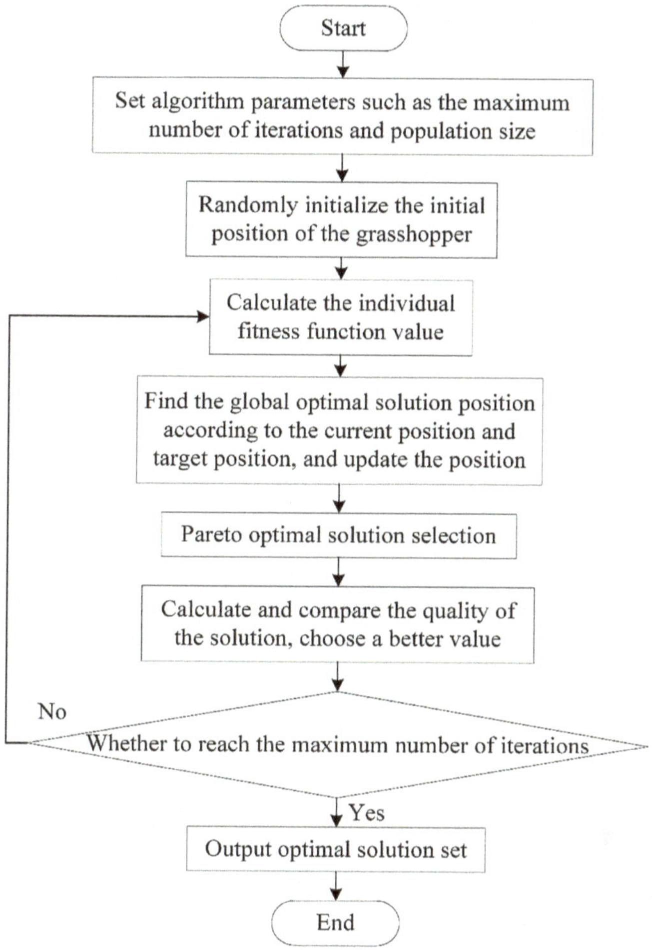

Figure 6.6 The flow chart of the MOGOA.

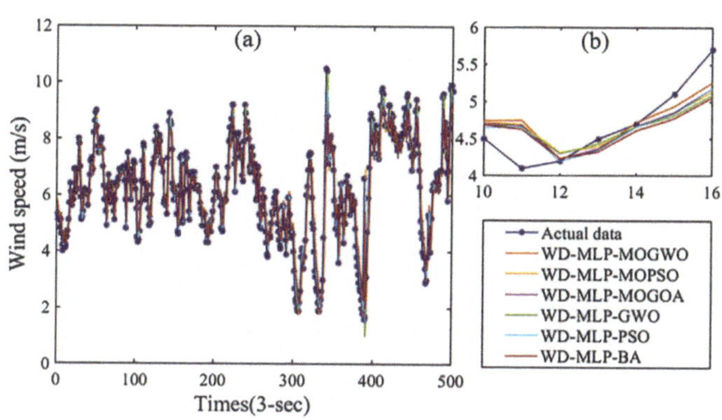

Figure 6.7 The 1-step prediction results of the optimization ensemble models: (A) prediction results of the entire test set, (B) partially enlarged view from 10 to 16.

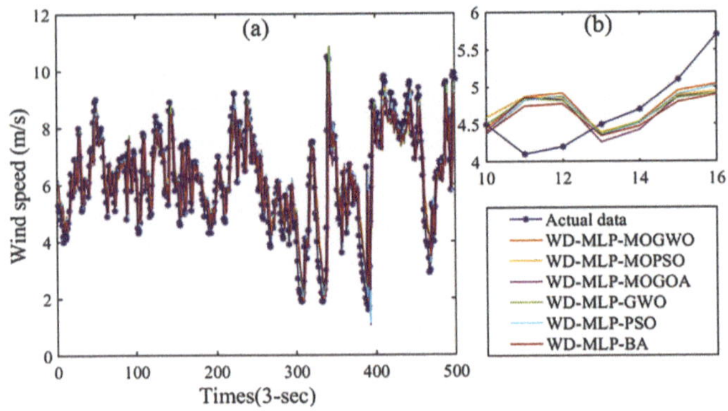

Figure 6.8 The 2-step prediction results of the optimization ensemble models: (A) prediction results of the entire test set, (B) partially enlarged view from 10 to 16.

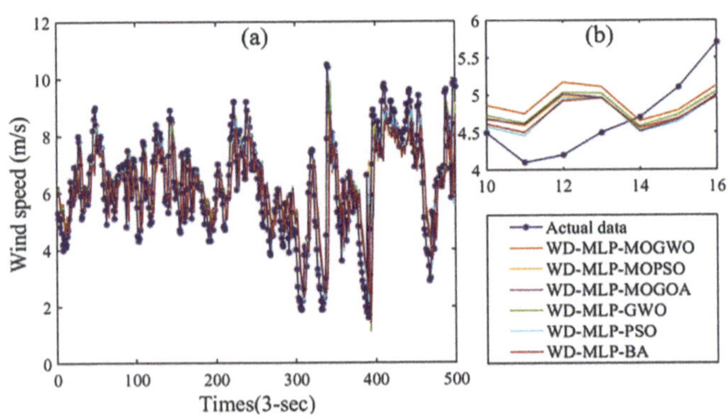

Figure 6.9 The 3-step prediction results of the optimization ensemble models: (A) prediction results of the entire test set, (B) partially enlarged view from 10 to 16.

selected for enlargement. And the specific prediction index results of the six ensemble models are shown in Tables 6.2–6.4.

By analyzing the information of these figures and tables, we can draw the following conclusions:

(1) Whether it is a single-objective ensemble model or a multi-objective ensemble model, good wind speed prediction results can all be obtained. It can be seen from Table 6.2 that the WD-MLP-BA model has the highest one-step prediction error among the six ensemble models, and the MAPE of it is only 9.662%. Therefore, the prediction results of the single-objective ensemble models can be better close to the real results.

Table 6.2 The 1-step forecasting performance of the ensemble models.

Ensemble models	MAE(m/s)	MAPE(%)	RMSE(m/s)
WD-MLP-MOGWO	0.498	9.110	0.711
WD-MLP-MOPSO	0.502	9.605	0.712
WD-MLP-MOGOA	0.481	8.970	0.694
WD-MLP-GWO	0.493	9.117	0.710
WD-MLP-PSO	0.485	8.812	0.702
WD-MLP-BA	0.518	9.662	0.728

Table 6.3 The 2-step forecasting performance of the ensemble models.

Ensemble models	MAE(m/s)	MAPE(%)	RMSE(m/s)
WD-MLP-MOGWO	0.807	14.517	1.075
WD-MLP-MOPSO	0.816	15.200	1.071
WD-MLP-MOGOA	0.815	14.831	1.082
WD-MLP-GWO	0.812	14.749	1.083
WD-MLP-PSO	0.805	14.648	1.072
WD-MLP-BA	0.804	14.696	1.067

Table 6.4 The 3-step forecasting performance of the ensemble models.

Ensemble models	MAE(m/s)	MAPE(%)	RMSE(m/s)
WD-MLP-MOGWO	0.987	18.360	1.289
WD-MLP-MOPSO	0.839	18.090	1.283
WD-MLP-MOGOA	0.983	18.024	1.283
WD-MLP-GWO	0.984	17.876	1.292
WD-MLP-PSO	0.974	17.637	1.277
WD-MLP-BA	0.975	17.707	1.288

(2) The larger the prediction step, the lower the prediction accuracy and the more obvious the lag. Taking WD-MLP-MOGWO as an example, from the perspective of MAPE, the model prediction accuracy of the 2-step ahead is 59.34% lower than that of the 1-step ahead, and the prediction accuracy of the 3-step ahead is 26.47% lower than that of the 2-step ahead.

(3) In the 1-step ahead prediction, when the selection of the base learner is consistent, the performance improvement effect of the multi-objective optimization algorithm on the forecasting model is generally better than that of the single-objective optimization algorithm. However, as the number of ahead prediction steps increases, the improvement effect gradually decreases. Even in the 3-step ahead prediction, from the perspective of MAPE, the prediction effects of the three multi-objective ensemble models used in this section are all worse than the three single-objective ensemble models.

6.3.4 Conclusions

This section proposes three multi-objective ensemble models and three single-objective ensemble models to predict wind speed data. It can be seen from the test results that

(a) Both multi-objective optimization algorithms and single-objective optimization algorithms can improve the forecasting performance of the base learner and achieve a better wind speed prediction effect.

(b) When the steps of ahead prediction increase, the forecasting accuracy of the objective ensemble model decreases, and the lag phenomenon becomes more obvious.

(c) In the 1-step ahead prediction, the multi-objective ensemble model can achieve better prediction results than the single-objective ensemble model. However, as the steps of ahead prediction increase, the improvement effect gradually decreases.

6.4 Single-point wind speed forecasting algorithm based on stacking

The Stacking algorithm is an ensemble algorithm, also known as Stacked generalization, proposed by Wolpert in 1992 [22]. The main idea of Stacking is to first use several base learners to learn the training samples, and then use the output results of the base learners as the training samples of the next layer of meta-learners to implement nonlinear weighting of the base learners, and the output result of the model is finally obtained. The Stacking algorithm does not need to consider the specific details of the base learners, has strong scalability, and can achieve better learning effects than a single model [23].

Stacking has been widely used in many fields for its better learning effect. For example, Hansen et al. applied the Stacking algorithm to the taxable sales prediction field, and compared it with autoregressive integrated moving average, radial basis function neural network, and other models [24]. The experimental results proved that the Stacking algorithm has the smallest prediction error. Wang et al. used the Stacking algorithm to predict the type of membrane protein and obtained a higher prediction success rate than Support Vector Machine (SVM) and instance-based learning [25]. And the Stacking algorithm has also been deeply studied by many scholars in the field of wind speed prediction, and has achieved good prediction results.

Therefore, this section uses the Stacking algorithm for single-point wind speed forecasting. The data decomposition algorithm WD is also combined with the base learner. In the Stacking algorithm, the choice of the base learner is of great significance and will greatly affect the forecasting accuracy of the model. In this section, a series of commonly used models are selected as base learners and meta-learners to study the influence of the combination of different types and different numbers of base learners on the accuracy of the Stacking model. Besides, in the multi-step forecasting of time series, recursive and Multiple-Input Multiple-Output (MIMO) strategies are two commonly used and effective forecasting strategies [26]. This section also further studies the accuracy changes of the Stacking ensemble model when the forecasting strategy is different.

This section uses the single-point wind speed data shown in Section 6.2 as the original sequence and divides it into three parts, namely D_1, D_2, and D_3. D_1 and D_2 contain the 1st−1500th data. The specific samples in D_1 and D_2 are randomly sampled to train the base learner and meta-learner. D_3 contains the 1501th−2000th data, which is used to test the forecasting performance. Finally, this section uses the error evaluation indicators mentioned in Section 6.2 to compare and analyze the prediction performance of different Stacking ensemble models.

6.4.1 Model framework

In the section of the Stacking prediction model, WD is used as the data decomposition algorithm, MIMO or recursive is used as forecasting strategy, different numbers of MLP and ENN are used as base learners, and SVM is used as the meta-learner. They are combined to form eight different Stacking ensemble models. The specific content is as follows:

- Model 1: Stacking-3-MLP-MIMO (3 MLPs as base learners)
- Model 2: Stacking-5-MLP-MIMO (5 MLPs as base learners)
- Model 3: Stacking-3-ENN-MIMO (3 ENNs as base learners)
- Model 4: Stacking-5-ENN-MIMO (5 ENNs as base learners)
- Model 5: Stacking-3-MLP-Recursive (3 MLPs as base learners)
- Model 6: Stacking-5-MLP-Recursive (5 MLPs as base learners)
- Model 7: Stacking-3-ENN-Recursive (3 ENNs as base learners)
- Model 8: Stacking-5-ENN-Recursive (5 ENNs as base learners)

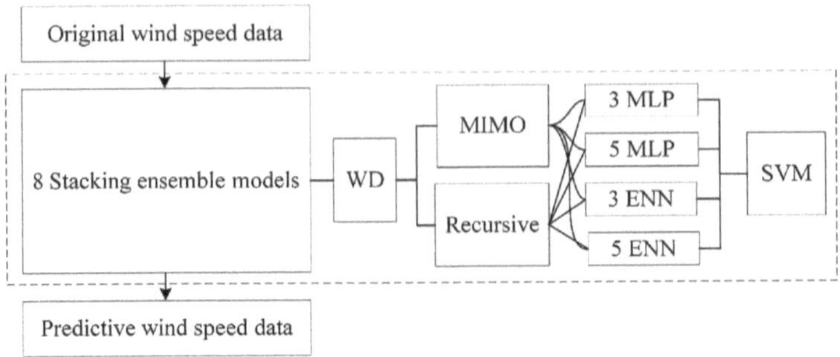

Figure 6.10 The model framework of Stacking ensemble.

The overall model framework of the Stacking ensemble is shown in Fig. 6.10.

6.4.2 Theoretical basis

The Stacking algorithm is an ensemble algorithm proposed by Wolpert in 1992 [22]. It can use a meta-learner to implement nonlinear weighting of different base learners. In the Stacking algorithm, the model is constructed by base learners and a meta-learner. The base learners are used as the primary learning models, and the meta-learner performs secondary learning on the learning results of the primary models [27]. The output of the base learners is used as a feature of the corresponding sample in the new dataset, and the data label is still the label of the sample in the original data.

6.4.3 Result analysis

To compare the influence of the two forecasting strategies on the wind speed prediction effect, the wind speed prediction results of different forecasting strategies and the same base learner are plotted in the same figure and compared with the original wind speed data, as shown in Figs. 6.11–6.14. To more clearly show the forecasting accuracy of the different models, the specific prediction error indicator results of the eight Stacking ensemble models are shown in Tables 6.5–6.7.

By analyzing the information of these figures and tables, we can draw the following conclusions:

(1) In the three multi-steps ahead prediction, all eight Stacking ensemble prediction models can predict wind speed well. But the more ahead prediction steps, the worse the prediction effect. Taking Stacking-3-MLP-MIMO as an example, from the perspective of MAPE, the

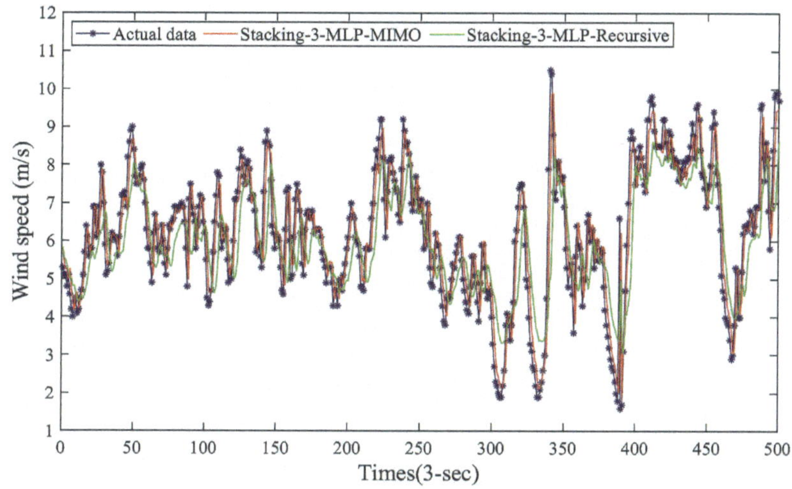

Figure 6.11 The 1-step prediction results of Stacking-3-MLP ensemble models.

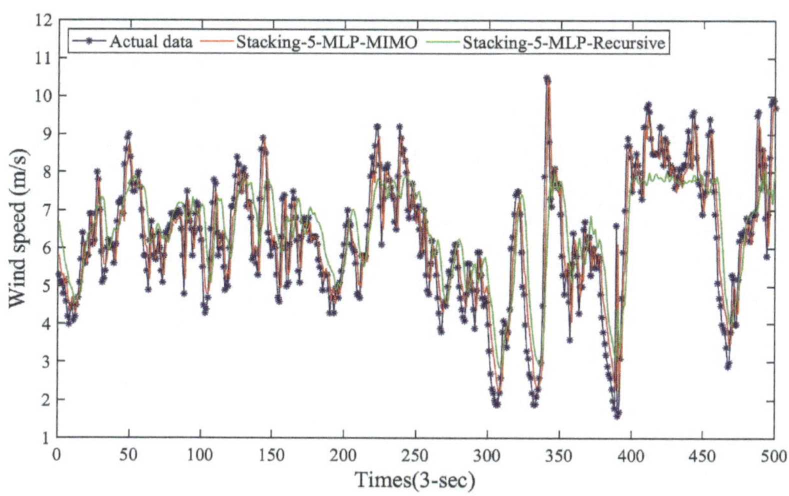

Figure 6.12 The 1-step prediction results of Stacking-5-MLP ensemble models.

model prediction accuracy of the 2-step ahead is 66.55% lower than that of the 1-step ahead, and the prediction accuracy of the 3-step ahead is 21.36% lower than that of the 2-step ahead.

(2) In the Stacking ensemble prediction, when the base learner is MLP, regardless of the number of ahead prediction steps, the MIMO strategy has a better prediction effect than the recursive strategy. When the base learner is ENN, as the number of ahead prediction steps changes, the

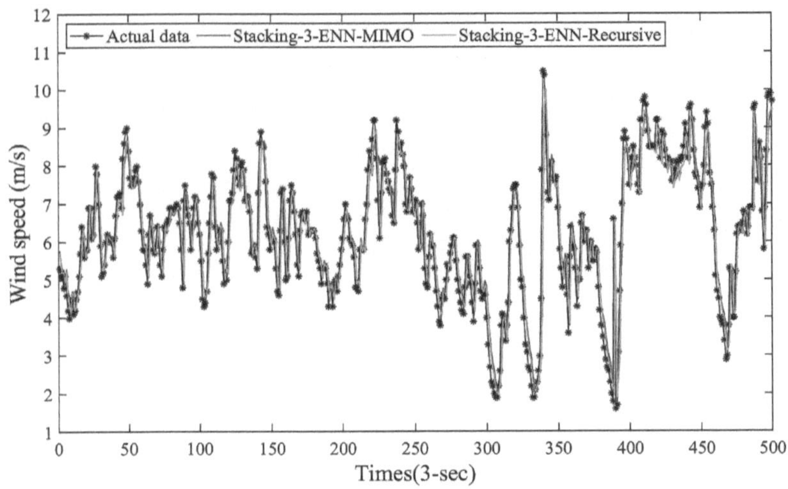

Figure 6.13 The 1-step prediction results of Stacking-3-ENN ensemble models.

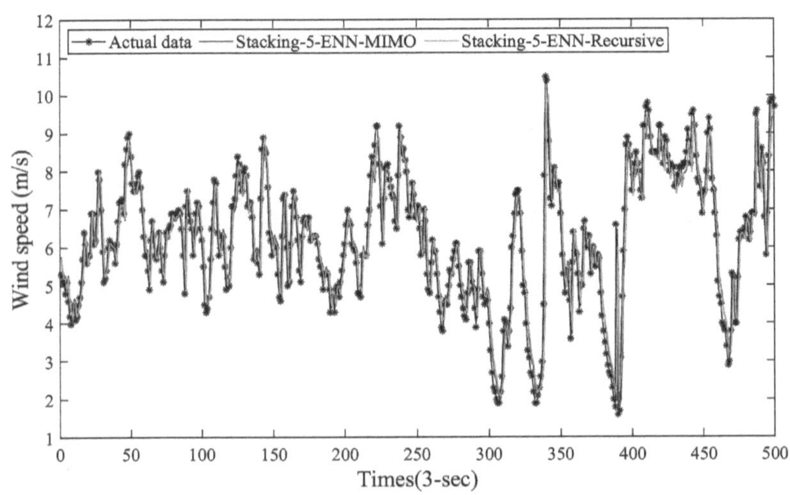

Figure 6.14 The 1-step prediction results of Stacking-5-ENN ensemble models.

Table 6.5 The 1-step forecasting performance of the Stacking ensemble models.

Stacking ensemble models	MAE(m/s)	MAPE(%)	RMSE(m/s)
Stacking-3-MLP-MIMO	0.491	8.944	0.699
Stacking-5-MLP-MIMO	0.521	9.978	0.726
Stacking-3-ENN-MIMO	0.499	9.284	0.706
Stacking-5-ENN-MIMO	0.487	8.951	0.701
Stacking-3-MLP-Recursive	0.804	14.434	1.054
Stacking-5-MLP-Recursive	0.7156	13.839	0.932
Stacking-3-ENN-Recursive	0.490	9.031	0.705
Stacking-5-ENN-Recursive	0.491	9.126	0.702

Table 6.6 The 2-step forecasting performance of the Stacking ensemble models.

Stacking ensemble models	MAE(m/s)	MAPE(%)	RMSE(m/s)
Stacking-3-MLP-MIMO	0.817	14.895	1.080
Stacking-5-MLP-MIMO	0.808	14.733	1.077
Stacking-3-ENN-MIMO	0.811	14.805	1.072
Stacking-5-ENN-MIMO	0.806	14.676	1.066
Stacking-3-MLP-Recursive	0.902	16.694	1.200
Stacking-5-MLP-Recursive	0.913	18.174	1.159
Stacking-3-ENN-Recursive	0.817	14.790	1.085
Stacking-5-ENN-Recursive	0.812	14.621	1.078

Table 6.7 The 3-step forecasting performance of the Stacking ensemble models.

Stacking ensemble models	MAE(m/s)	MAPE(%)	RMSE(m/s)
Stacking-3-MLP-MIMO	0.993	18.077	1.294
Stacking-5-MLP-MIMO	0.988	18.162	1.293
Stacking-3-ENN-MIMO	0.984	18.097	1.286
Stacking-5-ENN-MIMO	0.978	17.833	1.279
Stacking-3-MLP-Recursive	1.132	20.731	1.483
Stacking-5-MLP-Recursive	1.138	22.534	1.422
Stacking-3-ENN-Recursive	1.002	18.535	1.312
Stacking-5-ENN-Recursive	0.983	17.887	1.293

effects of the two prediction strategies are different. For example, from the perspective of MAPE, in the 2-step ahead prediction, the combination of ENN and recursive strategy can achieve better prediction results; and in the 3-step ahead prediction, the combination of ENN and MIMO strategy has better prediction results.

(3) In the 1-step ahead prediction, as the number of base learners increases, the Stacking ensemble prediction effect does not get better but worsens. In the 2-step ahead and 3-step ahead prediction, as the number of base learners increases, the prediction effect of some Stacking ensemble models is improved. For example, in the 2-step ahead prediction, when combined with the MIMO strategy, from the perspective of MAPE, the prediction accuracy of five MLPs as the base learner is 1.09% higher than that of the three MLPs as the base learner.

(4) In the Stacking ensemble prediction, when the meta-learner is SVM, regardless of the number of ahead prediction steps, choosing ENN as the base learner can achieve higher prediction accuracy than choosing MLP in most cases. For example, in the 1-step ahead forecast, from the perspective of MAPE, Stacking-3-ENN-Recursive has a 37.43% higher prediction accuracy than Stacking-3-MLP-Recursive. It shows that the combination of ENN and SVM in Stacking is better.

6.4.4 Conclusions

This section proposes eight Stacking ensemble models to predict wind speed data. It can be seen from the test results that

(a) Stacking ensemble forecasting model can achieve a better wind speed forecasting effect, but as the number of ahead forecasting steps increases, the forecasting effect gradually becomes worse.

(b) In Stacking ensemble prediction, the MIMO strategy prediction effect is better when MLP is used as the base learner, and when ENN is used as the base learner, the MIMO and recursive strategies are comparable.

(c) When the number of base learners increases, the prediction effect of the Stacking ensemble model will not necessarily improve, and there is no absolute positive correlation between the two.

6.5 Single-point wind direction forecasting algorithm based on boosting

The Boosting algorithm is an ensemble method that continuously trains a series of weak learners to obtain a strong learner with excellent performance. The Boosting algorithm was first proposed by Schapire in 1990 [28], and then improved by Freund in 1995 [29]. The Boosting algorithm was first proposed to solve the classification problem. In 1997, Freund and Schapire extended the Boosting algorithm to solve the regression problem and proposed the AdaBoost.R algorithm [30]. The Adaboost.R algorithm improves many shortcomings of the early Boosting algorithm, and improves the lower limit of the weak learning algorithm to adaptive. Subsequently, other scholars have successively proposed some improved AdaBoost algorithms, such as AdaBoost.RT algorithm, AdaBoost.MRT algorithm, Modified AdaBoost.RT algorithm, and so on.

When predicting the wind direction sequence, the hybrid model has great potential. Hybrid models generally combine decomposition algorithms, intelligent forecasting models, ensemble algorithms, etc. The original wind direction sequence has strong instability and randomness. The decomposition method can decompose the wind direction sequence into several more stable subsequences. Predicting each subsequence can effectively reduce the difficulty of prediction. This section uses WD as the data decomposition algorithm. Using WD, the original unsteady wind direction sequence is decomposed into a plurality of relatively stable wind direction subsequences, and a corresponding prediction model can be built according to the characteristics of each subsequence.

This section uses the wind direction data in Section 6.2 as the original sequence and divides it into two parts, namely D_1 and D_2. D_1 contains the 1st–1500th data as training data; D_2 contains the 1501th–2000th data as test data. This section uses MLP as the weak learner and uses different Boosting algorithms for ensemble prediction. The impact of using a data decomposition algorithm or not on the effect of multi-step wind direction prediction is also considered. Finally, we use the error evaluation indicators mentioned in Section 6.2 to compare and analyze the prediction performance.

6.5.1 Model framework

In this section, we choose MLP as the weak learner and combine data decomposition algorithm WD to comprehensively compare the following Boosting algorithms: AdaBoost.RT, AdaBoost.MRT, Modified AdaBoost.RT, and Gradient Boosting. Eight wind direction multi-step forecasting hybrid models are proposed. The specific model framework of Boosting ensemble models is shown in Fig. 6.15.

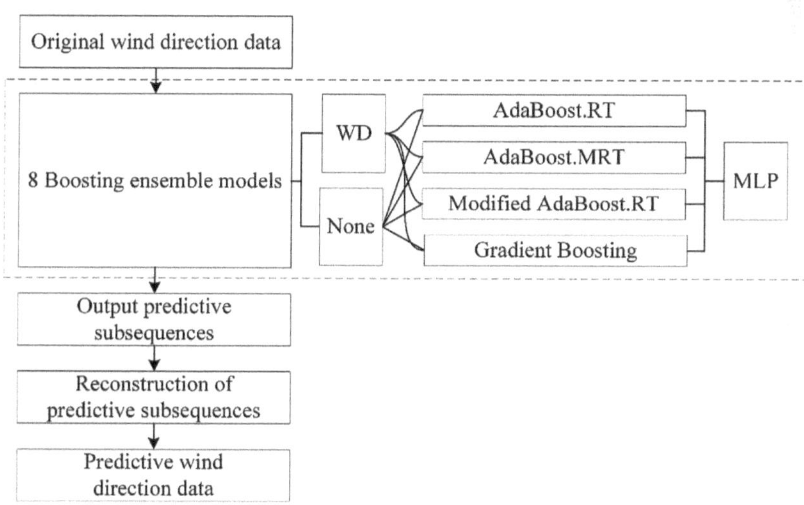

Figure 6.15 The model framework of boosting ensemble.

6.5.2 Theoretical basis

The Boosting algorithm is an ensemble method that has been extensively studied in wind prediction. The Boosting algorithm continuously trains a series of weak learners and dynamically changes the training set based on the training results. This makes the training sample that the previous weak learner judged incorrectly will receive more attention in the follow-up. The next weak learner is then trained based on the changed sample distribution. By repeating this process, a strong learner with excellent performance can be obtained [29].

Assuming $H = \{h_i(x)|i = 1, 2, ..., n\}$ are several weak learners, $W = \{w_i|i = 1, 2, ..., n\}$ are the weight coefficients of the weak learners, where n is the number of weak learners. Then the general structure of the Boosting algorithm is [29]

$$F(x) = \sum_{i=1}^{n} w_i h_i(x) \tag{6.2}$$

The purpose of the Boosting algorithm is to find the optimal parameter w that can minimize the loss and to improve the performance of the overall algorithm. In this section, we will conduct in-depth research on four commonly used Boosting algorithms in wind prediction, including AdaBoost.RT, AdaBoost.MRT, Modified AdaBoost.RT, and Gradient Boosting.

6.5.2.1 *AdaBoost.RT*

The AdaBoost.RT algorithm is an improved algorithm of the AdaBoost.R algorithm, which was proposed by Solomatine et al. in 2004 [31]. The RT stands for regression and threshold. The specific content of the AdaBoost.RT algorithm is as Algorithm 6.1 [31].

Algorithm 6.1. AdaBoost.RT

Input:

Sample sequence set $S = \{(x_m, y_m)|m = 1, 2, ..., M\}$

Weak learner algorithm WL_t

The number of iterations T

Threshold ϕ, used to divide and judge the correctness of the prediction instance

Number of sample instances M

(Continued)

Output:

Prediction function $f_{fin}(x) = \sum_t \left(\log\frac{1}{\beta_t}\right) f_t(x) \Big/ \sum_t \left(\log\frac{1}{\beta_t}\right)$

1: The initial distribution weight of all training sample instances is $D_t(i) = 1/M$

2: Initial error rate $\varepsilon_t = 0$

3: The number of initial iterations $t = 1$

4: **While** $t \leq T$

5: Call the WL_t to establish a regression model $f_t(x) \rightarrow y$

6: Calculate the error of each sample instance according to $E_t(i) = \left|\frac{f_t(x_i) - y_i}{y_i}\right|$

7: Calculate the error of the regression model $f_t(x)$ according to $\varepsilon_t = \sum_{E_t(i)} D_t(i)$

8: $\beta^t = \varepsilon_t^n$, where $n = 1$

9: At each iteration, update the distribution weight D_t according to $D_{t+1}(i) = \frac{D_t(i)}{Z_t} \times \begin{cases} \beta_t, E_t(i) \leq \phi \\ 1, otherwise \end{cases}$, where Z_t is the normalization factor

10: **End while**

6.5.2.2 AdaBoost.MRT

The AdaBoost.MRT algorithm is an improvement of the AdaBoost.RT algorithm, which eliminates the singularity in the misjudgment function and enhances the robustness of noise by adjusting the multivariable output [32]. The specific content of the AdaBoost.MRT algorithm is as Algorithm 6.2 [32].

Algorithm 6.2. AdaBoost.MRT

Input:

Sample sequence set $S = \{(x_m, y_m) | m = 1, 2, ..., M\}$

Weak learner algorithm WL_t

The number of iterations T

Threshold $\Phi = \{\phi^r | r = 1, ..., R\}$

Number of sample instances M

(Continued)

Output:

Prediction function $y(x) = \frac{\sum_t f_t(x)}{T}$

1: The initial distribution weight of all training sample instances is $D_t(i) = 1/M$

2: Initial error rate $\varepsilon_t^r = 0$

3: The number of initial iterations $t = 1$

4: The output error weight of the output variables of all training sample instances $D_{y,t}^r(i) = 1\big/M$

5: **While** $t \le T$

6: Take N samples from the sample sequence set M as an example, using sampling with replacement

7: The probability weight of each sample instance is D_t

8: Call the WL_t to establish a regression model $f_t(x) \rightarrow y$

9: Calculate the error of the output variable of each sample instance m according to the following equation: $Er_t^{(r)}(i) = \frac{|f_t(x_i) - y_{r,t}|}{\sigma_t^{(r)}}$, where $\sigma_t^{(r)}$ is the sampling standard deviation of $\left(f_t(x_i) - y_{r,t}\right)$

10: Calculate the error of the regression model according to $\varepsilon_t = \sum_{E_t(i)} D_t(i)$

11: $\varepsilon_t^{(r)} = \sum_{Er_t^{(r)}(i) > \phi_r} D_{y,t}^{(r)}(i)$, and $\beta_{t,r} = \left(\varepsilon_t^{(r)}\right)^n$, where n is 1

12: Update the distribution weight $D_{y,t}$ of the output error according to the following equation: $D_{y,t+1}(i) = \frac{D_{y,t}(i)}{Z_t} \times \begin{cases} \beta_{t,r}, E_t(i) \le \phi \\ 1, otherwise \end{cases}$, where Z_t is the normalization factor

13: Further update the distribution weight D_t of the sample instance according to the equation: $D_{t+1}(i) = \frac{1}{R} \sum_1^r D_{y,t}^{(r)}(i)$

14: **End while**

6.5.2.3 Modified AdaBoost.RT

In the aforementioned AdaBoost algorithms, the set of the initial threshold will affect the accuracy of the algorithm. Therefore, the Modified AdaBoost.RT algorithm is proposed, which can adaptively adjust the threshold according to the root mean squared error of the previous iteration. The specific steps of the Modified AdaBoost.RT algorithm are as Algorithm 6.3 [33].

Algorithm 6.3. Modified AdaBoost.RT

Input:

Sample sequence set $S = \{(x_m, y_m)|m = 1, 2, ..., M\}$

Weak learner algorithm WL_t

The number of iterations T

The threshold vector ϕ_t, the change rate of the specified threshold ϕ_t is r

Number of sample instances M

Output:

Prediction function $f_{fin}(x) = \sum_t \left(\log\frac{1}{\beta_t}\right) f_t(x) \Big/ \sum_t \left(\log\frac{1}{\beta_t}\right)$

1: The initial distribution weight of all training sample instances is $D_t(i) = 1/M$

2: Initial error rate $\varepsilon_t = 0$

3: The number of initial iterations $t = 1$

4: **While** $t \leq T$

5: The probability weight of each sample instance is D_t

6: Call the WL_t to establish a regression model $f_t(x) \rightarrow y$

7: Calculate the error of each sample instance according to $ARE_t(i) = \left|\frac{f_t(x_i) - y_i}{y_i}\right|$

8: Calculate the error of the regression model $f_t(x)$ according to $\varepsilon_t = \sum_{E_t(i)} D_t(i)$

9: $\beta^t = \varepsilon_t^n$, where $n = 1$

10: At each iteration, update the distribution weight D_t according to $D_{t+1}(i) = \frac{D_t(i)}{Z_t} \times \begin{cases} \beta_t, ARE_t(i) \leq \phi \\ 1, otherwise \end{cases}$, where Z_t is the normalization factor

11: Update the threshold ϕ_t according to the following equation: $\phi_{t+1}(i) = \phi_t(i) \times \begin{cases} (1 - \lambda), e_t < e_{t-1} \\ (1 + \lambda), otherwise \end{cases}$, where $\lambda = r \times \left|\frac{e_t - e_{t-1}}{e_t}\right|$, e is the root mean square error, and its calculation equation is as follows: $e_t = \sqrt{\frac{1}{N}\sum_{i=1}^{N}(f_t(x_t) - y_t)^2}$

12: **End while**

6.5.2.4 Gradient Boosting

Different from the previous AdaBoost algorithms, the idea of the Gradient Boosting algorithm is that on a suitable cost function, the Boosting algorithm can be interpreted as an optimization algorithm, which originated from the observation report of Breiman [34]. Subsequently, Mason et al. proposed a more general-purpose Gradient Boosting algorithm [35], and Friedman proposed an explicit regression Gradient Boosting algorithm [36,37].

The difference between the Gradient Boosting algorithm and the traditional Boosting algorithm is that each calculation of the Gradient Boosting algorithm is to reduce the last residual, and it is implemented by establishing a new model in the direction of the gradient of the residual reduction. This view of functional Gradient Boosting algorithm goes beyond simple regression and classification, and promotes the development of Boosting algorithms in many fields such as machine learning and mathematical statistics.

The specific steps of the Gradient Boosting algorithm are as Algorithm 6.4.

Algorithm 6.4. Gradient Boosting

Input:

Sample sequence set $S = \{(x_m, y_m)|m = 1, 2, ..., M\}$

Differentiable loss function $L(y, F(x))$

The number of iterations T

Number of sample instances M

Output:

Model $F_T(x)$ obtained in the Tth iteration

1: Fixed value initialization model: $F_0(x) = \text{argmin} \sum_{i=1}^{n} L(y_i, \gamma)$

2: The number of initial iterations $t = 1$

3: **While** $t \leq T$

4: Calculate pseudo-residuals: $\tau_{it} = -\left[\frac{\partial L(y_i, F(x_i))}{\partial F(x_i)}\right]_{F(x)=F_{t-1}(x)} \quad (i = 1, 2, ..., m)$

5: Combine the pseudo-residual with the base learner $h_t(x)$ to establish the training set $\{(x_i, \tau_{it})\}_{i=1}^{m}$

6: Obtain the multiplier γ^m by calculating the following one-dimensional

(Continued)

optimization problem $\gamma^m = \text{argmin} \sum_{i=1}^{n} L(y_i, F_{m-1}(x_i) + \gamma h_m(x_i))$

Update the model: $F_m(x) = F_{m-1}(x) + \gamma_m h_m(x)$

End while

6.5.3 Result analysis

This section proposes eight Boosting ensemble prediction models, all of which can complete the multi-step ahead wind direction prediction. To compare the influence of whether the decomposition algorithm is used or not on the prediction effect, combined with the 1501th–2000th sample points in the original wind direction sequence, the wind direction prediction results corresponding to the four Boosting algorithms are drawn as shown in Figs. 6.16–6.19. To more clearly show the forecasting accuracy of the different models, the specific prediction error indicator results of the eight Boosting ensemble models are shown in Tables 6.8–6.10.

By analyzing the information of these figures and tables, we can draw the following conclusions:

(1) The forecasting results of the eight Boosting ensemble prediction models are very close to the actual wind direction data. And the error evaluation indicator values of the three multi-steps ahead prediction

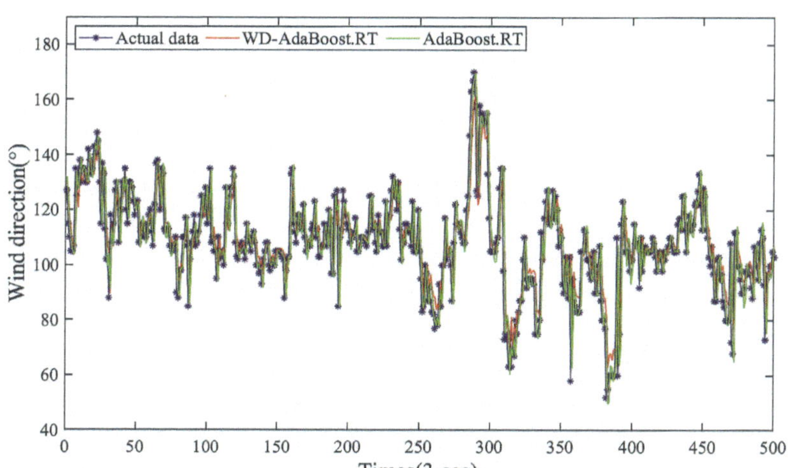

Figure 6.16 The 1-step prediction results of AdaBoost.RT ensemble models.

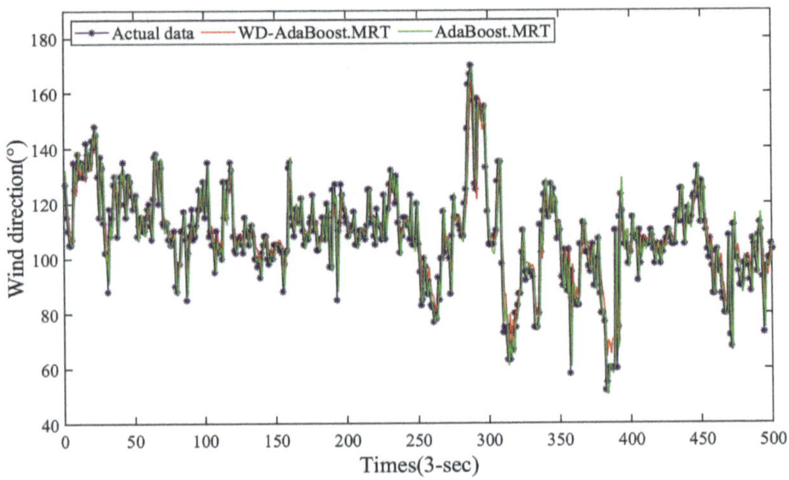

Figure 6.17 The 1-step prediction results of AdaBoost.MRT ensemble models.

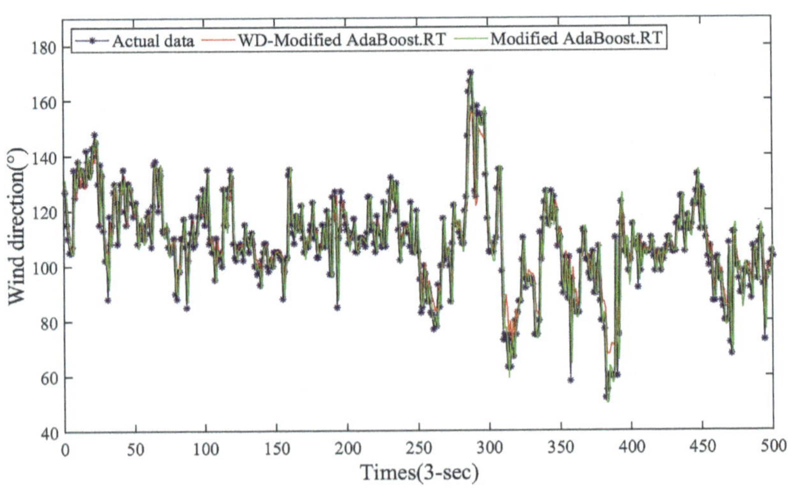

Figure 6.18 The 1-step prediction results of Modified AdaBoost.RT ensemble models.

results are all small, and the largest MAPE is only 12.362%. This shows that the four Boosting algorithms can achieve wind direction prediction well.

(2) The wind direction prediction accuracy is significantly reduced with large forecasting steps. Taking AdaBoost.RT-MLP as an example, from the perspective of MAPE, the model prediction accuracy of the 2-step ahead is 30.09% lower than that of the 1-step ahead, and the prediction accuracy of the 3-step ahead is 10.25% lower than that of the 2-step ahead.

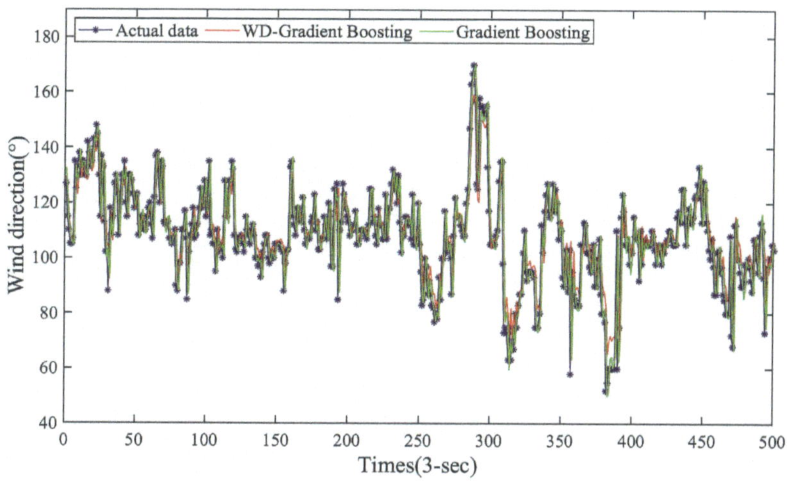

Figure 6.19 The 1-step prediction results of Gradient Boosting ensemble models.

Table 6.8 The 1-step forecasting performance of the boosting ensemble models.

Boosting ensemble models	MAE(°)	MAPE(%)	RMSE(°)
WD-AdaBoost.RT-MLP	7.581	7.334	10.891
WD-AdaBoost.MRT-MLP	7.574	7.315	10.953
WD-Modified AdaBoost.RT-MLP	7.526	7.282	10.806
WD-Gradient Boosting-MLP	7.540	7.300	10.858
AdaBoost.RT-MLP	7.361	7.235	10.207
AdaBoost.MRT-MLP	7.398	7.249	10.272
Modified AdaBoost.RT-MLP	7.431	7.335	10.341
Gradient Boosting-MLP	7.454	7.373	10.399

Table 6.9 The 2-step forecasting performance of the boosting ensemble models.

Boosting ensemble models	MAE(°)	MAPE(%)	RMSE(°)
WD-AdaBoost.RT-MLP	11.052	10.624	14.763
WD-AdaBoost.MRT-MLP	11.204	10.786	14.893
WD-Modified AdaBoost.RT-MLP	10.997	10.531	14.785
WD-Gradient Boosting-MLP	11.084	10.627	14.867
AdaBoost.RT-MLP	9.543	9.412	12.589
AdaBoost.MRT-MLP	9.465	9.277	12.618
Modified AdaBoost.RT-MLP	9.637	9.535	12.767
Gradient Boosting-MLP	9.690	9.615	12.734

Table 6.10 The 3-step forecasting performance of the boosting ensemble models.

Boosting ensemble models	MAE(°)	MAPE(%)	RMSE(°)
WD-AdaBoost.RT-MLP	12.664	12.298	16.929
WD-AdaBoost.MRT-MLP	12.750	12.362	16.889
WD-Modified AdaBoost.RT-MLP	12.393	11.998	16.655
WD-Gradient Boosting-MLP	12.397	11.998	16.577
AdaBoost.RT-MLP	10.427	10.377	13.827
AdaBoost.MRT-MLP	10.339	10.197	13.799
Modified AdaBoost.RT-MLP	10.512	10.477	13.965
Gradient Boosting-MLP	10.575	10.579	13.944

(3) The four Boosting algorithms, AdaBoost.RT, AdaBoost.MRT, Modified AdaBoost.RT, and Gradient Boosting, have little difference in the prediction effect when combined with MLP for wind direction prediction. In the 2-step and 3-step ahead prediction, it can be known from the error indicator data that the combination of AdaBoost.MRT algorithm and MLP for wind direction prediction is relatively good. For example, in the 2-step ahead prediction, from the perspective of MAPE, the wind direction prediction accuracy of AdaBoost.MRT-MLP is improved by 3.51% than Gradient Boosting-MLP.

(4) In the 1-step ahead prediction, the use of the decomposition algorithm WD can help to improve the accuracy of wind direction prediction. For example, from the perspective of MAPE, the ahead prediction accuracy of WD-Gradient Boosting-MLP is improved by 0.99% compared with the Gradient Boosting-MLP ensemble model. In the 2-step and 3-step ahead forecast, the advantage of WD is not obvious.

6.5.4 Conclusions

This section proposes eight Boosting ensemble models to predict wind direction. It can be seen from the test results that

(a) The Boosting ensemble forecasting model can achieve a good wind direction forecasting effect, but as the number of ahead forecasting steps increases, the forecasting effect decreases.

(b) When combined with MLP for wind direction prediction, four Boosting algorithms, AdaBoost.RT, AdaBoost.MRT, Modified AdaBoost.RT, and Gradient Boosting, can all achieve high prediction accuracy, and no specific algorithm has obvious advantages.

(c) In the 1-step ahead prediction, the use of the decomposition algorithm WD can help to improve the accuracy of wind direction prediction. However, the advantage of WD is not obvious in the 2-step and 3-step ahead prediction.

6.6 Summary and outlook

This chapter introduces the ensemble learning algorithm in wind prediction in detail. Ensemble learning combines multiple individual learners for data training, which can achieve better prediction results than a single learner. This chapter mainly introduces the three ensemble learning methods of the multi-objective ensemble, Stacking, and Boosting. The real wind data along the strong wind railway is predicted, and the three error evaluation indicators of MAE, MAPE, and RMSE are used for evaluation.

In Section 6.3, the relevant content of the multi-objective ensemble algorithm is introduced in detail, and its application in wind speed prediction is studied. The multi-objective ensemble algorithm combines multiple individual learners, and uses the multi-objective optimization algorithm to calculate the combined weights of the individual learners. This section uses WD as the decomposition algorithm and MLP as the intelligent learner, and compares the multi-objective optimization algorithms MOPSO, MOGWO, and MOGOA with the single-objective optimization algorithms PSO, GWO, and BA. The test results show that (1) Single-point wind speed prediction using multi-objective ensemble algorithm can achieve better prediction results. (2) The prediction accuracy of the ensemble model decreases with large forecasting steps. (3) In the 1-step ahead prediction, compared with the single-objective optimization algorithm, the multi-objective optimization algorithm improves the model prediction accuracy more obviously.

In Section 6.4, the relevant content of the Stacking ensemble algorithm is introduced in detail, and its application in single-point wind speed prediction is studied. The Stacking algorithm first uses several base learners to learn the training samples, and then uses the meta-learner to implement nonlinear weighting of different base learners. This section uses WD as the decomposition algorithm, different numbers of MLP or ENN as the base learners, SVM as the meta-learner, and MIMO or recursive as the forecasting strategy. A total of eight Stacking ensemble models are proposed. The test results show that (1) Stacking ensemble prediction model can achieve better wind speed prediction effect, but as the number of ahead prediction steps increases, the forecasting accuracy decreases. (2) In Stacking ensemble prediction, when MLP is used as the base learner, the MIMO strategy has obvious advantages, and when ENN is used as the base learner, the prediction effects of MIMO and recursive strategies are similar. (3) When the number of base learners increases, the prediction effect of the Stacking ensemble model will not necessarily improve.

In Section 6.5, the relevant content of the Boosting ensemble algorithm is introduced in detail, and its application in single-point wind direction prediction is studied. The Boosting algorithm can train multiple weak learners and further change the distribution of the training set to obtain a strong learner with excellent performance. In this section, four Boosting algorithms AdaBoost.RT, AdaBoost.MRT, Modified AdaBoost.RT, and Gradient Boosting are used, and MLP is used as the weak learner. The influence of whether the WD algorithm is used or not on the wind direction prediction results is also compared. The test results show that (1) The Boosting ensemble prediction model can achieve a good wind direction prediction effect, but the prediction accuracy will decrease as the number of ahead prediction steps increases. (2) When combined with MLP for wind direction prediction, the prediction effects of the four Boosting algorithms are not very different. (3) In the 1-step ahead prediction, the use of the decomposition algorithm WD has an effect on improving the accuracy of wind direction prediction.

Ensemble learning has great potential in wind prediction by combining the advantages of multiple learners. In the future, in addition to the related ensemble learning algorithms mentioned in this chapter, other algorithms can also be further studied in wind prediction. And deep learning, reinforcement learning, and other algorithms can also be combined with ensemble learning to explore the potential of ensemble learning.

References

[1] T.G. Dietterich, Ensemble Learning. The Handbook of Brain Theory and Neural Networks, 2002, pp. 110−125, 2.

[2] J.D. Wichard, M. Ogorzalek, Time series prediction with ensemble models, in: 2004 IEEE International Joint Conference on Neural Networks, vol. 2, 2004, pp. 1625−1630.

[3] Y. Li, H. Shi, F. Han, et al., Smart wind speed forecasting approach using various boosting algorithms, big multi-step forecasting strategy, Renew. Energy 135 (2019) 540−553.

[4] Z. Qu, K. Zhang, W. Mao, et al., Research and application of ensemble forecasting based on a novel multi-objective optimization algorithm for wind-speed forecasting, Energy Convers. Manag. 154 (2017) 440−454.

[5] H. Liu, C. Chen, Multi-objective data-ensemble wind speed forecasting model with stacked sparse autoencoder and adaptive decomposition-based error correction, Appl. Energy 254 (2019) 113686.

[6] K. Deb, Multi-objective optimization, in: Search Methodologies, Springer, 2014, pp. 403−449.

[7] Y. Censor, Pareto optimality in multiobjective problems, Appl. Math. Optim. 4 (1977) 41−59.

[8] H. Liu, S. Yin, C. Chen, et al., Data multi-scale decomposition strategies for air pollution forecasting: a comprehensive review, J. Clean. Prod. 277 (2020) 124023.
[9] I. Daubechies, The wavelet transform, time-frequency localization and signal analysis, IEEE Trans. Inform. Theory 36 (1990) 961–1005.
[10] S.G. Mallat, A theory for multiresolution signal decomposition: the wavelet representation, IEEE Trans. Pattern Anal. Machine Intell. 11 (1989) 674–693.
[11] C.M. Bishop, Neural Networks for Pattern Recognition, Oxford university press, 1995.
[12] S. Mirjalili, S.M. Mirjalili, A. Lewis, Grey wolf optimizer, Adv. Eng. Software 69 (2014) 46–61.
[13] J. Kennedy, R. Eberhart, Particle swarm optimization, in: Proceedings of ICNN'95-International Conference on Neural Networks, vol. 4, 1995, pp. 1942–1948.
[14] U. Baumgartner, C. Magele, W. Renhart, Pareto optimality and particle swarm optimization, IEEE Trans. Magn. 40 (2004) 1172–1175.
[15] X.-S. Yang, A new metaheuristic bat-inspired algorithm, in: Nature Inspired Cooperative Strategies for Optimization (NICSO 2010), Springer, 2010, pp. 65–74.
[16] X.-S. Yang, X. He, Bat algorithm: literature review and applications, Int. J. Bio-Inspired Comput. 5 (2013) 141–149.
[17] S. Mirjalili, S. Saremi, S.M. Mirjalili, et al., Multi-objective grey wolf optimizer: a novel algorithm for multi-criterion optimization, Expert Syst. Appl. 47 (2016) 106–119.
[18] C.C. Coello, M.S.L. MOPSO, A proposal for multiple objective particle swarm optimization, in: Proceedings of the 2002 Congress on Evolutionary Computation CEC'02 (Cat No 02TH8600), vol. 2, 2002, pp. 1051–1056.
[19] C.a.C. Coello, G.T. Pulido, M.S. Lechuga, Handling multiple objectives with particle swarm optimization, IEEE Trans. Evol. Comput. 8 (2004) 256–279.
[20] S.Z. Mirjalili, S. Mirjalili, S. Saremi, et al., Grasshopper optimization algorithm for multi-objective optimization problems, Appl. Intell. 48 (2018) 805–820.
[21] S. Saremi, S. Mirjalili, A. Lewis, Grasshopper optimisation algorithm: theory and application, Adv. Eng. Software 105 (2017) 30–47.
[22] D.H. Wolpert, Stacked generalization, Neural Network. 5 (1992) 241–259.
[23] S. Džeroski, B. Ženko, Is combining classifiers with stacking better than selecting the best one? Mach. Learn. 54 (2004) 255–273.
[24] J.V. Hansen, R.D. Nelson, Data mining of time series using stacked generalizers, Neurocomputing 43 (2002) 173–184.
[25] S.-Q. Wang, J. Yang, K.-C. Chou, Using stacked generalization to predict membrane protein types based on pseudo-amino acid composition, J. Theor. Biol. 242 (2006) 941–946.
[26] S.B. Taieb, G. Bontempi, A.F. Atiya, et al., A review and comparison of strategies for multi-step ahead time series forecasting based on the NN5 forecasting competition, Expert Syst. Appl. 39 (2012) 7067–7083.
[27] C.Y. Low, J. Park, A.B.J. Teoh, Stacking-based deep neural network: deep analytic network for pattern classification, IEEE Trans. Cybernetic. 50 (2020) 5021–5034.
[28] R.E. Schapire, The strength of weak learnability, Mach. Learn. 5 (1990) 197–227.
[29] Y. Freund, Boosting a weak learning algorithm by majority, Inf. Comput. 121 (1995) 256–285.
[30] Y. Freund, R.E. Schapire, A decision-theoretic generalization of on-line learning and an application to boosting, J. Comput. Syst. Sci. 55 (1997) 119–139.
[31] D.P. Solomatine, D.L. Shrestha, R.T. AdaBoost, A boosting algorithm for regression problems, in: 2004 IEEE International Joint Conference on Neural Networks (IEEE Cat No 04CH37541) vol. 2, 2004, pp. 1163–1168.

[32] N. Kummer, H. Najjaran, M.R.T. Adaboost, Boosting regression for multivariate estimation, Artif. Intell. Res. 3 (2014) 64–76.

[33] H.-X. Tian, Z.-Z. Mao, An ensemble ELM based on modified AdaBoost.RT algorithm for predicting the temperature of molten steel in ladle furnace, IEEE Trans. Autom. Sci. Eng. 7 (2010) 73–80.

[34] L. Breiman, Arcing the edge.Technical Report 486, Statistics Department, University of California at Berkeley, 1997.

[35] L. Mason, J. Baxter, P.L. Bartlett, et al., Boosting algorithms as gradient descent, in: Advances in Neural Information Processing Systems, 2000, pp. 512–518.

[36] J.H. Friedman, Greedy function approximation: a gradient boosting machine, Ann. Stat. (2001) 1189–1232.

[37] J.H. Friedman, Stochastic gradient boosting, Comput. Stat. Data Anal. 38 (2002) 367–378.

CHAPTER 7

Description methods of spatial wind along railways

Contents

7.1 Introduction

The railway is the main part of national transportation and the basic guarantee for economic development. Strong wind has become a huge threat to safe operation, which often endangers the normal operation of trains [1]. Under the action of strong lateral wind, especially on superlarge bridges, high embankments, or special lines, the lateral aerodynamic force received by the train may lead the train to swing beyond the limit, fall off

Wind Forecasting in Railway Engineering
ISBN 978-0-12-823706-9
https://doi.org/10.1016/B978-0-12-823706-9.00007-7

the track, and even roll over and cause casualties [2,3]. The spatial wind speed description along the railway can analyze the wind speed spatial correlation and provide a basis for the construction of an accurate wind speed prediction model.

7.2 Spatial wind correlation analysis

7.2.1 Wind analysis methods and data collection

To study the spatial correlation between wind speeds measured at different measuring points, 10 measuring points were selected from the measuring stations in strong wind area, as shown in Fig. 7.1. The grid in Fig. 7.1 is 0.1×0.1 degree. These data are collected from National Renewable Energy Laboratory (NREL), the website is https://www.nrel.gov/.

7.2.2 Cross-correlation analysis by MI

7.2.2.1 Theory basis

The Mutual Information (MI) is the measurement of the interdependence of variables and works as the transformation method. The mutual information method is a filtering method commonly used in feature selection,

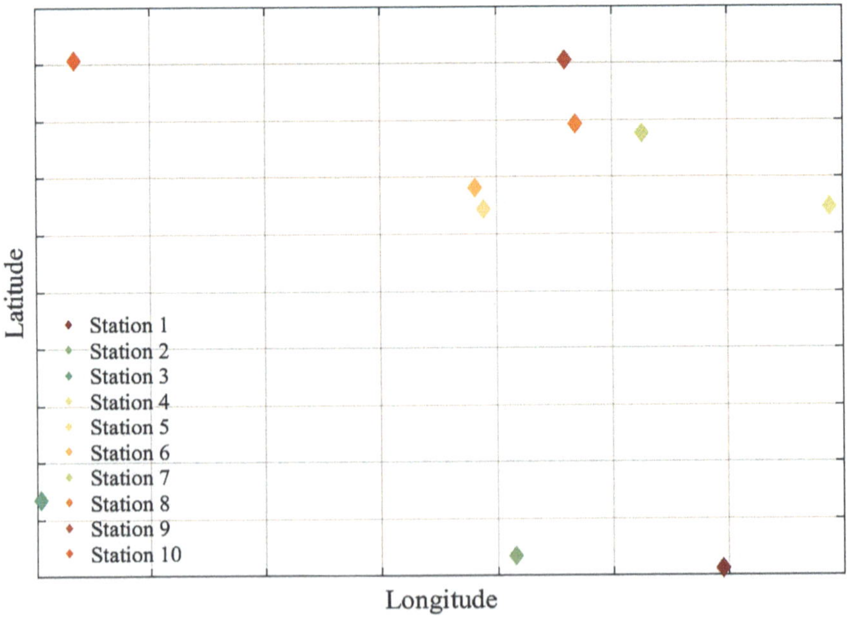

Figure 7.1 Locations of the wind monitoring stations in strong wind area.

which measures the dependence between two variables. The MI is defined as the following form [4]:

$$\mathrm{MI}(X,\ Y)=\sum_{x}\sum_{y}p(x,y)\log\frac{p(x,y)}{p(x)p(y)} \tag{7.1}$$

where X and Y represent the variables, $p(x)$, $p(y)$ are the probability density of X, Y, respectively; $p(x,y)$ is the joint probability density of X and Y.

For continuous values, the integral equation of MI is as follows [5]:

$$\mathrm{MI}(X,\ Y)=\int_{Y}\int_{X}p(x,\ y)\log\left(\frac{p(x,\ y)}{p(x)p(y)}\right)dxdy \tag{7.2}$$

where $p(x)$, $p(y)$ are the probability distribution of X, Y, respectively; $p(x,y)$ is the joint probability distribution of X and Y.

When X and Y are independent random variables, the mutual information of them will be zero, that is, $\mathrm{MI}(X,\ Y)=0$. By statistical research, it can be considered that the joint probability distribution could be $p(x,\ y)=p(x)p(y)$.

7.2.2.2 Cross-correlation of the wind locations

The results from the MI for variable correlation analysis are shown below. Figs. 7.2 and 7.3 are heat maps drawn based on MI values of wind speed and wind direction, and Tables 7.1 and 7.2 are the MI values among all variables.

It can be concluded that the strong correlation is among the wind information of the measuring stations, and the variables have weak independence. In particular, the changing trend of X_{L7}–X_{L10} are almost the same, and the variables are deeply correlated. The correlations between other variables are not so important as before.

7.2.3 Cross-correlation analysis by Pearson coefficient

7.2.3.1 Theory basis

The Pearson coefficient is a method used to present the linear correlation. It is mainly related to the linear relationship between random variables, and its value is between −1 and 1. Table 7.3 is the absolute value of Pearson correlation coefficient of each correlation grad. The Pearson correlation coefficient is suitable for two groups of continuous data that maintain a linear relationship and the overall obey a normal distribution, and each pair of measured values are independent of each other. The formula is as follows [6]:

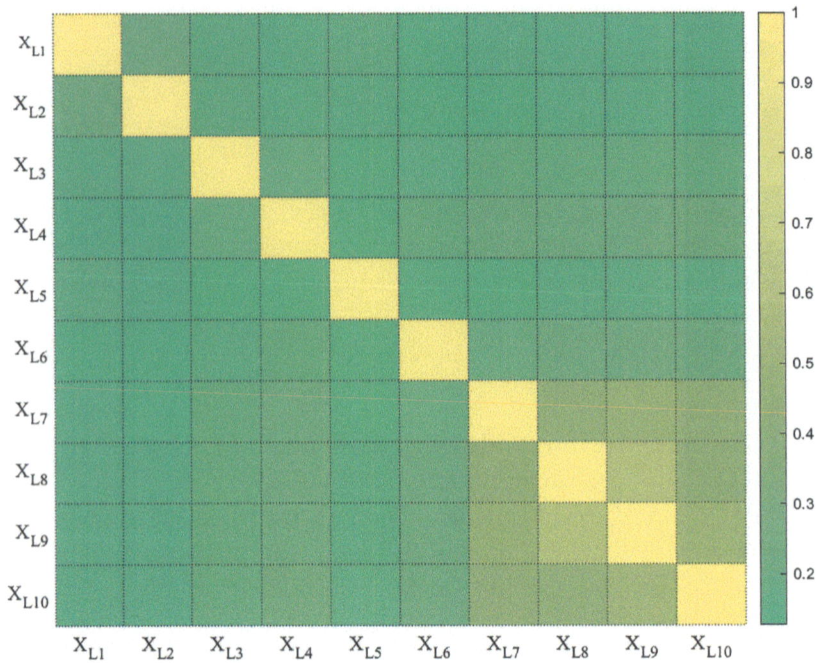

Figure 7.2 Heat map of cross-correlation result based on MI for wind speed.

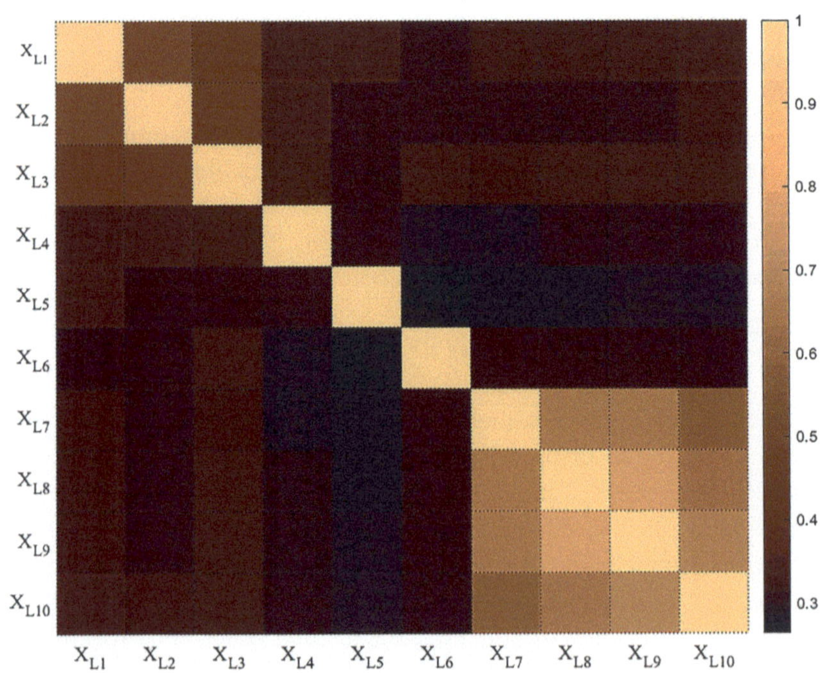

Figure 7.3 Heat map of cross-correlation result based on MI for wind direction.

Table 7.1 The cross-correlation coefficient based on MI for wind speed.

	X_{D1}	X_{D2}	X_{D3}	X_{D4}	X_{D5}	X_{D6}	X_{D7}	X_{D8}	X_{D9}	X_{D10}
X_{D1}	1	0.2447	0.1742	0.1536	0.1757	0.1343	0.1483	0.1541	0.1503	0.1536
X_{D2}	0.2447	1	0.1695	0.1458	0.1485	0.1409	0.1396	0.1461	0.1452	0.1537
X_{D3}	0.1742	0.1695	1	0.2002	0.1347	0.1649	0.1836	0.1836	0.1832	0.1899
X_{D4}	0.1536	0.1458	0.2002	1	0.1380	0.1862	0.2132	0.2156	0.2168	0.2261
X_{D5}	0.1757	0.1485	0.1347	0.1380	1	0.1277	0.1401	0.1447	0.1437	0.1550
X_{D6}	0.1343	0.1409	0.1649	0.1862	0.1277	1	0.2015	0.2187	0.2161	0.2197
X_{D7}	0.1483	0.1396	0.1836	0.2132	0.1401	0.2015	1	0.4175	0.4466	0.4152
X_{D8}	0.1541	0.1461	0.1836	0.2156	0.1447	0.2187	0.4175	1	0.5803	0.4266
X_{D9}	0.1503	0.1452	0.1832	0.2168	0.1437	0.2161	0.4466	0.5803	1	0.4842
X_{D10}	0.1536	0.1537	0.1899	0.2261	0.1550	0.2197	0.4152	0.4266	0.4842	1

Table 7.2 The cross-correlation coefficient based on MI for wind direction.

1	X_{D1}	X_{D2}	X_{D3}	X_{D4}	X_{D5}	X_{D6}	X_{D7}	X_{D8}	X_{D9}	X_{D10}
X_{D1}	1	0.5070	0.4534	0.3496	0.3464	0.3213	0.3300	0.3308	0.3317	0.3496
X_{D2}	0.5070	1	0.4697	0.3485	0.3129	0.3148	0.3114	0.3178	0.3206	0.3217
X_{D3}	0.4534	0.4697	1	0.3333	0.3121	0.3336	0.3272	0.3407	0.3419	0.3395
X_{D4}	0.3496	0.3485	0.3333	1	0.3043	0.2923	0.2950	0.2996	0.3002	0.3058
X_{D5}	0.3464	0.3129	0.3121	0.3043	1	0.2639	0.2803	0.2862	0.2877	0.2978
X_{D6}	0.3213	0.3148	0.3336	0.2923	0.2639	1	0.3116	0.3140	0.3136	0.3084
X_{D7}	0.3300	0.3114	0.3272	0.2950	0.2803	0.3116	1	0.6510	0.6550	0.5948
X_{D8}	0.3308	0.3178	0.3407	0.2996	0.2862	0.3140	0.6510	1	0.7759	0.6348
X_{D9}	0.3317	0.3206	0.3419	0.3002	0.2877	0.3136	0.6550	0.7759	1	0.6860
X_{D10}	0.3496	0.3217	0.3395	0.3058	0.2978	0.3084	0.5948	0.6348	0.6860	1

$$P_{x,y} = \frac{\sum x \cdot y - \frac{\sum x \cdot \sum y}{n}}{\sqrt{\left(\sum x^2 - \frac{\left(\sum x\right)^2}{n}\right) \cdot \left(\sum y^2 - \frac{\left(\sum y\right)^2}{n}\right)}} \tag{7.3}$$

where x and y represent the variables and n represents the number of the values. Under the normal circumstances, the relative strength of variables can be judged by the value range.

7.2.3.2 Cross-correlation of wind locations

The results from the Pearson methods for correlation analysis are shown as follows. Figs. 7.4 and 7.5 are heat maps drawn based on Pearson methods of wind speed and wind direction, and Tables 7.4 and 7.5 are the Pearson methods values among all variables.

It can be concluded that the strong correlation is between the wind information of the measuring stations with weak independence of variables. In particular, the changing trend of $X_{L1}-X_{L2}$ and $X_{L7}-X_{L10}$ are almost the

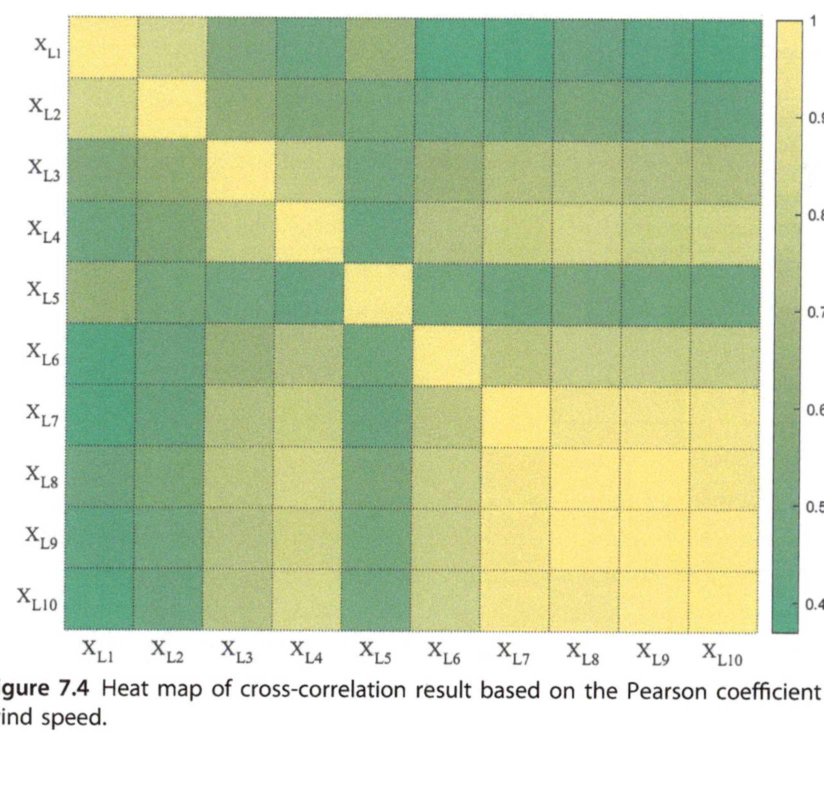

Figure 7.4 Heat map of cross-correlation result based on the Pearson coefficient for wind speed.

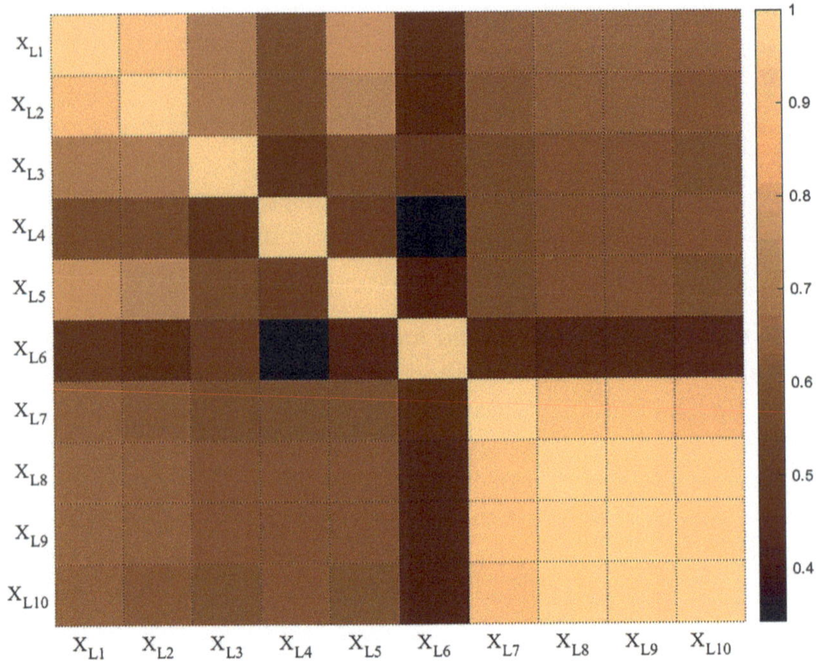

Figure 7.5 Heat map of cross-correlation result based on the Pearson coefficient for wind direction.

Table 7.3 Absolute value of correlation coefficient and correlation grad.

The absolute value of the correlation coefficient	Correlation grad
0.8–1.0	Highly linear correlation
0.6–0.8	Strong linear correlation
0.4–0.6	Medium linear correlation
0.2–0.4	Weak linear correlation
0.0–0.2	Weak or no correlation

same, and the variables are deeply correlated, which can contribute to feature extraction in the feature selection process. The correlations between other variables are less significant. Different from other methods, the Pearson coefficient can be negative in the results, but the information here obviously presents the positive correlation between variables, which has a certain reference for the next selection of different locations.

Table 7.4 The cross-correlation coefficient based on the Pearson coefficient for wind speed.

	X_{D1}	X_{D2}	X_{D3}	X_{D4}	X_{D5}	X_{D6}	X_{D7}	X_{D8}	X_{D9}	X_{D10}
X_{D1}	1	0.8341	0.5560	0.4620	0.6434	0.3697	0.3780	0.4498	0.4007	0.3766
X_{D2}	0.8341	1	0.6213	0.5372	0.5025	0.4574	0.4395	0.5180	0.4715	0.4465
X_{D3}	0.5560	0.6213	1	0.8009	0.4956	0.6547	0.7325	0.7561	0.7437	0.7246
X_{D4}	0.4620	0.5372	0.8009	1	0.4349	0.7354	0.8165	0.8328	0.8294	0.8350
X_{D5}	0.6434	0.5025	0.4956	0.4349	1	0.4627	0.4406	0.5118	0.4869	0.4754
X_{D6}	0.3697	0.4574	0.6547	0.7354	0.4627	1	0.7631	0.7923	0.7901	0.7929
X_{D7}	0.3780	0.4395	0.7325	0.8165	0.4406	0.7631	1	0.9645	0.9731	0.9673
X_{D8}	0.4498	0.5180	0.7561	0.8328	0.5118	0.7923	0.9645	1	0.9920	0.9699
X_{D9}	0.4007	0.4715	0.7437	0.8294	0.4869	0.7901	0.9731	0.9920	1	0.9828
X_{D10}	0.3766	0.4465	0.7246	0.8350	0.4754	0.7929	0.9673	0.9699	0.9828	1

Table 7.5 The cross-correlation coefficient based on the Pearson coefficient for wind direction.

	X_{D1}	X_{D2}	X_{D3}	X_{D4}	X_{D5}	X_{D6}	X_{D7}	X_{D8}	X_{D9}	X_{D10}
X_{D1}	1	0.9259	0.7158	0.5906	0.7817	0.4910	0.6505	0.6680	0.6636	0.6478
X_{D2}	0.9259	1	0.7127	0.5959	0.7480	0.4648	0.6326	0.6458	0.6405	0.6247
X_{D3}	0.7158	0.7127	1	0.4903	0.5975	0.5466	0.5931	0.6188	0.6138	0.6013
X_{D4}	0.5906	0.5959	0.4903	1	0.5428	0.3426	0.6047	0.6126	0.6132	0.6128
X_{D5}	0.7817	0.7480	0.5975	0.5428	1	0.4440	0.6085	0.6135	0.6113	0.6080
X_{D6}	0.4910	0.4648	0.5466	0.3426	0.4440	1	0.4609	0.4513	0.4463	0.4435
X_{D7}	0.6505	0.6326	0.5931	0.6047	0.6085	0.4609	1	0.9201	0.9224	0.9053
X_{D8}	0.6680	0.6458	0.6188	0.6126	0.6135	0.4513	0.9201	1	0.9847	0.9713
X_{D9}	0.6636	0.6405	0.6138	0.6132	0.6113	0.4463	0.9224	0.9847	1	0.9791
X_{D10}	0.6478	0.6247	0.6013	0.6128	0.6080	0.4435	0.9053	0.9713	0.9791	1

7.2.4 Cross-correlation analysis by Kendall coefficient

7.2.4.1 Theory basis

Kendall correlation coefficient, also known as the Kendall rank correlation coefficient, is used to expound the statistical correlation. Its value range is also between −1 and 1. The value of 1 represents a positive correlation and the value of −1 shows a negative correlation. The value of 0 represents two random variables that are mutually independent [7]. The calculation is as follows [7]:

$$K_{x,\,y} = \frac{A - B}{\sqrt{(n_3 - n_1)\cdot(n_3 - n_2)}} \tag{7.4}$$

$$\begin{cases} n_3 = n\cdot(n-1)/2 \\ n_1 = \sum_{i=1}^{m} P_i\cdot(P_i - 1)/2 \\ n_2 = \sum_{j=1}^{t} Q_j\cdot(Q_j - 1)/2 \end{cases} \tag{7.5}$$

where A represents the number of concordant pairs in x or y, B represents the number of discordant pairs in x or y. The n represents the number of variables, n_1 and n_2 are calculated for sets x and y respectively. The same elements in x are combined into a small set, m is the number of all small sets, P_i represents the number of elements in the ith small set, and t and Q_j are based on the variable y.

7.2.4.2 Cross-correlation of wind locations

The results from the Kendall method are shown as follows. Figs. 7.6 and 7.7 are heat maps drawn based on the Kendall correlation coefficient of wind speed and wind direction, and Tables 7.6 and 7.7 are the Kendall correlation coefficient values among all variables.

It can be concluded that the results by Kendall methods are similar to those by Pearson correlation analysis. In particular, the changing trend of X_{L1}−X_{L2} and X_{L7}−X_{L10} are similar, and the variables are completely positively correlated, which can be used for feature extraction.

7.2.5 Cross-correlation analysis by Spearman coefficient

7.2.5.1 Theory basis

The Spearman coefficient, also known as the Spearman rank coefficient, is a monotonic function to expresses the relationship between two variables

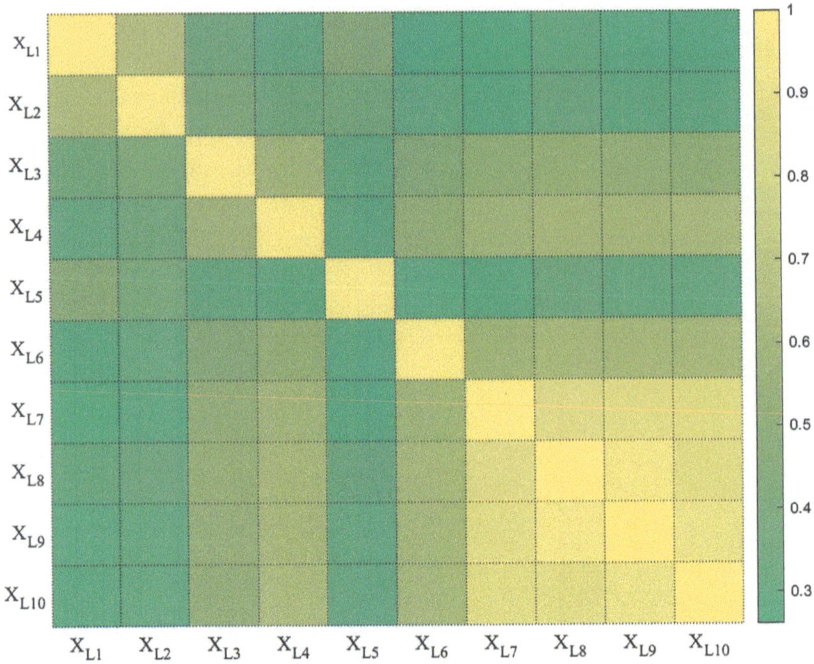

Figure 7.6 Heat map of cross-correlation result based on the Kendall coefficient for wind speed.

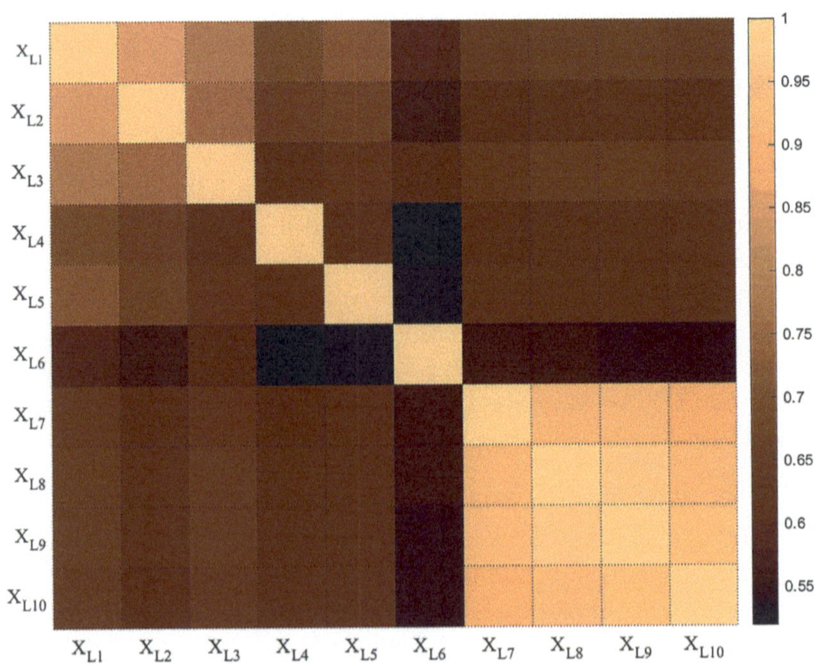

Figure 7.7 Heat map of cross-correlation result based on the Kendall coefficient for wind direction.

Table 7.6 The cross-correlation coefficient based on the Kendall coefficient for wind speed.

	X_{D1}	X_{D2}	X_{D3}	X_{D4}	X_{D5}	X_{D6}	X_{D7}	X_{D8}	X_{D9}	X_{D10}
X_{D1}	1	0.6589	0.3728	0.3240	0.4856	0.2753	0.2615	0.3191	0.2881	0.2840
X_{D2}	0.6589	1	0.4142	0.3550	0.3917	0.3241	0.2768	0.3366	0.3061	0.2984
X_{D3}	0.3728	0.4142	1	0.5975	0.3170	0.4719	0.5358	0.5504	0.5439	0.5342
X_{D4}	0.3240	0.3550	0.5975	1	0.3000	0.5423	0.6067	0.6247	0.6219	0.6319
X_{D5}	0.4856	0.3917	0.3170	0.3000	1	0.2930	0.2823	0.3371	0.3194	0.3201
X_{D6}	0.2753	0.3241	0.4719	0.5423	0.2930	1	0.5917	0.6186	0.6132	0.6155
X_{D7}	0.2615	0.2768	0.5358	0.6067	0.2823	0.5917	1	0.8564	0.8779	0.8645
X_{D8}	0.3191	0.3366	0.5504	0.6247	0.3371	0.6186	0.8564	1	0.9322	0.8678
X_{D9}	0.2881	0.3061	0.5439	0.6219	0.3194	0.6132	0.8779	0.9322	1	0.9003
X_{D10}	0.2840	0.2984	0.5342	0.6319	0.3201	0.6155	0.8645	0.8678	0.9003	1

Table 7.7 The cross-correlation coefficient based on the Kendall coefficient for wind direction.

	X_{D1}	X_{D2}	X_{D3}	X_{D4}	X_{D5}	X_{D6}	X_{D7}	X_{D8}	X_{D9}	X_{D10}
X_{D1}	1	0.8622	0.7831	0.6923	0.7224	0.5944	0.6491	0.6556	0.6552	0.6541
X_{D2}	0.8622	1	0.7654	0.6698	0.6777	0.5678	0.6132	0.6204	0.6191	0.6167
X_{D3}	0.7831	0.7654	1	0.6310	0.6467	0.6116	0.6456	0.6564	0.6555	0.6508
X_{D4}	0.6923	0.6698	0.6310	1	0.6258	0.5194	0.6141	0.6191	0.6205	0.6261
X_{D5}	0.7224	0.6777	0.6467	0.6258	1	0.5315	0.6192	0.6226	0.6232	0.6295
X_{D6}	0.5944	0.5678	0.6116	0.5194	0.5315	1	0.5588	0.5574	0.5541	0.5522
X_{D7}	0.6491	0.6132	0.6456	0.6141	0.6192	0.5588	1	0.9243	0.9259	0.8989
X_{D8}	0.6556	0.6204	0.6564	0.6191	0.6226	0.5574	0.9243	1	0.9701	0.9263
X_{D9}	0.6552	0.6191	0.6555	0.6205	0.6232	0.5541	0.9259	0.9701	1	0.9425
X_{D10}	0.6541	0.6167	0.6508	0.6261	0.6295	0.5522	0.8989	0.9263	0.9425	1

and to solve the problem according to the sort position of the original data [8]. The calculation process is as follows. Assuming two sets of random x and y, the number of the elements is n, then x_i and y_i are the sorted i^{th} elements in x and y. The calculation formula of Spearman correlation coefficient is as follows [8]:

$$S_{x,y} = 1 - \frac{6\sum_{i=1}^{n}(x_i - y_i)^2}{n\cdot(n^2-1)} \tag{7.6}$$

The Spearman correlation coefficient requires the fact that the data of the two sets are in a pairwise correspondence or the corresponding quantity obtained by continuous measurement. The Spearman correlation coefficient can be used in different overall distributions or different quantities for analysis calculations [8].

7.2.5.2 Cross-correlation of wind locations

The results from the Spearman method for variable correlation analysis are displayed as follows. Figs. 7.8 and 7.9 are heat maps drawn based on Spearman values of wind speed and wind direction, and Tables 7.8 and 7.9 are the Spearman correlation coefficient values among all variables.

It can be concluded with Fig. 7.9 and Table 7.9 that the strong correlation is among the wind information of the measuring stations, and the variables have weak independence. In particular, the changing trend of $X_{L1}-X_{L2}$ and $X_{L6}-X_{L10}$ are similar, and the variables are positively correlated when others are not so important.

7.2.6 Analysis of correlation results

To compare variance, the scatter plot of wind speed correlations and wind direction correlations are shown in Fig. 7.10. In Fig. 7.10, the correlation values of different correlation coefficients are sorted in ascent. From Fig. 7.10, it can be observed that the mutual information correlations have the smallest values in the wind speed and direction. The Pearson coefficient is similar to the Spearman coefficient in the wind speed correlations, while the Pearson coefficient is significantly smaller than the Spearman coefficient in the wind direction correlations.

To analyze whether the distance affects the correlations, the scatter plots between the distance and correlations are shown in Fig. 7.11. The linear goodness of fit are presented in Table 7.10. The distance values are

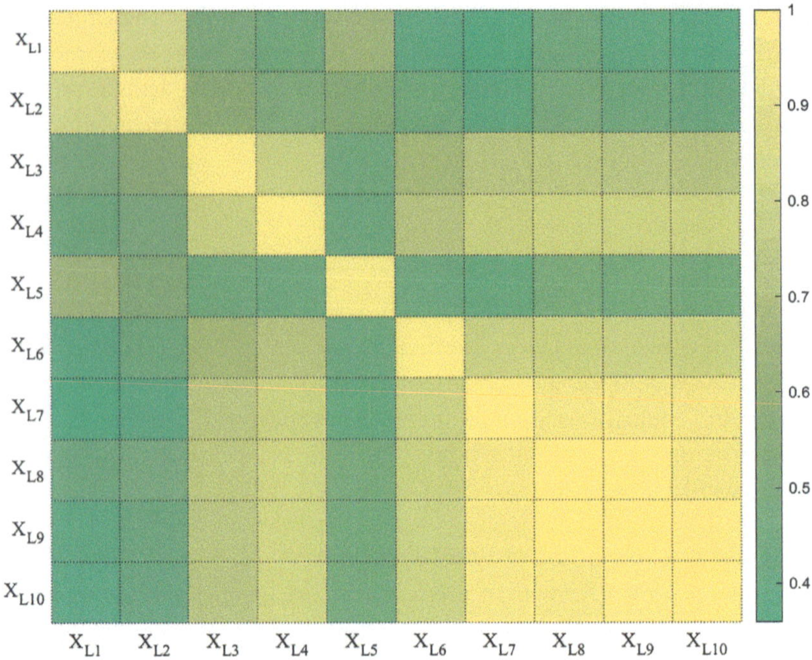

Figure 7.8 Heat map of cross-correlation result based on the Spearman coefficient for wind speed.

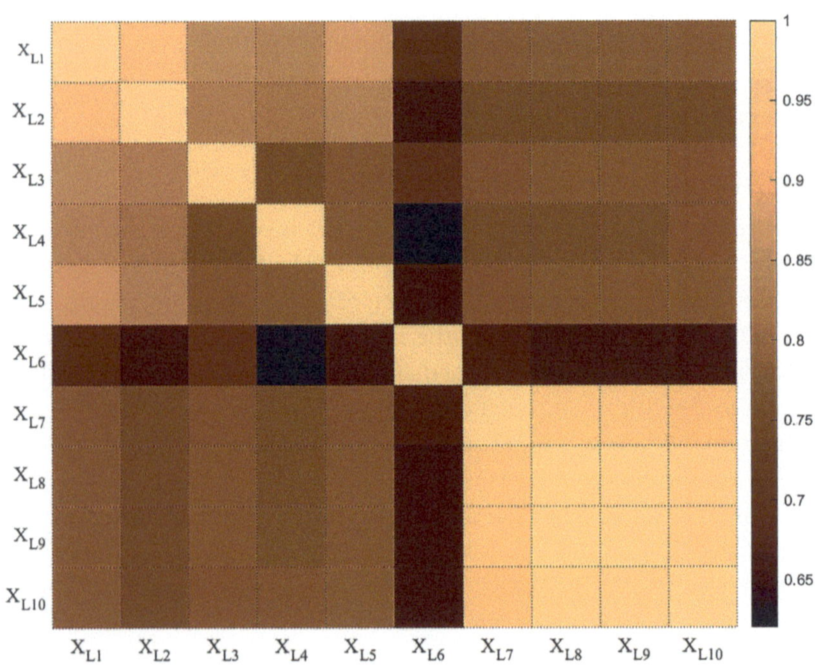

Figure 7.9 Heat map of cross-correlation result based on the Spearman coefficient for wind direction.

Table 7.8 The cross-correlation coefficient based on the Spearman coefficient for wind speed.

	X_{D1}	X_{D2}	X_{D3}	X_{D4}	X_{D5}	X_{D6}	X_{D7}	X_{D8}	X_{D9}	X_{D10}
X_{D1}	1	0.8414	0.5060	0.4502	0.6535	0.3922	0.3595	0.4446	0.3977	0.3890
X_{D2}	0.8414	1	0.5720	0.5070	0.5467	0.4593	0.3945	0.4807	0.4366	0.4224
X_{D3}	0.5060	0.5720	1	0.7991	0.4455	0.6642	0.7402	0.7566	0.7493	0.7389
X_{D4}	0.4502	0.5070	0.7991	1	0.4212	0.7368	0.8105	0.8237	0.8209	0.8294
X_{D5}	0.6535	0.5467	0.4455	0.4212	1	0.4230	0.4020	0.4776	0.4539	0.4523
X_{D6}	0.3922	0.4593	0.6642	0.7368	0.4230	1	0.7913	0.8133	0.8080	0.8099
X_{D7}	0.3595	0.3945	0.7402	0.8105	0.4020	0.7913	1	0.9698	0.9780	0.9748
X_{D8}	0.4446	0.4807	0.7566	0.8237	0.4776	0.8133	0.9698	1	0.9928	0.9764
X_{D9}	0.3977	0.4366	0.7493	0.8209	0.4539	0.8080	0.9780	0.9928	1	0.9861
X_{D10}	0.3890	0.4224	0.7389	0.8294	0.4523	0.8099	0.9748	0.9764	0.9861	1

Table 7.9 The cross-correlation coefficient based on the Spearman coefficient for wind direction.

	X_{D1}	X_{D2}	X_{D3}	X_{D4}	X_{D5}	X_{D6}	X_{D7}	X_{D8}	X_{D9}	X_{D10}
X_{D1}	1	0.9555	0.8519	0.8402	0.8785	0.6906	0.7865	0.7910	0.7913	0.7906
X_{D2}	0.9555	1	0.8346	0.8188	0.8454	0.6594	0.7576	0.7621	0.7624	0.7616
X_{D3}	0.8519	0.8346	1	0.7592	0.7850	0.7110	0.7757	0.7812	0.7807	0.7777
X_{D4}	0.8402	0.8188	0.7592	1	0.7903	0.6206	0.7654	0.7678	0.7708	0.7760
X_{D5}	0.8785	0.8454	0.7850	0.7903	1	0.6675	0.7802	0.7817	0.7828	0.7880
X_{D6}	0.6906	0.6594	0.7110	0.6206	0.6675	1	0.6681	0.6637	0.6623	0.6635
X_{D7}	0.7865	0.7576	0.7757	0.7654	0.7802	0.6681	1	0.9681	0.9699	0.9574
X_{D8}	0.7910	0.7621	0.7812	0.7678	0.7817	0.6637	0.9681	1	0.9945	0.9839
X_{D9}	0.7913	0.7624	0.7807	0.7708	0.7828	0.6623	0.9699	0.9945	1	0.9881
X_{D10}	0.7906	0.7616	0.7777	0.7760	0.7880	0.6635	0.9574	0.9839	0.9881	1

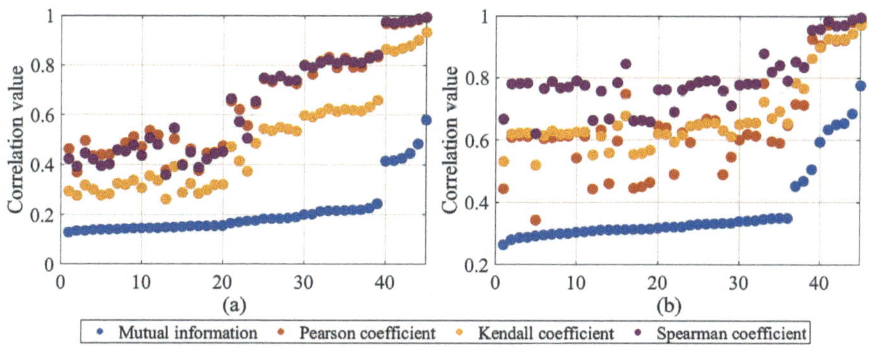

Figure 7.10 The correlation values of different coefficients. (A) wind speed correlation, (B) wind direction correlation.

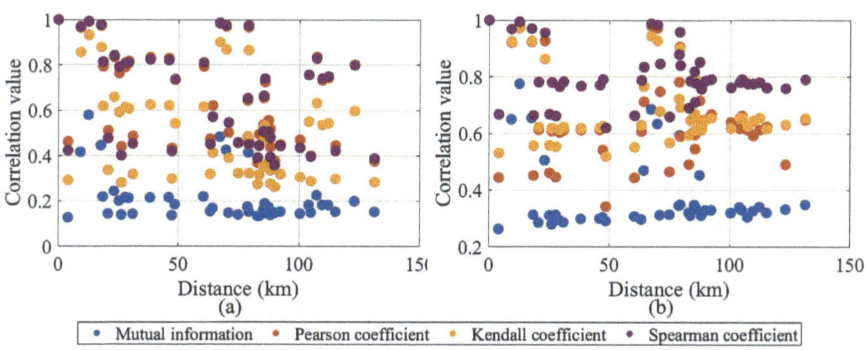

Figure 7.11 The relationship between distances and correlation values. (A) wind speed correlation, (B) wind direction correlation.

Table 7.10 The goodness of fit between distance and correlations.

Correlation coefficients	Wind speed	Wind direction
Mutual information	0.5442	0.4551
Pearson coefficient	0.2829	0.1953
Kendall coefficient	0.3747	0.3653
Spearman coefficient	0.2672	0.1756

calculated without considering the terrain. From Fig. 7.11 and Table 7.10, it can be observed that the correlation values have a bare correlation with the distance. This phenomenon can be explained by the complex terrain.

To research the impact of the wind direction correlation to the wind speed correlation, the scatter plots of the wind speed correlation and wind

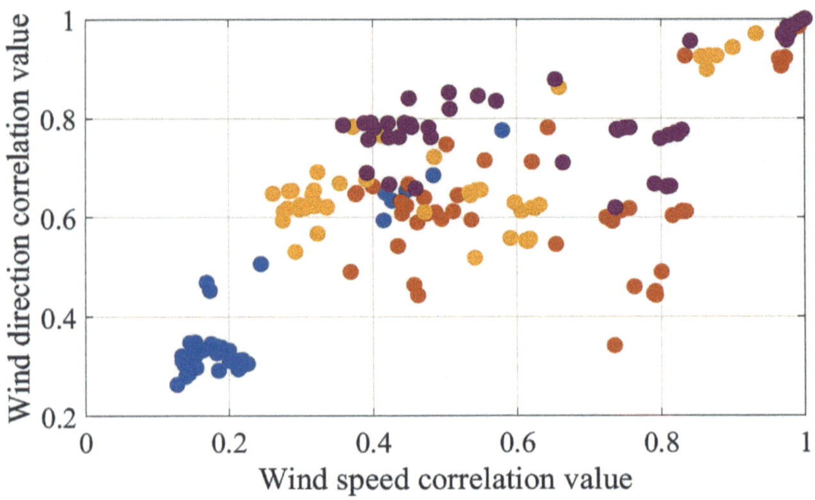

Figure 7.12 The relationship between wind speed and wind direction correlation values.

direction correlation are shown in Fig. 7.12. From Fig. 7.12, it can be seen that the wind speed correlation has a strong positive relationship with wind direction correlation.

7.3 Spatial wind description based on WRF

Mesoscale meteorological models provide meteorological field data and basic grids for emission source models and air quality models [9]. At present, the most commonly used weather field model is Weather Research and Forecasting Model (WRF). The WRF is a mesoscale weather forecasting model, which is a unified development centered on the National Center for Atmospheric Research (NCAR) and other American scientific research institutions [10]. The weather mode WRF includes multiple regions, flexible resolutions ranging from several kilometers to thousands of kilometers, multiple nested networks, and a coordinated three-position variational assimilation system [11]. The WRF numerical model adopts highly modular, parallel, and hierarchical design technology, and integrates a huge number of mesoscale research results [12].

The WRF model has wide applications, ranging from small and medium scale to global scale numerical prediction and simulation [12]. It can be used for operational numerical weather prediction and atmospheric numerical simulation research, including the study of data assimilation, the study of physical process parameterization, regional climate simulation, air quality simulation, and ideal experimental simulation [13].

7.3.1 Main structures

The WRF model is a fully compressible and nonstatic mode and it is characterized by flexibility, easy maintenance, scalability, efficiency, and a wide range of applied computing platforms [14]. The main advantages rely on the advanced data assimilation technology, the powerful nesting capabilities, and the advanced physical processes, especially in convection and mesoscale precipitation processing capabilities [15]. The whole WRF model is mainly composed of the system, the basic software framework, and postprocessing [16].

Both Advanced Research WRF (ARW) and Non-hydrostatic Mesoscale Model (NMM) are included in the WRF basic software framework. In addition to the difference in dynamic solution methods, they share the same WRF model system framework and physical process modules [17]. ARW is developed for research based on NCAR's mesoscale model version 5 and NMM is developed based on NCAR's Eta model [17,18].

The WRF model can not only be used for case simulation of real weather but also can be used as a theoretical basis for the discussion of basic physical processes [19]. The WRF model system has many features such as portability, good maintenance and scalability, high efficiency, and convenience [20]. It is an advanced mesoscale numerical weather model and it is widely used in weather forecasting operations and related business departments and scientific research units [19,20]. The WRF numerical model has become the main tool for daily numerical weather forecasting by meteorological bureaus at all levels, and an important method for researchers to carry out simulation studies of special weather phenomena [21].

7.3.2 WRF modeling along the railway

The WRF model is designed with double nested grids. The National Center for Environmental Prediction (NCEP) Global Data Assimilation System (GDAS) Final Analysis data are used as input (https://rda.ucar.edu/datasets/ds083.3/). In this study, the boundary data were taken from analysis field data with a resolution of 0.25×0.25 degrees once every 6 h. The mode is nested in 2 layers and divided into 34 layers vertically. The nested grid is shown in Fig. 7.13, which aims at the two domains around the target area. The domains are generated with WRF Domain Wizard (https://esrl.noaa.gov/gsd/wrfportal/DomainWizardForLAPS.html).

After the geogrid.exe operation, the geographic information of the two domains is obtained. Taking altitude as an example, the altitude of domain

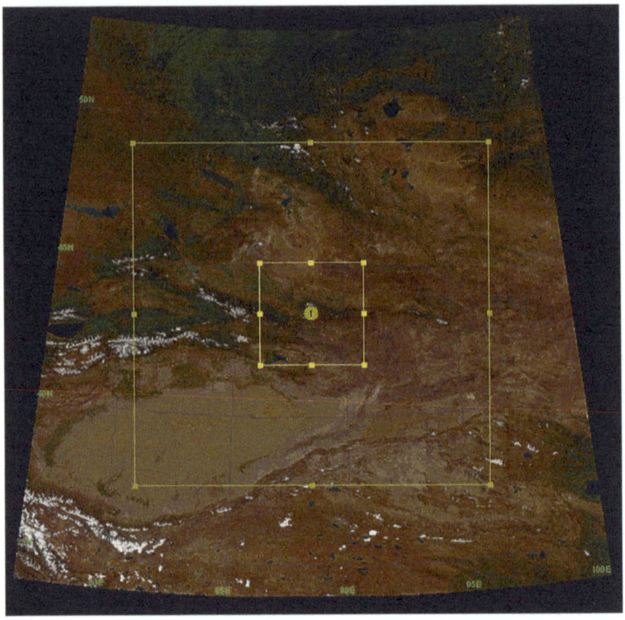

Figure 7.13 The target area of the domain 1 and domain 2.

1 and domain 2 are shown in Figs. 7.14 and 7.15. These plots are generated with Panoply (https://www.giss.nasa.gov/tools/panoply/).

After the unbrib.exe and metgrib.exe calculations, the horizontal and vertical component diagram and the vector diagram of wind speed at 2020-10-03 00:00:00 UTC (domain1) are obtained as Figs. 7.16–7.18. These plots are generated with Panoply.

After the unbrib.exe and metgrib.exe calculations, the horizontal and vertical component diagram and the vector diagram of wind speed at 2020-10-03 00:00:00 UTC (domain2) are obtained as Figs. 7.19–7.21. These plots are generated with Panoply.

After the calculation of real.exe and wrf.exe, taking the more refined domain 2 as an example, the horizontal and vertical component diagram and the vector diagram of wind speed at 2020-10-03 06:00:00 UTC (domain2) are obtained as Figs. 7.22–7.24. These plots are generated with Panoply.

After comparison calculation, the difference between the described values and the actual value at 2020-10-03 06:00:00 UTC are obtained as Figs. 7.25 and 7.26. These plots are generated with Panoply.

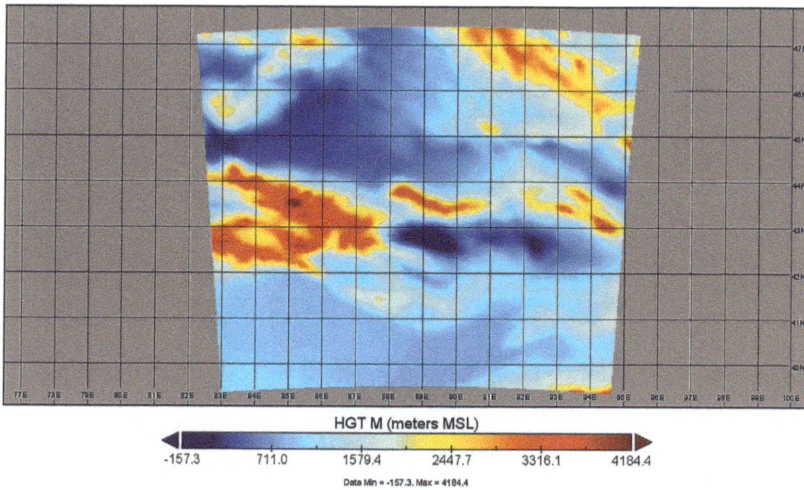

Figure 7.14 The altitude of the domain 1.

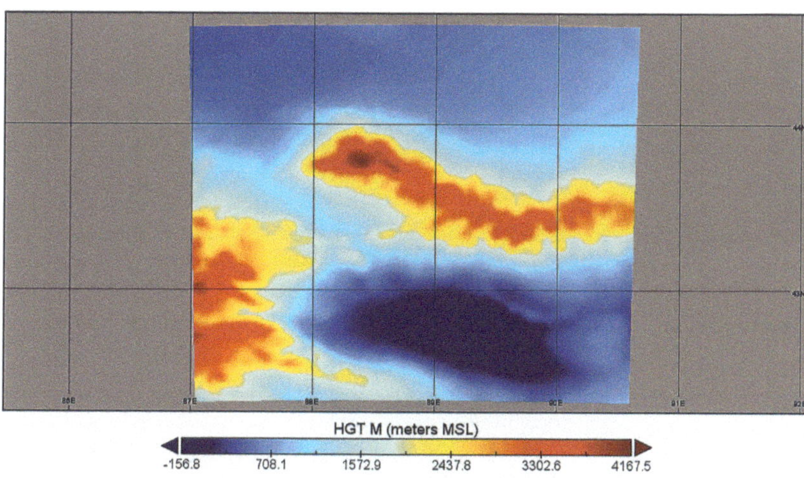

Figure 7.15 The altitude of the domain 2.

7.3.3 WRF future development trends

Since the state of the WRF model is only an approximation of the real state, it cannot accurately reflect the complex motions of the atmosphere and the ocean [22]. It can only approximate the motion characteristics of the atmosphere and the ocean in a certain area [22,23]. This requires the complementary fusion of observational data and model information, to

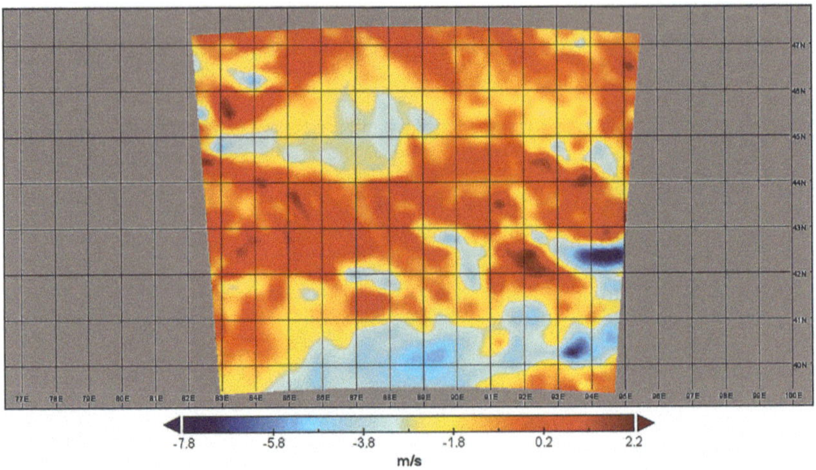

Figure 7.16 The horizontal component diagram of wind speed in the domain 1.

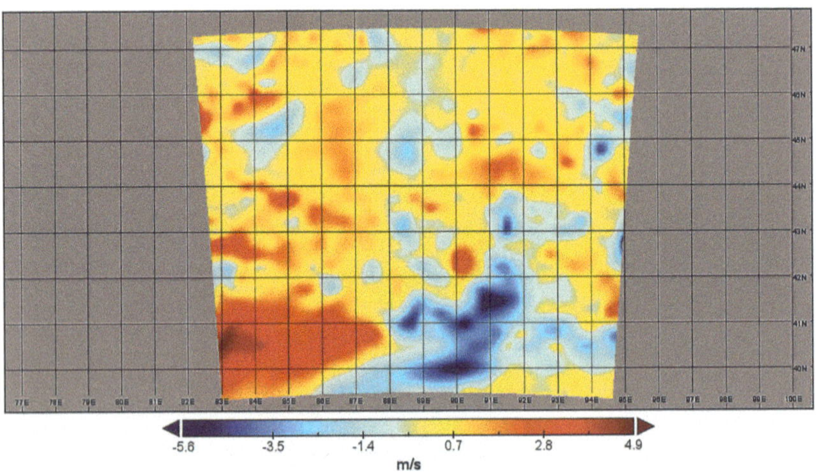

Figure 7.17 The vertical component diagram of wind speed in the domain 1.

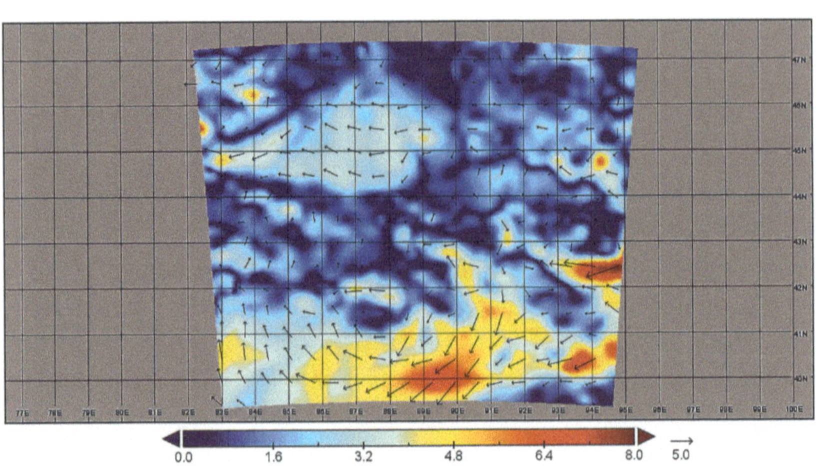

Figure 7.18 The wind speed vector diagram in the domain 1.

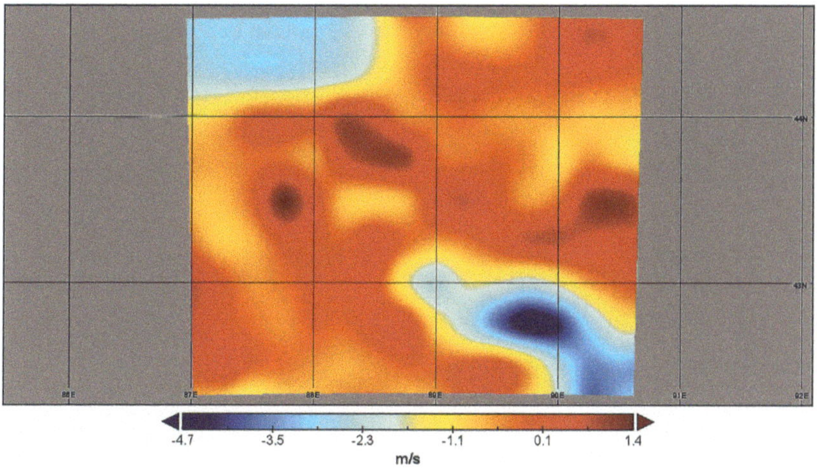

Figure 7.19 The horizontal component diagram of wind speed in the domain 2 (2020-10-03 00:00:00 UTC).

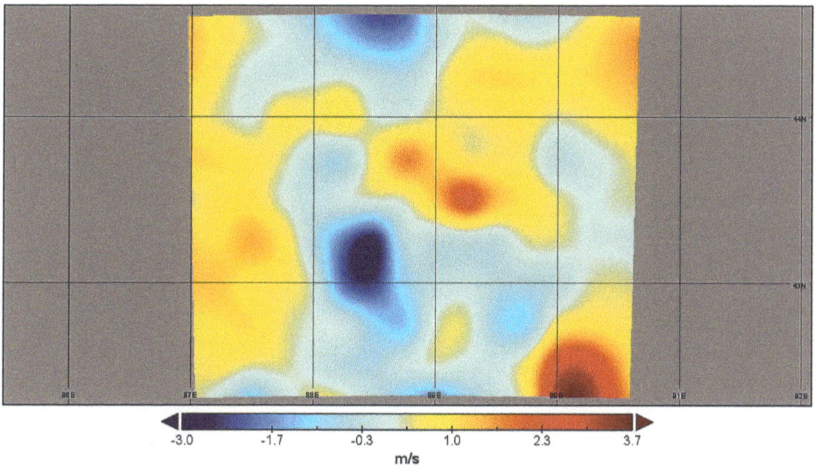

Figure 7.20 The vertical component diagram of wind speed in the domain 2 (2020-10-03 00:00:00 UTC).

describe us a weather picture that is close to the real state and contains internal physical processes. Therefore, it is necessary to appropriately adjust, process, and objectively analyze these meteorological data through data assimilation techniques, and adopt approximation and hypothetical methods to make the forecast field of the WRF model closer to the real natural weather field [23].

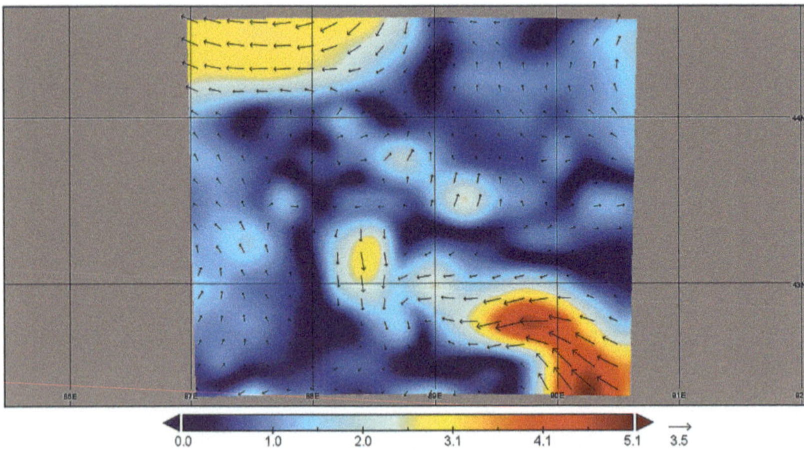

Figure 7.21 The wind speed vector diagram in the domain 2 (2020-10-03 00:00:00 UTC).

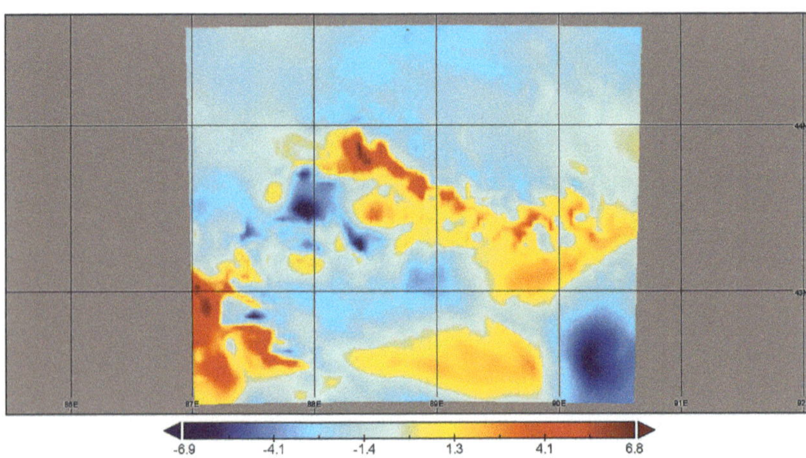

Figure 7.22 The horizontal component diagram of wind speed in the domain 2 (2020-10-03 06:00:00 UTC).

By the data preprocessing, some large meteorological data are corrected according to the processing of abnormal values. However, these wind speed data may be collected under extreme weather, which is real and has a certain meaning. In-depth research can be also conducted on the data sample classification method to make the data more regular [24]. For short-term wind speed under complex terrain conditions, the research area could be expanded. More different terrain forecast experiments can be carried out for

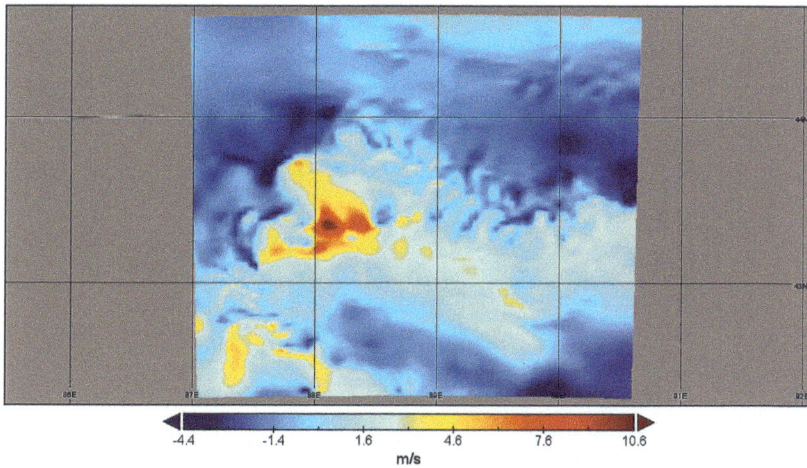

Figure 7.23 The vertical component diagram of wind speed in the domain 2 (2020-10-03 06:00:00 UTC).

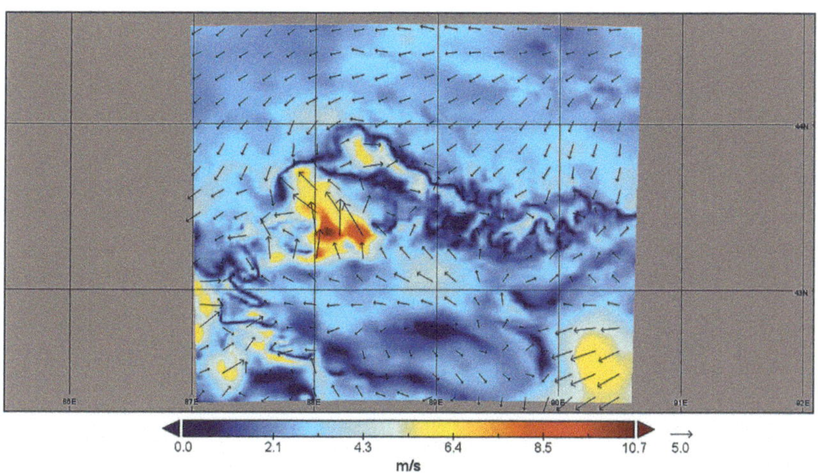

Figure 7.24 The wind speed vector diagram in the domain 2 (2020-10-03 06:00:00 UTC).

suitable solutions in complex terrain conditions, and a wind energy forecast system suitable for various national terrains could be established [25].

Increasing the calculation speed of the WRF model can further improve the degradation resolution of the WRF model and obtain more accurate forecast results for long-term wind resource assessment. Methods such as radiosonde data and observational data from equivalent technology can be applied to increase the accuracy of the mesoscale WRF model and make the system more efficient [26].

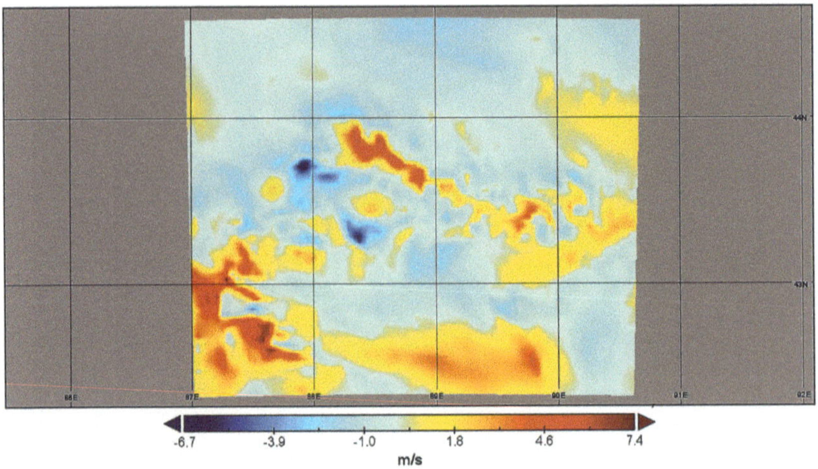

Figure 7.25 Difference of the horizontal component of actual value in the domain 2 (2020-10-03 06:00:00 UTC).

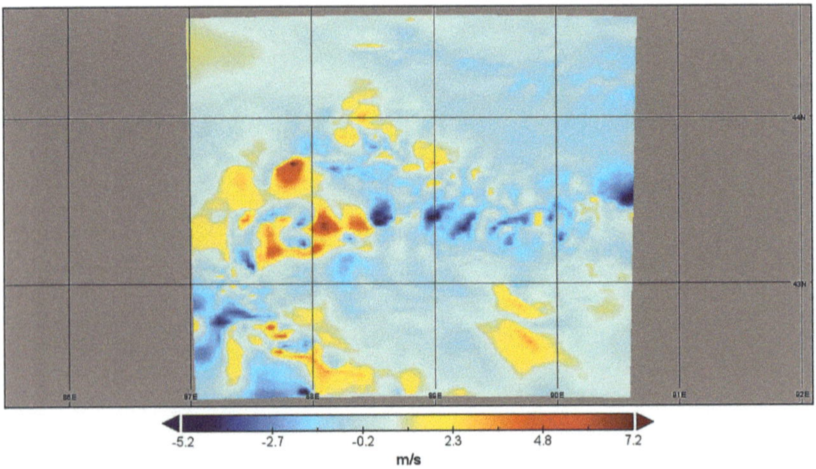

Figure 7.26 Difference of the vertical component of the actual value in the domain 2 (2020-10-03 06:00:00 UTC).

However, to obtain the optimal integrated forecasting method and better wind speed forecasting effects are the focus of the next step of the research. For example, some intelligent algorithms can be used to establish a hybrid model for wind speed forecasting [27]. Through further data mining and analysis of the structural characteristics of deep learning, combined with the characteristics of wind speed in wind farms, these new methods and new rules can be used to increase the forecasting accuracy of wind speed [28].

7.4 Description accuracy evaluation indicators

The spatial correlation of wind speed refers to a certain correlation between the wind speed in a certain place and the wind speed of various places around its geographic space. The correlation of wind speed includes the temporal correlation of the same spatial position and the spatial correlation between different spatial positions [29]. The spatial correlation analysis of wind speed is based on the time-series correlation analysis of wind speed at each spatial point. Using this spatial correlation, the wind speed and direction in the surrounding area can be used to improve the description effect along the railway [30]. One of the methods is to fully explore the characteristics and laws of wind speed correlation from historical observations, and use optimization models to obtain more accurate forecast results [29].

Three evaluation indicators Spatial Mean Absolute Error (SMAE), Spatial Mean Absolute Percentage Error (SMAPE), and Spatial Root Mean Square Error (SRMSE) can be used to evaluate the accuracy of the spatial description results. The smaller their value, the better the description accuracy. Their specific calculation formulas are as follows:

$$\mathrm{SMAE} = \frac{1}{n}\sum_{i=1}^{n}|y_i - f_i| \tag{7.7}$$

$$\mathrm{SMAPE} = \frac{100\%}{n}\sum_{i=1}^{n}\left|\frac{y_i - f_i}{y_i}\right| \tag{7.8}$$

$$\mathrm{SRMSE} = \sqrt{\frac{1}{n}\sum_{i=1}^{n}(y_i - f_i)^2} \tag{7.9}$$

where y_i is the true value of spatial wind speed data, f_i represents the description value of different models, and n is the total number of samples in different test sets.

Besides, the skewness and kurtosis of the spatial wind data can be calculated, and the two indicators of the actual data and description data can be compared to evaluate the description effectiveness from the spatial wind data. The calculation formulas for skewness and kurtosis are as follows [31]:

$$\mathrm{Skewness} = \frac{1}{n}\sum_{i=1}^{n}\left[\left(\frac{X_i - \mu}{\sigma}\right)^3\right] \tag{7.10}$$

$$\text{Kurtosis} = \frac{1}{n}\sum_{i=1}^{n}\left[\left(\frac{X_i - \mu}{\sigma}\right)^4\right] \tag{7.11}$$

where X_i is the true value or description value from spatial wind speed data, μ represents the mean of spatial wind speed data, and σ represents the standard deviation of spatial wind speed data.

7.5 Summary and outlook

Wind description technology is a cross-application discipline of meteorology, physical engineering, statistics, and computing science. From the analysis of the above sections, the spatial wind correlation research methods contribute to the precise description of the wind data, including the MI, the Pearson coefficient, the Kendall coefficient, and the Spearman coefficient. The WRF model also provides a wide range of wind descriptions with the integration of mesoscale research results, which further improves the description accuracy.

Continuously improving description accuracy still requires decision-making under the common constraints of basic science, technology, and cost. The development of big data, energy internet, high-performance computing, and other theories and technologies will provide a broader data foundation and more powerful calculation tools for wind prediction dynamics and statistical modeling methods.

(a) Use higher resolution, more detailed description of the terrain and underlying surface attributes, detailed description of the environmental factors in and around the wind area. To more accurately describe the difference in wind speed at different locations in the wind field, and adopt the temporal—spatial resolution data of wind area will be verified and analyzed in more detail.

(b) An appropriate boundary layer parameterization scheme can increase the vertical resolution accordingly, and even adopt a collection of multiple boundary layer parameterization schemes to provide wind speed description at different heights.

(c) It can be combined with statistical methods such as time series processing methods, random process processing methods, etc., to describe with a shorter time scale.

(d) Use the big data platform to count massive historical data, and integrate various influencing factors to make possible changes in the future period.

References

[1] C.S. Yang, W.P. Zhou, Q. Wang, Design of meteorological monitoring system for high-speed railway, in: Advanced Materials Research, vol. 655, 2013, pp. 777–780.

[2] H. Xia, W.W. Guo, N. Zhang, et al., Dynamic analysis of a train–bridge system under wind action, Comput. Struct. 86 (2008) 1845–1855.

[3] M.A. Barcala, J. Meseguer, Visualization study of the influence of parapets on the flow around a train vehicle under cross winds, WIT Trans. Built Environ. 103 (2008) 797–806.

[4] S. Barman, Y.-K. Kwon, A novel mutual information-based Boolean network inference method from time-series gene expression data, PLoS One 12 (2017) e0171097.

[5] Y. Koizumi, K. Niwa, Y. Hioka, et al., Informative acoustic feature selection to maximize mutual information for collecting target sources, IEEE/ACM Trans. Audio Speech & Lang. Process. 25 (2017) 768–779.

[6] Y. Mu, X. Liu, L. Wang, A Pearson's correlation coefficient based decision tree and its parallel implementation, Inf. Sci. 435 (2018) 40–58.

[7] J. Van Doorn, A. Ly, M. Marsman, et al., Bayesian inference for Kendall's rank correlation coefficient, Am. Stat. 72 (2018) 303–308.

[8] F. Dikbas, A new two-dimensional rank correlation coefficient, Water Resour. Manag. 32 (2018) 1539–1553.

[9] A.B. Gilliland, C. Hogrefe, R.W. Pinder, et al., Dynamic evaluation of regional air quality models: assessing changes in O_3 stemming from changes in emissions and meteorology, Atmos. Environ. 42 (2008) 5110–5123.

[10] Y. Zhang, V. Dulière, P.W. Mote, et al., Evaluation of WRF and HadRM mesoscale climate simulations over the US Pacific Northwest, J. Clim. 22 (2009) 5511–5526.

[11] F. Zhang, Y. Yang, C. Wang, The effects of assimilating conventional and ATOVS data on forecasted near-surface wind with WRF-3DVAR, Mon. Weather Rev. 143 (2015) 153–164.

[12] F. Chen, H. Kusaka, R. Bornstein, et al., The integrated WRF/urban modelling system: development, evaluation, and applications to urban environmental problems, Int. J. Climatol. 31 (2011) 273–288.

[13] W.C. Skamarock, J.B. Klemp, A time-split nonhydrostatic atmospheric model for weather research and forecasting applications, J. Comput. Phys. 227 (2008) 3465–3485.

[14] L. Oana, M. Frincu, Benchmarking the WRF model on Bluegene/P, cluster, and cloud platforms and accelerating model setup through parallel genetic algorithms, in: 2017 16th International Symposium on Parallel and Distributed Computing (ISPDC), 2017, pp. 78–84.

[15] X. Qie, R. Zhu, T. Yuan, et al., Application of total-lightning data assimilation in a mesoscale convective system based on the WRF model, Atmos. Res. 145 (2014) 255–266.

[16] S. Yu, R. Mathur, J. Pleim, et al., Comparative evaluation of the impact of WRF–NMM and WRF–ARW meteorology on CMAQ simulations for O_3 and related species during the 2006 TexAQS/GoMACCS campaign, Atmos. Pollut. Res. 3 (2012) 149–162.

[17] P. Zannetti, Air Pollution Modeling: Theories, Computational Methods and Available Software, Springer Science & Business Media, 2013.

[18] D.B. Rao, V. Tallapragada, Tropical cyclone prediction over Bay of Bengal: a comparison of the performance of NCEP operational HWRF, NCAR ARW, and MM5 models, Nat. Hazards 63 (2012) 1393–1411.

[19] K. Darmenova, I.N. Sokolik, Y. Shao, et al., Development of a physically based dust emission module within the Weather Research and Forecasting (WRF) model:

assessment of dust emission parameterizations and input parameters for source regions in Central and East Asia, J. Geophys. Res. Atmos. 114 (2009).
[20] J. Ploski, G. Scherp, T.I. Petroliagis, et al., Grid-based deployment and performance measurement of the Weather Research & Forecasting model, Future Generat. Comput. Syst. 25 (2009) 346–350.
[21] Y. Huang, Y. Liu, Y. Liu, et al., Mechanisms for a record-breaking rainfall in the coastal metropolitan city of Guangzhou, China: observation analysis and nested very large eddy simulation with the WRF model, J. Geophys. Res. Atmos. 124 (2019) 1370–1391.
[22] D. Carvalho, A. Rocha, M. Gómez-Gesteira, et al., A sensitivity study of the WRF model in wind simulation for an area of high wind energy, Environ. Model. Software 33 (2012) 23–34.
[23] G. Skok, J. Tribbia, J. Rakovec, Object-based analysis and verification of WRF model precipitation in the low-and midlatitude Pacific Ocean, Mon. Weather Rev. 138 (2010) 4561–4575.
[24] C. Zhang, H. Lin, M. Chen, et al., Scale matching of multiscale digital elevation model (DEM) data and the Weather Research and Forecasting (WRF) model: a case study of meteorological simulation in Hong Kong, Arab. J. Geosci. 7 (2014) 2215–2223.
[25] H. Zhang, Z. Pu, X. Zhang, Examination of errors in near-surface temperature and wind from WRF numerical simulations in regions of complex terrain, Weather Forecast. 28 (2013) 893–914.
[26] H. Liu, J. Anderson, Y.-H. Kuo, et al., Evaluation of a nonlocal quasi-phase observation operator in assimilation of CHAMP radio occultation refractivity with WRF, Mon. Weather Rev. 136 (2008) 242–256.
[27] H. Liu, H. Tian, X. Liang, et al., Wind speed forecasting approach using secondary decomposition algorithm and Elman neural networks, Appl. Energy 157 (2015) 183–194.
[28] H. Liu, X. Mi, Y. Li, Smart multi-step deep learning model for wind speed forecasting based on variational mode decomposition, singular spectrum analysis, LSTM network and ELM, Energy Convers. Manag. 159 (2018) 54–64.
[29] M.C. Alexiadis, P.S. Dokopoulos, H.S. Sahsamanoglou, Wind speed and power forecasting based on spatial correlation models, IEEE Trans. Energy Convers. 14 (1999) 836–842.
[30] T.G. Barbounis, J.B. Theocharis, A locally recurrent fuzzy neural network with application to the wind speed prediction using spatial correlation, Neurocomputing 70 (2007) 1525–1542.
[31] B. Hasche, General statistics of geographically dispersed wind power, Wind Energy 13 (2010) 773–784.

CHAPTER 8

Data-driven spatial wind forecasting methods along railways

Contents

Wind Forecasting in Railway Engineering
ISBN 978-0-12-823706-9
https://doi.org/10.1016/B978-0-12-823706-9.00008-9

8.1 Introduction

Due to the unique topography along the railway, wind speed usually has a significant spatial correlation. The wind speed values between different spatial points will be influenced by each other. Therefore, wind speed time series using single-point forecasting cannot satisfy the requirement of actual application in some cases. Researchers have been dedicated to establishing spatial wind forecasting models. Different from nonspatial models, the spatial models consider wind speed data not only in target sites but also in adjacent monitoring sites.

The available studies on spatial wind forecasting are relatively rare. Yu et al. constructed spatiotemporal features by mapping the collected data to the plane [1]. And then a deep convolutional network was proposed to forecast the wind. The results indicated that the proposed deep learning spatial forecasting model significantly outperformed the existing methods. Jiang et al. adopted gray correlation analysis to choose features from adjacent sites and then trained v-Support Vector Machine (v-SVM) [2]. Pourhabib et al. used data from multiple turbines to construct an ensemble-like predictor [3]. They found that spatial data benefit short-term wind speed forecasting. Zhu et al. proposed a Predictive Spatio-Temporal Network (PSTN) that combines a Long Short-Term Memory (LSTM) and a Convolutional Neural Network (CNN) [4]. CNN is adopted to extract spatial features by treating the data collected from several monitoring sites as figures. And the LSTM is used to learn temporal dependencies. Velázquez et al. analyzed the correlation coefficient of wind speed obtained from 22 weather stations located in Spain and fed the spatial data into Artificial Neural Networks (ANNs) [5]. The proposed spatial model can reduce the single-point prediction error by 75%. Noorollahi et al. used Geographic Information System (GIS) software to analyze the circular terrain information with a radius of 5 km around the wind station and input the data into ANNs [6]. The proposed model generated forecasting results with about 2.6% error.

According to the above literature review, the existing spatial forecasting models have different characteristics. They consider spatial characteristics from different aspects. In this study, the spatial features are analyzed from the perspective of data science. The relationship between target sites and adjacent sites is analyzed by statistical, intelligent, and deep learning methods, respectively. Then, the extracted features are directly fed into forecasting models. This chapter applies different statistical, intelligent, and deep learning methods and introduces their methodologies. For statistical spatial forecasting models, mutual information is proposed to select spatially related sites. The forecasting performances of the Autoregressive Integrated Moving Average with Extra Input (ARIMAX) and Generalized Linear Regression (GLR) model are tested. For intelligent spatial forecasting models, four binary optimization algorithms including Binary Grey Wolf Optimization (BGWO), Binary Harris Hawk Optimization (BHHO), Binary Particle Swarm Optimization (BPSO), and Binary Differential Evolution (BDE) are investigated. For deep learning spatial forecasting models, Sparse AutoEncoder (SAE) is constructed to extract spatial features. Moreover, three deep learning models are applied, including LSTM, Bidirectional Long Short-Term Memory (BILSTM), and Deep Echo State Network (DeepESN). Finally, the forecasting performances of these spatial forecasting models are comprehensively evaluated by the actual wind speed dataset.

8.2 Wind data description

To analyze spatial wind speed forecasting models, a wind speed dataset that was collected from 415 adjacent monitoring sites in strong wind areas is adopted in the study. These data are collected from National Renewable Energy Laboratory (NREL), the website is https://www.nrel.gov/. The wind speed data are collected with a 5-min resolution. As a result, each monitoring site has 9216 observations. Four target sites are randomly selected among all sites. The wind speed time series of these four target sites are shown in Fig. 8.1. They are separated into three parts using nested cross-validation to avoid data leaks, including training set, validation set, and testing set. The percentages of the training set, validation set, and testing set are 60%, 20%, and 20%, respectively. Hence, the 1st−5490th observations are training set, the 5491th−7373rd observations are validation set, and the remaining 1843 observations are testing set.

The statistical characteristics of the wind speed series in four target sites are listed in Table 8.1. According to Table 8.1, the fluctuations in target site #1 are more dramatic than other target sites. The standard deviation and maximum of target site #1 are 6.4915 and 28.0064, respectively, which are much higher than other sites. It demonstrates a greater degree of dispersion. Besides, the kurtosis values of target sites #1−4 are 2.7656, 2.3559, 2.2231, and 2.7018, respectively. High kurtosis of target site #1 represents that it is more outlier-prone than the data distribution in other target sites.

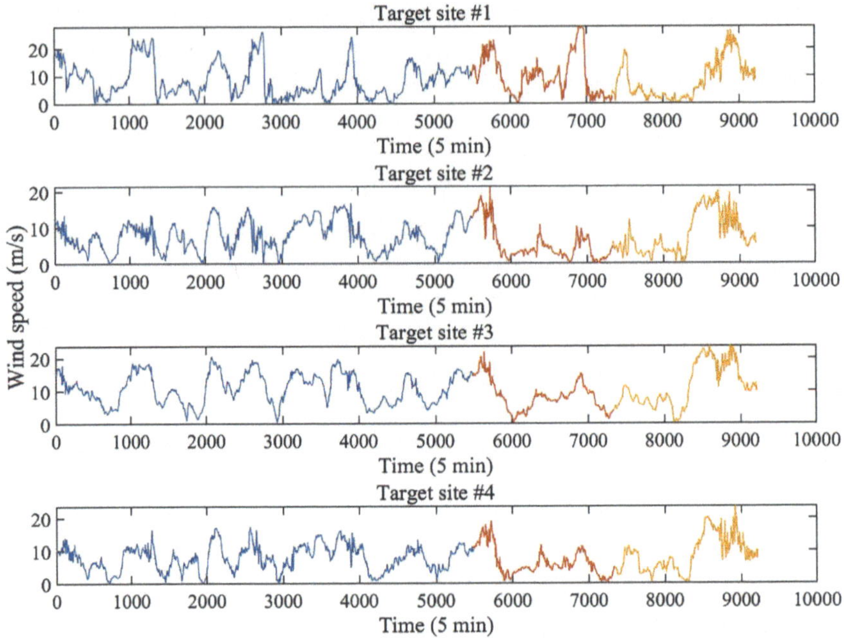

Figure 8.1 Description and separation of four wind speed series in target sites.

Table 8.1 Statistical characteristics of wind speed series in four target sites.

Target site	Maximum(m/s)	Minimum(m/s)	Mean(m/s)	Standard deviation(m/s)	Kurtosis
#1	28.006	0.083	8.848	6.492	2.766
#2	21.393	0.061	7.303	4.544	2.356
#3	23.756	0.101	10.442	5.153	2.223
#4	23.701	0.043	7.735	4.305	2.702

8.3 Spatial wind forecasting algorithm based on statistical model

8.3.1 Theoretical basis

8.3.1.1 Spatial feature selection based on mutual information

In the study, Mutual Information (MI) is adopted to analyze the spatial correlation of different wind speed monitoring sites and select correlated sites that contain sufficient information about target sites. The basic concept is to calculate the MI between the target site and other adjacent sites, and then select adjacent sites with relatively high MI values. To avoid data leaks and simulate actual application, the calculation of MI is only implemented by the training set and validation set. And then, the adjacent sites with the top 30% MI values are selected to form spatial features. The utilization of spatially related data can enhance the interpretability and forecasting performance of the model.

8.3.1.2 Generalized linear regression

GLR model is an extension of a linear model. It connects the response variables and the forecasting variable of the linear combination by link function. It is calculated as follows:

$$\eta_i = g(\mu_i) \tag{8.1}$$

where μ_i is the expectation of the dependent variable Y_i. g denotes the link function. η_i is the weighted combination of the independent variable X_i.

8.3.2 Model framework

Fig. 8.2 provides the framework of statistical spatial wind speed forecasting models.

The modeling steps are described as follows:

(a) Divide wind speed series of the target site and adjacent sites into the training set, validation set, and testing set.

(b) Utilize MI to analyze the correlation between wind speed series at target sites and adjacent sites. The adjacent sites with the top 30% MI values are selected to form spatial features. Note that only a training set is used during this process so that data leaks can be properly avoided.

(c) Normalize the wind speed series at target sites and spatially related sites.

(d) Use the normalized data in the training set to train two statistical forecasting models, including ARIMAX and GLR.

(e) Apply the trained ARIMAX and GLR model on the testing set. Generate forecasting results.

(f) Evaluate the forecasting results of two statistical spatial models by calculating MAPE, MAE, RMSE. Finally, compare different statistical forecasting models and conclude.

8.3.3 Analysis of statistical spatial forecasting models

8.3.3.1 Spatial analysis of monitoring sites

MI is adopted to evaluate the spatial correlation between the target sites and adjacent sites. To eliminate the influence of the target site itself, the MI value between the target site and itself is set as the mean of all MI values, so

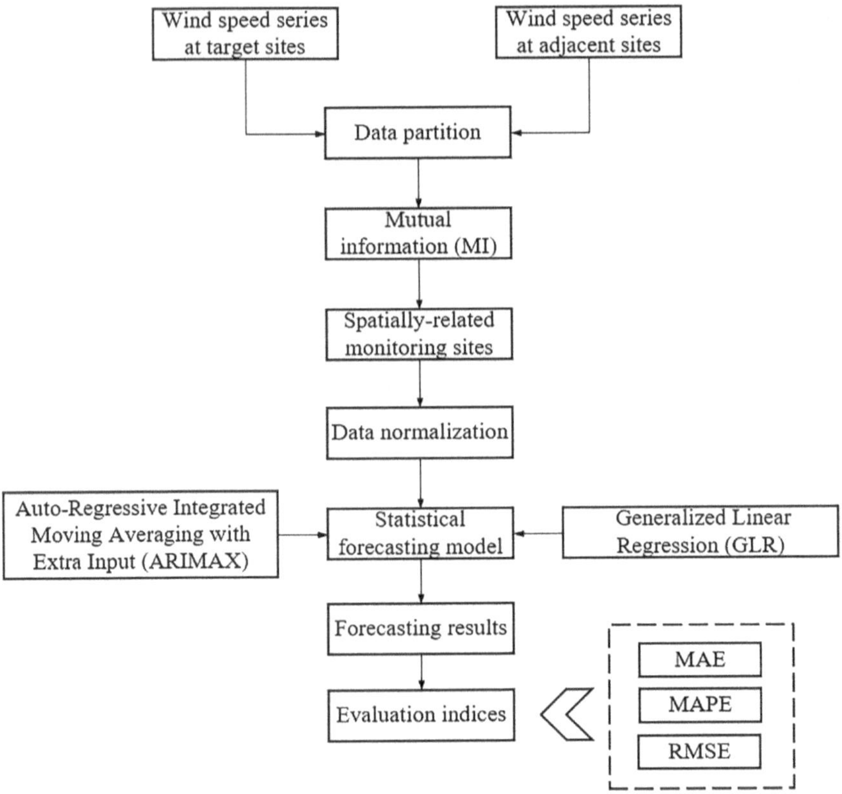

Figure 8.2 Framework of statistical spatial wind speed forecasting models.

that the top 30% MI values are still representative. The evaluation results of four target sites are shown in Fig. 8.3. For better comparison and selection of relevant sites, the evaluation results are normalized and sorted as shown in Fig. 8.4.

It can be observed that only a few adjacent sites have relatively high normalized MI (NMI) values, while the NMI values of the most adjacent sites are lower than 0.5. The sorted NMI values in four target sites show a similar trend: rapidly decrease in the beginning and slowly decrease afterward. The result is helpful to further construction of spatial forecasting models because a few but representative features are appropriate and reasonable. A mass of data may be counterproductive and result in the curse of dimensionality.

Since the threshold of this study is set as selecting sites with top 30% MI values, the sites whose MI values are higher than 0.7 are chosen. As a result,

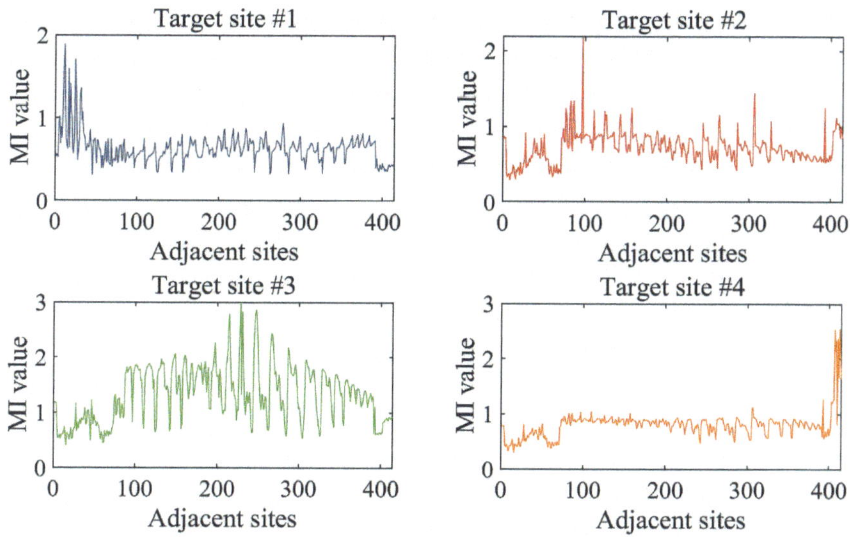

Figure 8.3 Evaluation results of MI values between adjacent sites and four target sites.

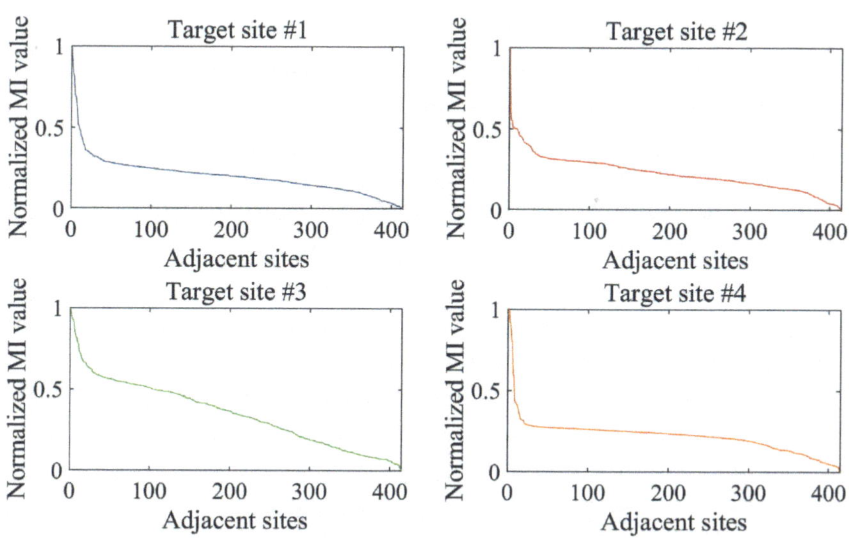

Figure 8.4 Normalized and sorted MI values between adjacent sites and four target sites.

the serial numbers of selected monitoring sites for four targets are listed in Table 8.2. The number of selected monitoring sites for four target sites is different. The number of selected monitoring sites for target sites #1−4 is 5, 1, 14, and 6, respectively. Target site #3 has the largest number of selected

Table 8.2 The serial numbers of selected monitoring sites for four targets.

Target site	Selected monitoring sites
#1	12 25 11 17 26
#2	97
#3	229 248 231 247 215 249 214 246 228 267 268 266 250 197
#4	413 407 411 409 408 414

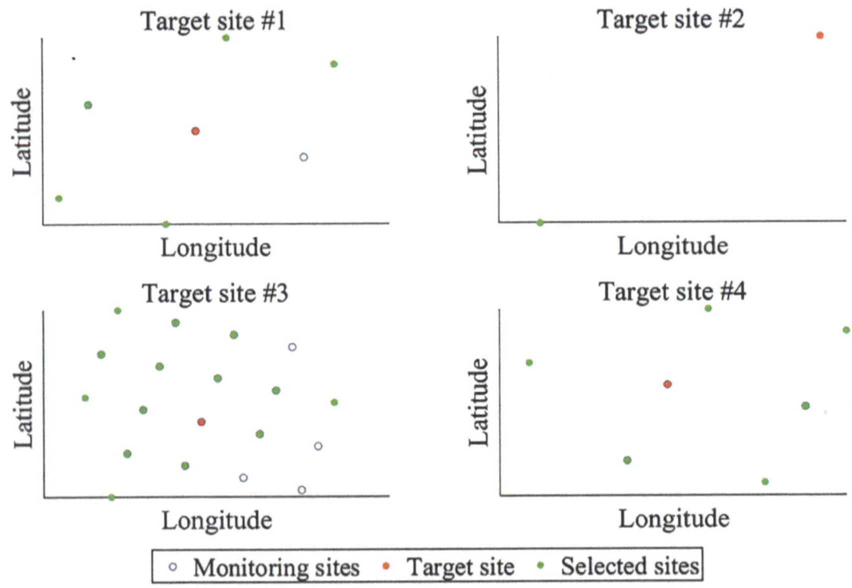

Figure 8.5 Locations of selected sites and target sites.

monitoring sites, while target site #2 has only one selected monitoring site. The results are mainly caused by the density of site distribution around the target sites. Target site #3 has a lot of adjacent monitoring sites, while target site #2 has only a few adjacent monitoring sites. Therefore, the proposed statistical method can automatically select spatially related monitoring sites. The selection results correspond to the actual situation.

Furthermore, Fig. 8.5 analyzes the locations of selected sites and target sites. The x-axis and y-axis represent the longitude and latitude of monitoring sites, respectively. To show the relative location of the target site, the other monitoring sites located within the range of selected sites and target sites are also provided. Hence, it can be observed that the selected sites are always the sites closest to the target sites. The phenomenon is reasonable because the most adjacent sites are usually more related to the targets than other sites. The spatially related data can help the forecasting model learn the spatial characteristics of wind speed in the specific zone.

8.3.3.2 Results of statistical spatial forecasting models

To evaluate the proposed statistical spatial analysis method, the forecasting performances of ARIMAX and GLR are compared. They utilize the wind speed data at the target site and selected adjacent sites to achieve spatial forecasting. The 1-step ahead forecasting results of the MI-ARIMAX and MI-GLR models for targets #1–#4 are shown in Figs. 8.6–8.9. The top subgraph represents the overall forecasting results in the whole testing set. The bottom subgraph represents the local forecasting results at the 1451th–1550th observations in the testing set. According to the figures, conclusions can be made as follows:

(a) Both MI-ARIMAX and MI-GLR models have graphically satisfactory forecasting performances as for the whole testing set. The 1-step ahead forecasting results of the two models are consistent with the actual wind speed series in the general trend.

(b) The MI-GLR model generates extreme values in some peaks and troughs, which significantly deviate from the actual wind speed values. The results may be caused by the iterated forecasting process of the

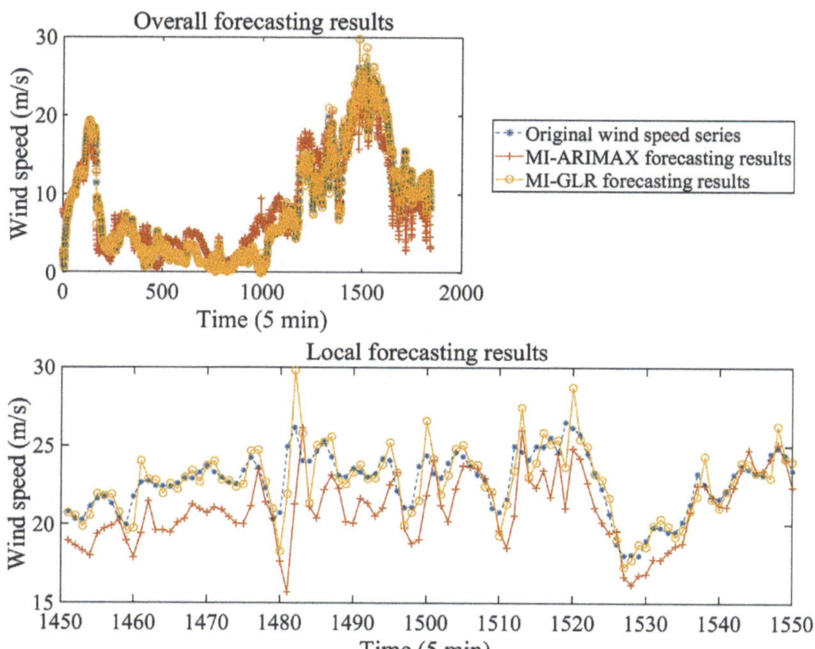

Figure 8.6 The 1-step ahead results of statistical spatial forecasting models for target site #1.

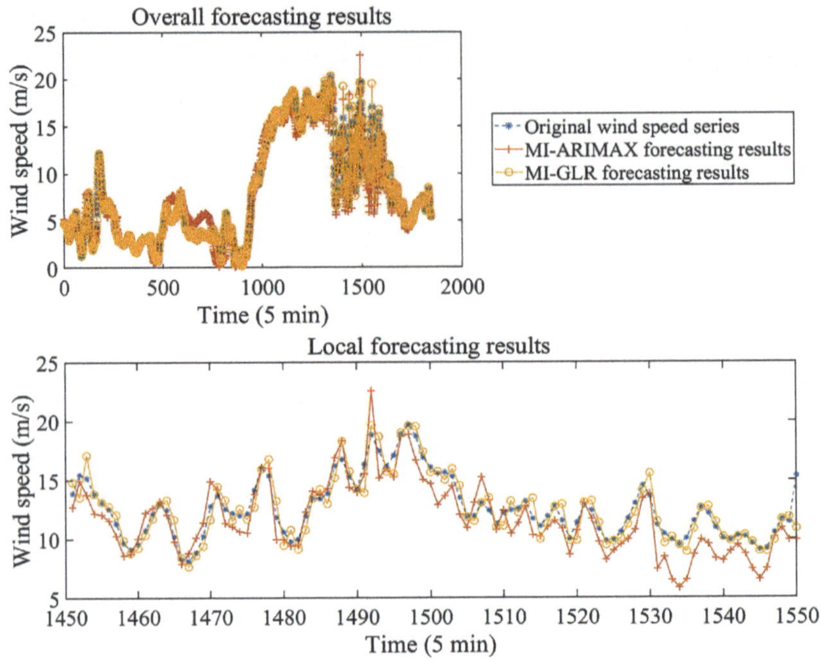

Figure 8.7 The 1-step ahead results of statistical spatial forecasting models for target site #2.

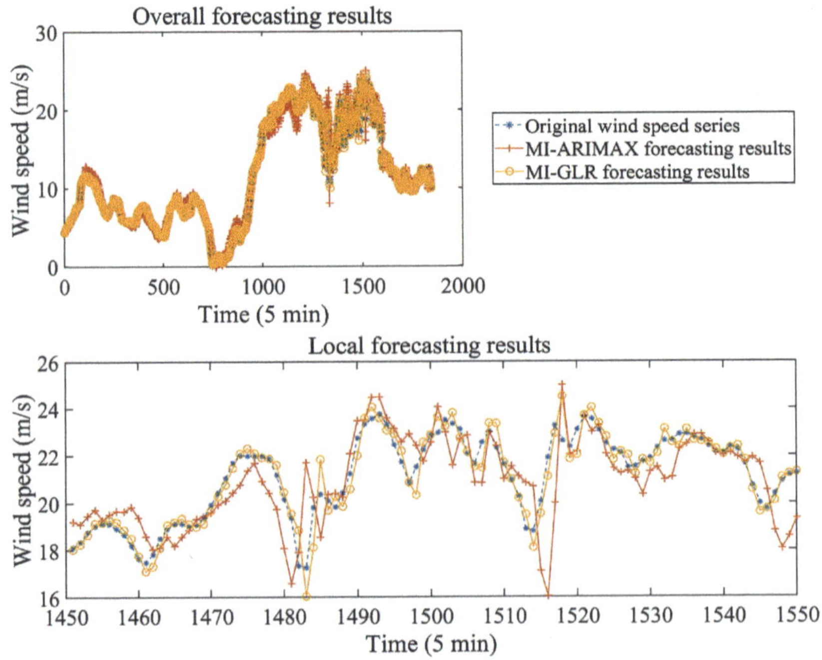

Figure 8.8 The 1-step ahead results of statistical spatial forecasting models for target site #3.

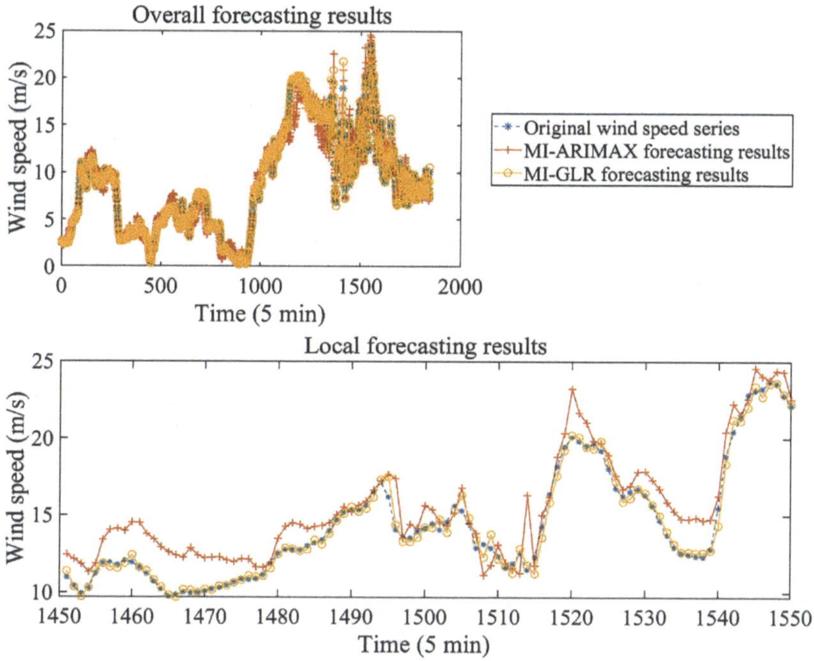

Figure 8.9 The 1-step ahead results of statistical spatial forecasting models for target site #4.

GLR model. In the multi-output design of the GLR model, an iterated strategy is utilized, which absorbs the forecasting output as the input of the next step. It could result in large iteration errors in multi-step ahead prediction.

(c) The MI-GLR model significantly outperforms the MI-ARIMAX model in local forecasting results. Taking the 1-step ahead result of statistical spatial forecasting models for target site #3 as an example, the MI-ARIMAX model shows an obvious delay phenomenon. This is because the theory of ARIMAX itself is to use the historical data of the target site, rather than make good use of the spatial data.

(d) Both MI-ARIMAX and MI-GLR models perform poorly in the local forecasting results for target #1. This is because the wind speed series collected from target site #1 is more discrete than that from other target sites, which brings extra difficulty to prediction. The reflection in the forecasting result is that the forecasting value deviates greatly from the actual value at some local observations.

Table 8.3 Evaluation indices of statistical spatial forecasting models.

Target site	Model	Step	MAE(m/s)	MAPE(%)	RMSE(m/s)
#1	MI-ARIMAX	1-step	2.131	72.210	2.626
		2-step	2.138	72.351	2.620
		3-step	2.145	72.465	2.622
	MI-GLR	1-step	0.205	4.374	0.417
		2-step	0.472	9.663	0.937
		3-step	0.751	15.013	1.481
#2	MI-ARIMAX	1-step	0.892	17.308	1.238
		2-step	0.927	17.289	1.317
		3-step	0.992	17.801	1.449
	MI-GLR	1-step	0.163	2.461	0.398
		2-step	0.374	5.613	0.798
		3-step	0.573	8.701	1.149
#3	MI-ARIMAX	1-step	0.726	11.592	0.974
		2-step	0.752	11.986	1.032
		3-step	0.797	12.611	1.115
	MI-GLR	1-step	0.114	1.328	0.233
		2-step	0.256	2.901	0.506
		3-step	0.407	4.460	0.796
#4	MI-ARIMAX	1-step	0.893	14.238	1.266
		2-step	0.951	15.029	1.353
		3-step	1.012	15.838	1.439
	MI-GLR	1-step	0.137	1.815	0.343
		2-step	0.345	4.518	0.845
		3-step	0.571	7.483	1.322

To quantitatively analyze the forecasting performance of statistical spatial forecasting models, the evaluation indices are provided in Table 8.3. According to Table 8.3, the following conclusions can be obtained:

(a) The MI-GLR model significantly outperforms the MI-ARIMAX model for all target sites and forecasting steps. The 3-step MAE, MAPE, and RMSE of MI-GLR model for target site #1 are 0.892 m/s, 17.308%, and 1.238 m/s, respectively. The 3-step MAE, MAPE, and RMSE of MI-GLR model for target site #2 are 0.573 m/s, 8.701%, and 1.149 m/s, respectively. The 3-step MAE, MAPE, and RMSE of MI-GLR model for target site #3 are 0.407 m/s, 4.460%, and 0.796 m/s, respectively. The 3-step MAE, MAPE, and RMSE of MI-GLR model for target site #1 are 0.571 m/s, 7.483%, and 1.322 m/s, respectively. The error indices of the MI-GLR model are much smaller than the MI-ARIMAX model.

These results are consistent with the local forecasting results, demonstrating the GLR model is more suitable than the ARIMAX model in this forecasting task.

(b) Regarding all target sites, the MI-ARIMAX and MI-GLR models have the worst forecasting performance in target site #1. And the MI-ARIMAX and MI-GLR models have better forecasting performance in target site #3 than the other target sites. The 3-step MAPEs of MI-ARIMAX model for target sites #1, #2, #3, #4 are 72.465%, 17.801%, 12.611%, and 15.838%, respectively. The 3-step MAPEs of MI-GLR model for target sites #1, #2, #3, #4 are 15.013%, 8.701%, 4.460%, and 7.483%, respectively. This is because target site #1 has the highest volatility and standard deviation, while target site #3 is much smoother.

8.4 Spatial wind forecasting algorithm based on intelligent model

8.4.1 Theoretical basis

8.4.1.1 Spatial feature selection based on binary optimization algorithms

In the study, four binary optimization algorithms are utilized to achieve spatial feature selection, namely BPSO [7], BDE [8], BGWO [9], and BHHO [10]. They are typical metaheuristic optimization algorithms, following the same algorithm structure and having similar modeling steps. The core concept of metaheuristic optimization algorithms is to utilize a series of agents to search for the optimal solution. During the searching process, the search direction is adjusted according to the fitness of agents. The general framework of metaheuristic optimization algorithms is as the following steps:

(a) Set the initial solutions.

(b) Utilize local search algorithms to enhance the performance of solutions.

(c) Iterate the above steps until the termination condition is met.

Different metaheuristic optimization algorithms obey the above modeling steps, but their behaviors are quite different. These algorithms simulate biological behaviors in nature. The BPSO graphically simulates the graceful and unpredictable movement of a flock of birds. The BDE simulates the processes of biological evolution. The BGWO simulates the hunting and preying processes of gray wolves. The BHHO simulates the prey, pounce, and attack strategies of hawks.

To fairly compare different optimization algorithms, the population size, and the maximum iterations of all algorithms are set as the same values. They are 50 and 100, respectively. The fitness is evaluated by the MAE of forecasting results.

8.4.1.2 Outlier robust extreme learning machine

Extreme Learning Machine (ELM) is a machine learning model based on the topological structure of Multi-Layer Perceptron (MLP). It differs from the MLP in that the ELM randomly initializes weights and biases. Besides, the final output is obtained by linear combination. To improve the robustness of outliers, an outlier robust extreme learning machine (ORELM) is proposed [11]. ORELM aims to minimize the following function:

$$\begin{gathered} \mathbf{min}\|\varepsilon\|_1 + \frac{1}{p}\|\beta\|_2^2 \\ s.t.\ \varepsilon = \mathrm{T} - \mathrm{H}\beta \end{gathered} \tag{8.2}$$

where ε denotes the error of output, p is the regularization parameter, β denotes the weights of output, T is the target vector of training output, and H is the output matrix.

The parameters of ORELM that needs to be set include the regularization parameter and the size of hidden neurons. The size of hidden neurons is set as 25 after trial and error. The regularization parameter is set as 2^{-20} by default.

8.4.2 Model framework

Fig. 8.10 shows the framework of intelligent spatial wind speed forecasting models. The modeling steps are described as follow:

(a) Divide wind speed series in the target site and adjacent sites into the training set, validation set, and testing set.

(b) Utilize MI to analyze the correlation between wind speed series at target sites and adjacent sites. Select the adjacent sites with the top 30% MI values are selected to form spatial features. Note that only a training set is used during this process so that data leaks can be properly avoided.

(c) Normalize the wind speed series at target sites and spatially related sites.

(d) Adopt binary optimization algorithms to further select lagged spatial features. Four binary optimization algorithms are used and compared,

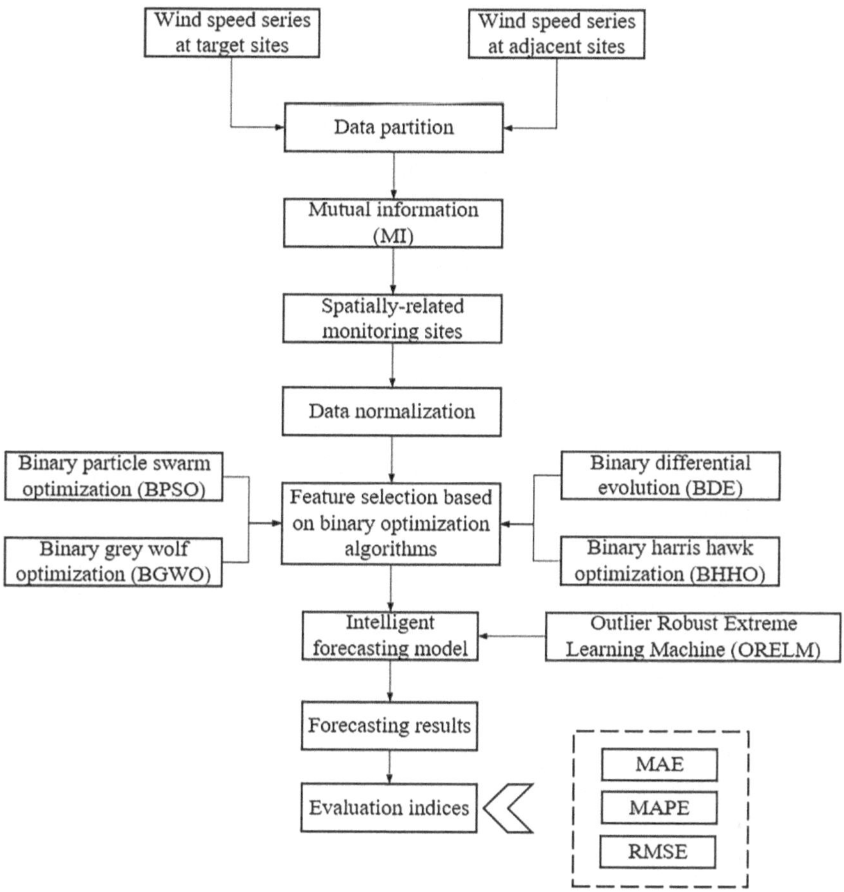

Figure 8.10 Framework of intelligent spatial wind speed forecasting models.

including BPSO, BDE, BGWO, and BHHO. The fitness values are calculated based on the forecasting results in the validation set.

(e) Use the selected spatial data in the training set to train the ORELM model, which forms four models: BPSO-ORELM, BDE-ORELM, BGWO-ORELM, and BHHO-ORELM.

(f) Apply the trained ORELM model on the testing set. Generate forecasting results.

(g) Evaluate the forecasting results of four intelligent spatial models by calculating MAPE, MAE, and RMSE. Finally, compare different binary optimization algorithms and conclude.

8.4.3 Analysis of intelligent spatial forecasting models

8.4.3.1 Spatial feature selection results

In the proposed intelligent spatial forecasting models, the binary optimization algorithms are utilized to further select time-lag features of spatially related sites. The binary optimization algorithms have an iteration process. Their convergence performances greatly affect the effectiveness of feature selection. To analyze the convergence performances of four different binary optimization algorithms, Fig. 8.11 shows the average fitness values (MAE in the study) of all search agents over the whole iteration process. In the study, fitness is evaluated by the MAE of spatial forecasting models. The selected spatial features of target sites #1–#4 are shown in Figs. 8.12–8.15. The x-axis represents the serial numbers of monitoring sites. The first column represents the target site and the other columns represent the spatially related sites. The y-axis represents time lags from *t-12* to *t-1*. 0 and 1

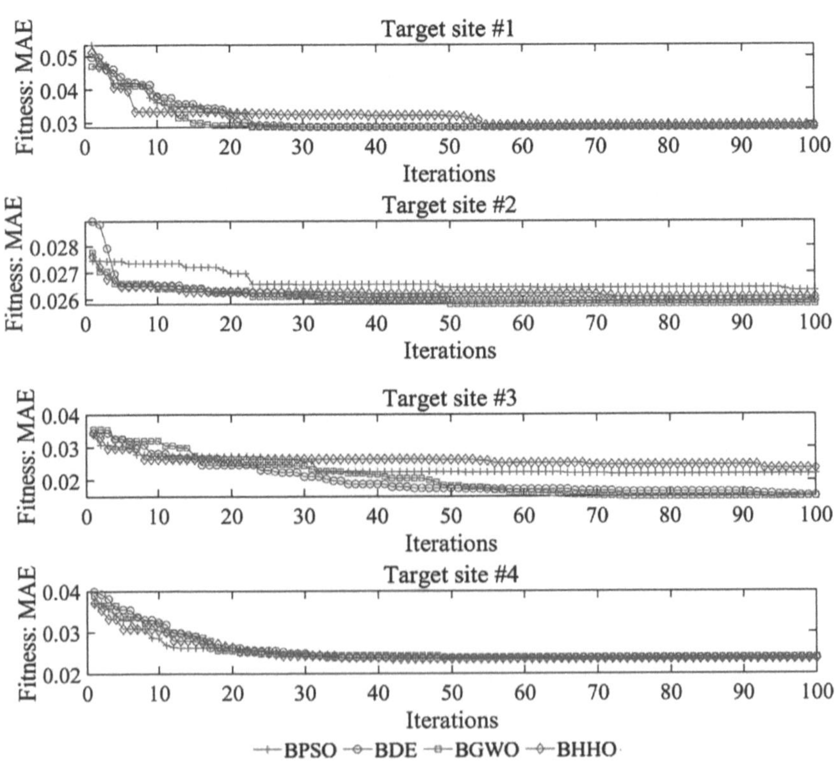

Figure 8.11 Average fitness values of all search agents over the whole iteration process.

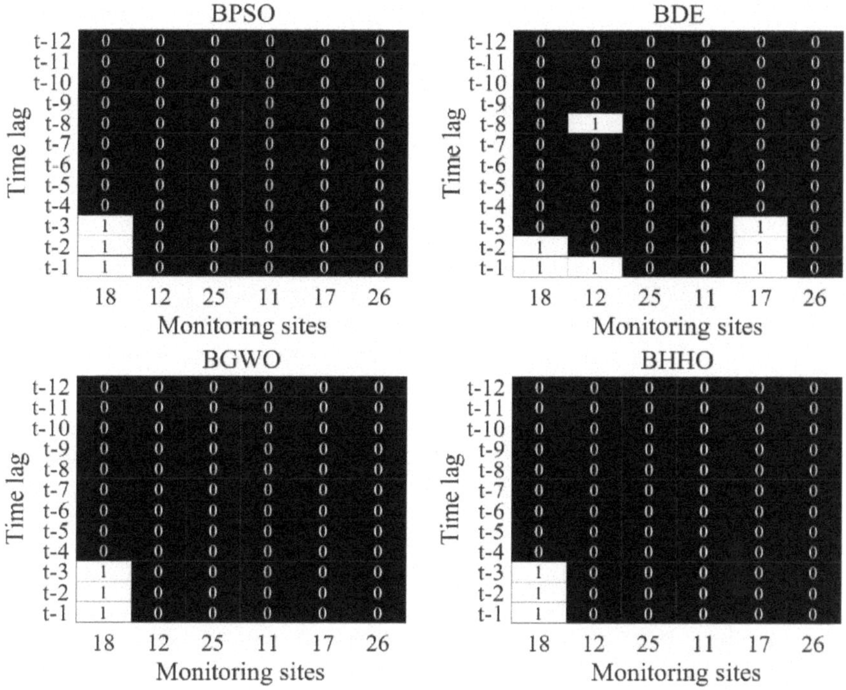

Figure 8.12 Spatial features of target sites #1 selected by binary optimization algorithms.

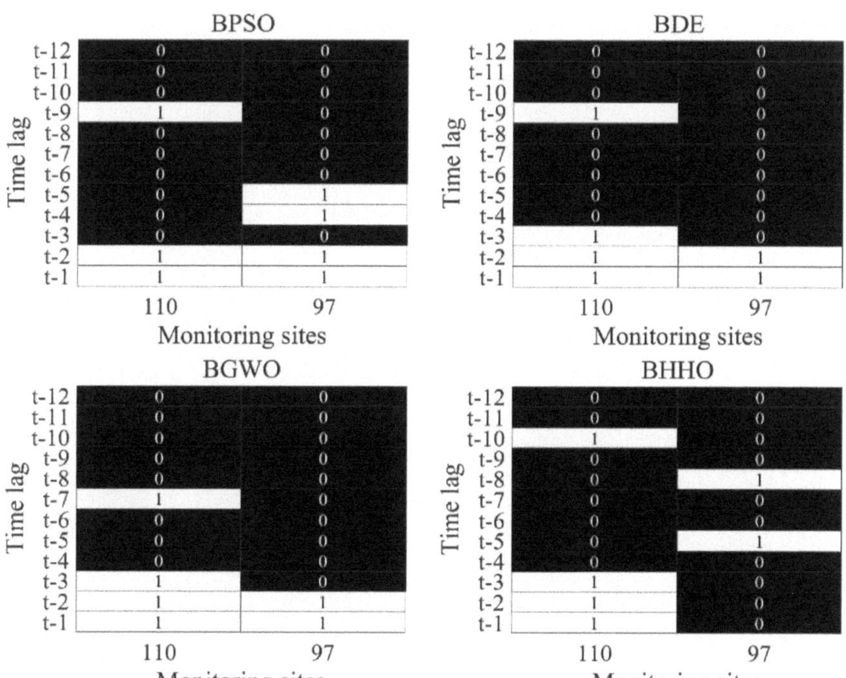

Figure 8.13 Spatial features of target sites #2 selected by binary optimization algorithms.

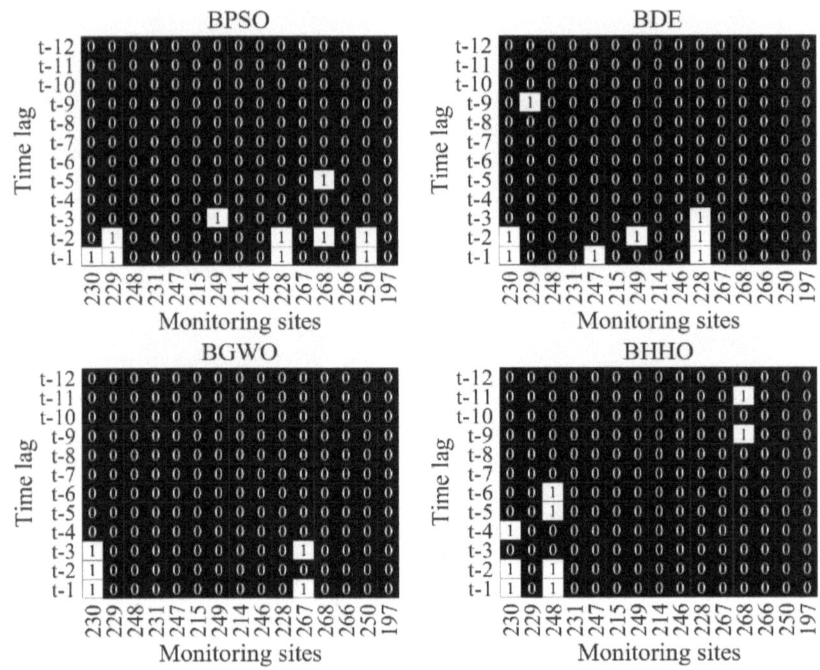

Figure 8.14 Spatial features of target sites #3 selected by binary optimization algorithms.

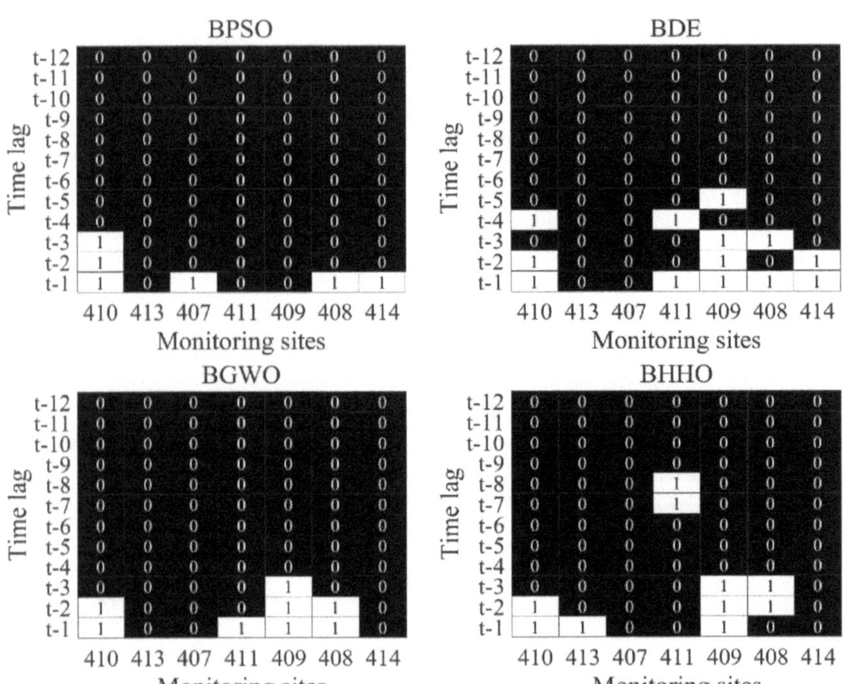

Figure 8.15 Spatial features of target sites #4 selected by binary optimization algorithms.

indicate that the features are discarded and selected, respectively. According to Figs. 8.11–8.15, conclusions can be made as follows:

(a) All binary optimization algorithms have satisfactory convergence performances. Their fitness values greatly decrease in the beginning and remain stable in the end. The results demonstrate that the MAE of the spatial forecasting model is reduced after several iterations. Since MAE evaluates the forecasting error, lower MAE represents a smaller error and better accuracy. In target site #2, the final fitness values of BPSO and BHHO are slightly higher than the other two algorithms. In target sites #1, #3, and #4, four binary optimization algorithms converge to a similar fitness value. Therefore, four binary optimization algorithms do not have many differences in reducing the MAE of spatial forecasting models.

(b) As for the rate of convergence, BGWO significantly outperforms the other three algorithms. In target sites #1 and #4, BGWO successfully converged to the minimum fitness after about 20 iterations. The superiority is mainly because of the mechanisms of the algorithm itself. The BGWO has excellent exploring ability and can rapidly search for optimal values.

(c) The selected features concentrate on the target site and short-term time lags. The *t-3*, *t-2*, and *t-1* features are frequently selected. The results are reasonable because the features of the target site are usually helpful to prediction. Besides, due to the characteristics of time series, wind speed series tends to have a short-term dependency. Short-term time lags are more significant than relatively long-term time lags. Therefore, the binary optimization algorithms can adaptively determine proper features for the spatial forecasting model. The dimensionality of spatial features is significantly reduced, and the optimal features are selected.

8.4.3.2 Results of intelligent spatial forecasting models

To evaluate the proposed intelligent spatial analysis method, the forecasting performances of MI-BPSO-ORLEM, MI-BDE-ORELM, MI-BGWO-ORELM, and MI-BHHO-ORELM are compared. They utilize the wind speed data at the target site and adjacent sites selected by MI to achieve spatial forecasting. The time lag features of selected sites are further processed by BPSO, BDE, BGWO, BHHO, respectively. To evaluate the effectiveness of intelligent algorithms, the MI-ORELM model is included as a comparison. The 1-step ahead forecasting results of the above models for targets #1–#4 are shown in Figs. 8.16–8.19. The top subgraph

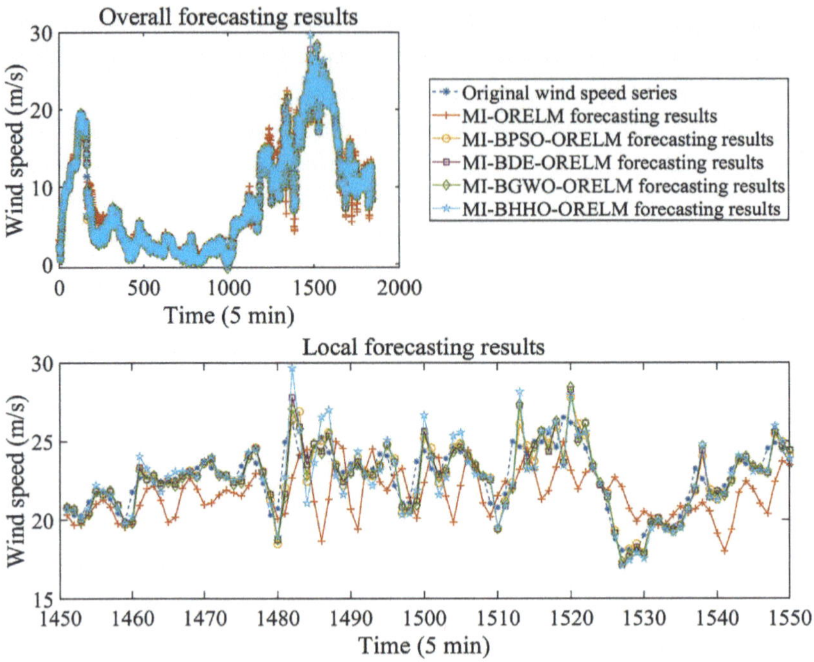

Figure 8.16 The 1-step ahead results of intelligent spatial forecasting models for target site #1.

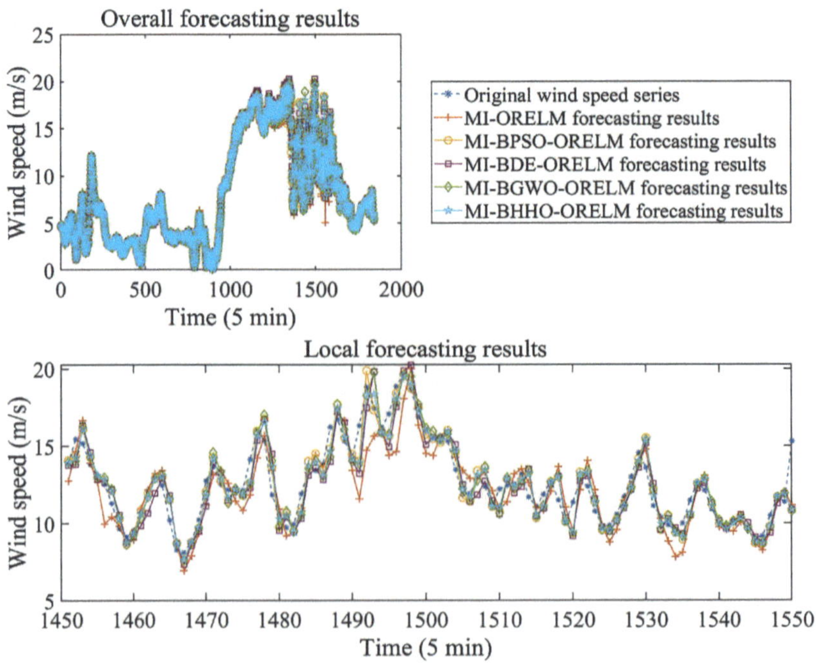

Figure 8.17 The 1-step ahead results of intelligent spatial forecasting models for target site #2.

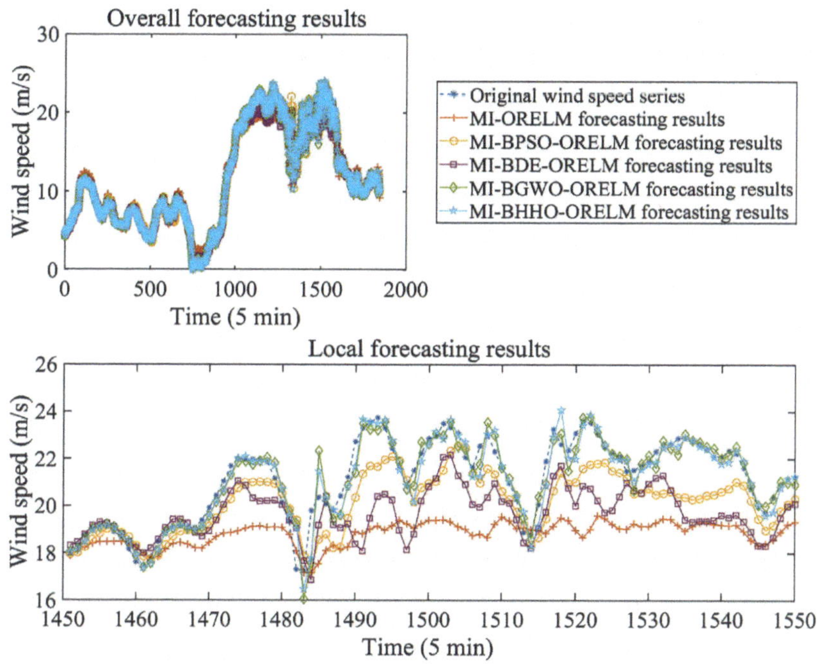

Figure 8.18 The 1-step ahead results of intelligent spatial forecasting models for target site #3.

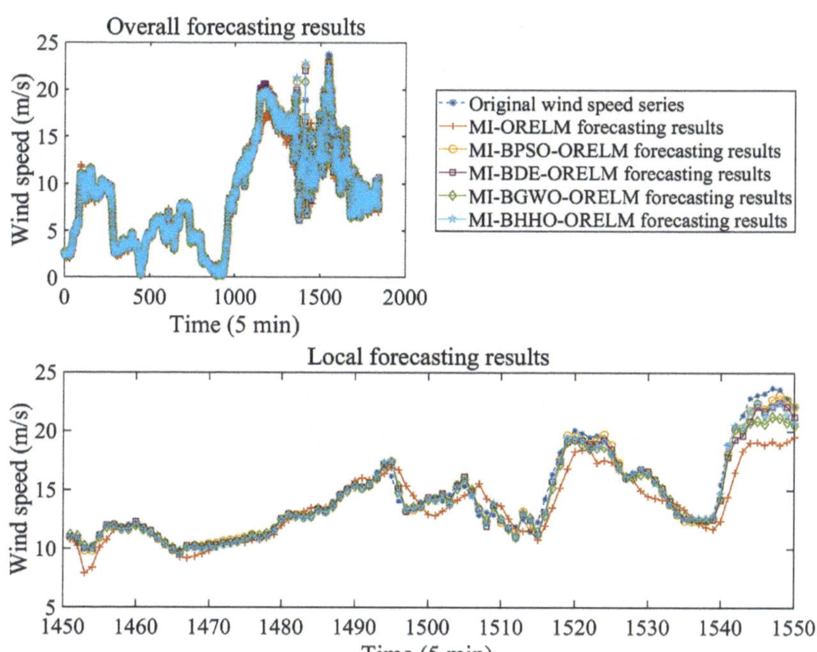

Figure 8.19 The 1-step ahead results of intelligent spatial forecasting models for target site #4.

represents the overall forecasting results in the whole testing set. The bottom subgraph represents the local forecasting results at the 1451th–1550th observations in the testing set. According to the figures, conclusions can be made as follows:

(a) All intelligent forecasting models have graphically satisfactory forecasting performances as for the whole testing set. The 1-step ahead forecasting results of the five models are consistent with the actual wind speed series in the general trend.

(b) As for the local forecasting results, the MI-ORELM model sometimes does not have satisfactory performances, especially in target sites #1 and #3. This is jointly determined by the dimension of features and the learning ability of ORELM. When the dimension of features is large and the learning ability of ORELM is insufficient, the forecasting model would be underfitting. Therefore, feature selection is significant to further reduce the dimensionality of spatial features.

(c) The forecasting performances of MI-BPSO-ORELM and MI-BDE-ORELM models are not stable when tackling large search space. For target site #3, there is a big difference between the 1-step forecasting results and the real wind speed. According to Table 8.2, the number of selected spatially related monitoring sites for target site #3 is 14, which results in $(1+14) \times 12 = 180$ features when considering the target site itself and 12-time lags. When applying optimization algorithms to achieve feature selection, the dimension of variables corresponds to the dimension of features. Therefore, such a large search space brings extra difficulty to the algorithms. In this study, BPSO and BDE perform significantly worse than the other two algorithms.

(d) The MI-BPSO-ORLEM, MI-BDE-ORELM, MI-BGWO-ORELM, and MI-BHHO-ORELM models significantly outperform the MI-ORELM model in local forecasting results. Taking the 1-step ahead forecasting result for target site #1 as an example, the MI-ORELM model significantly deviates from actual wind speed. However, the intelligent spatial forecasting models which utilize optimization algorithms perfectly track the tendency of wind speed series.

To quantitatively analyze the forecasting performance of intelligent spatial forecasting models, the evaluation indices of the MI-ORELM model are provided in Table 8.4 as a comparison. And the evaluation indices are provided in Table 8.5. According to Tables 8.4–8.5, the following conclusions can be summarized:

Table 8.4 Evaluation indices of the MI-ORELM model.

Target site	Model	Step	MAE(m/s)	MAPE(%)	RMSE(m/s)
#1	MI-ORELM	1-step	0.926	32.515	1.268
		2-step	1.007	34.134	1.381
		3-step	1.101	35.898	1.515
#2	MI-ORELM	1-step	0.318	7.023	0.532
		2-step	0.465	9.176	0.805
		3-step	0.607	11.443	1.045
#3	MI-ORELM	1-step	1.129	24.831	1.618
		2-step	1.154	25.384	1.633
		3-step	1.189	26.127	1.662
#4	MI-ORELM	1-step	0.653	12.153	1.078
		2-step	0.752	13.302	1.202
		3-step	0.846	14.565	1.333

Table 8.5 Evaluation indices of intelligent spatial forecasting models with binary optimization algorithms.

Target site	Model	Step	MAE(m/s)	MAPE(%)	RMSE (m/s)
#1	MI-BPSO-ORELM	1-step	0.189	3.678	0.387
		2-step	0.406	7.991	0.752
		3-step	0.614	12.075	1.063
	MI-BDE-ORELM	1-step	0.191	3.806	0.393
		2-step	0.400	8.050	0.745
		3-step	0.594	12.000	1.025
	MI-BGWO-ORELM	1-step	0.194	3.979	0.397
		2-step	0.410	8.404	0.757
		3-step	0.610	12.484	1.037
	MI-BHHO-ORELM	1-step	0.200	3.855	0.413
		2-step	0.429	8.191	0.814
		3-step	0.634	12.317	1.119
#2	MI-BPSO-ORELM	1-step	0.177	2.656	0.376
		2-step	0.392	5.835	0.763
		3-step	0.562	8.862	1.035
	MI-BDE-ORELM	1-step	0.174	2.384	0.412
		2-step	0.375	5.233	0.778
		3-step	0.537	7.906	1.006
	MI-BGWO-ORELM	1-step	0.172	2.447	0.390
		2-step	0.381	5.541	0.770
		3-step	0.562	8.436	1.029
	MI-BHHO-ORELM	1-step	0.166	2.502	0.360
		2-step	0.360	5.596	0.711
		3-step	0.533	8.556	0.978

Continued

Table 8.5 Evaluation indices of intelligent spatial forecasting models with binary optimization algorithms.—cont'd

Target site	Model	Step	MAE(m/s)	MAPE(%)	RMSE (m/s)
#3	MI-BPSO-ORELM	1-step	0.408	8.028	0.612
		2-step	0.462	9.240	0.686
		3-step	0.523	10.402	0.786
	MI-BDE-ORELM	1-step	0.531	10.456	0.817
		2-step	0.606	11.514	0.907
		3-step	0.674	12.336	0.993
	MI-BGWO-ORELM	1-step	0.122	1.473	0.249
		2-step	0.238	2.637	0.459
		3-step	0.346	3.730	0.632
	MI-BHHO-ORELM	1-step	0.114	1.237	0.231
		2-step	0.245	2.756	0.462
		3-step	0.377	4.134	0.659
#4	MI-BPSO-ORELM	1-step	0.147	1.874	0.342
		2-step	0.337	4.180	0.716
		3-step	0.514	6.436	0.970
	MI-BDE-ORELM	1-step	0.152	1.837	0.355
		2-step	0.331	4.051	0.715
		3-step	0.495	6.212	0.962
	MI-BGWO-ORELM	1-step	0.179	2.140	0.378
		2-step	0.333	4.228	0.686
		3-step	0.477	6.303	0.907
	MI-BHHO-ORELM	1-step	0.159	1.936	0.371
		2-step	0.331	4.226	0.729
		3-step	0.488	6.538	0.958

(a) The four intelligent spatial forecasting models with optimization algorithms significantly outperform the MI-ORELM model for all target sites and forecasting steps. The 3-step MAPE, MAE, RMSE of MI-BPSO-ORELM model for target site #1 are 12.075%, 0.614 m/s, 1.063 m/s, respectively. The 3-step MAPE, MAE, RMSE of MI-BDE-ORELM model for target site #1 are 12.000%, 0.594 m/s, 1.025 m/s, respectively. The 3-step MAPE, MAE, RMSE of MI-BGWO-ORELM model for target site #1 are 12.484%, 0.610 m/s, 1.037 m/s, respectively. The 3-step MAPE, MAE, RMSE of MI-BHHO-ORELM model for target site #1 are 12.317%, 0.634 m/s, 1.119 m/s, respectively. However, the 3-step MAPE, MAE, RMSE of MI-ORELM model for target site #1 are 35.898%, 1.101 m/s,

1.515 m/s, respectively. The error indices of the MI-ORELM model are much bigger than the other models. The results are consistent with the local forecasting results. Using binary optimization algorithms to reduce the dimensionality of features is significant.

(b) After a horizontal comparison of four optimization algorithms, no significant difference was found in the forecasting models except target site #3. In most cases, the forecasting results of MI-BPSO-ORLEM, MI-BDE-ORELM, MI-BGWO-ORELM, and MI-BHHO-ORELM are similar. When the wind speed dataset has a high degree of dispersion, the four models show different performances. Taking target site #3 as an example, the 3-step MAPEs of MI-BPSO-ORELM and MI-BDE-ORELM is 10.402% and 12.336%, respectively. While the 3-step MAPEs of MI-BGWO-ORELM and MI-BHHO-ORELM are 3.730% and 4.134%, respectively.

(c) Regarding all target sites, the intelligent spatial forecasting models have the worst forecasting performance in target site #1. The 3-step MAPEs of MI-ORELM model for target sites #1, #2, #3, and #4 are 32.515%, 11.443%, 26.127%, and 14.565%, respectively. The results correspond to Figs. 8.16–8.19. The forecasting results of intelligent spatial models in Fig. 8.16 deviate from the actual wind speed series more than other datasets.

8.5 Spatial wind forecasting algorithm based on deep learning model

8.5.1 The theoretical basis of deep learning spatial forecasting models

8.5.1.1 Spatial feature selection based on sparse autoencoder

Autoencoder is an artificial neural network utilized to achieve representation learning of input data. By extracting useful features from input data, autoencoder can be used to reduce the dimensionality of features. An autoencoder contains two parts, namely encoder and decoder. Its core concept is to encode the input data and minimize the error of decoding results. Therefore, although the dimension of encoding results is smaller than input data, the original data can be also represented.

8.5.1.2 Deep Echo State Network (DeepESN)

Echo State Network (ESN) belongs to the recurrent neural network. It is made up of an input layer, reservoir, and output layer. Different from the traditional MLP model, the reservoir of ESN contains a mass of neurons to learn the dynamic characteristics of the dataset [12]. Deep Echo State Network (DeepESN) stacks several reservoir layers to form a deep learning model. The outputs of reservoir layers are used as the inputs of subsequent reservoir layers, which improves the learning ability of the forecasting model.

The number of reservoir units and reservoir layers of DeepESN are two important hyper-parameters that need to be predetermined. In the study, the number of reservoir units is set as 10 and the number of reservoir layers is validated within $[1, 10]$. The number of reservoir layers with the smallest MAE on the validation set is selected as the final parameter.

8.5.2 Model framework

Fig. 8.20 shows the framework of deep learning spatial wind speed forecasting models. The modeling steps are described as follows:

(a) Divide wind speed series of the target site and adjacent sites into the training set, validation set, and testing set.
(b) Utilize MI and select the adjacent sites with the top 30% MI values to form spatial features. Note that only a training set is used during this process so that data leaks can be properly avoided.
(c) Normalize the wind speed series at target sites and spatially related sites.
(d) Adopt a sparse autoencoder to further select the lagged spatial features.
(e) Use the selected spatial data in the training set to train deep learning models, including LSTM, BILSTM, and DeepESN.
(f) Apply the trained deep learning model on the testing set.
(g) Evaluate the prediction results of three deep learning spatial models by calculating MAPE, MAE, and RMSE. Finally, compare different deep learning models and conclude.

8.5.3 Analysis of deep learning spatial forecasting models

8.5.3.1 The convergence of deep learning models

Deep learning models usually require massive datasets and complicated training processes to make them effective. Therefore, it is necessary to analyze the convergence performance of deep learning models. In the study, the convergence of SAE is evaluated by the mean squared error of reconstruction results during the training process. High reconstruction errors demonstrate poor representation ability of the built SAE model, and vice versa. The mean squared error of SAE during the training process of

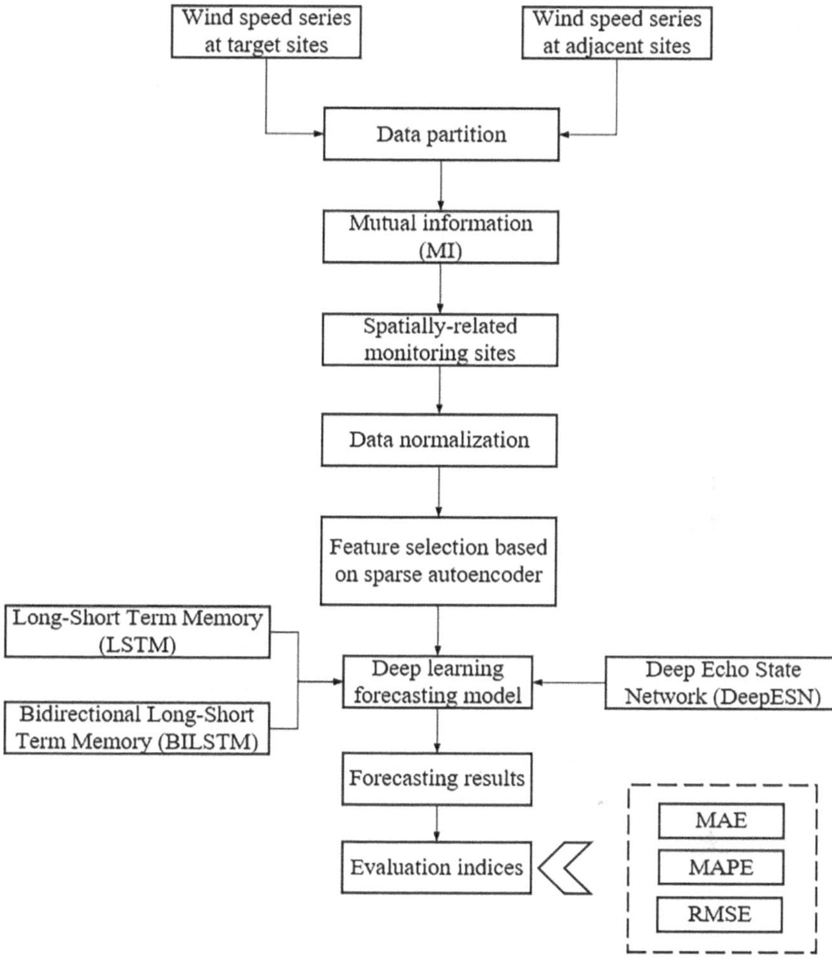

Figure 8.20 Framework of deep learning spatial wind speed forecasting models.

four target sites is shown in Fig. 8.21. The mean squared error rapidly decreased during the initial epochs and stabilized after about 200 epochs. Therefore, the SAE has a satisfactory convergence performance. Considering four target sites, the built SAEs have similar convergence performances. Therefore, it can converge on different datasets.

The convergence performances of LSTM and BILSTM are evaluated by training loss and validation loss together. The loss function is used to measure the forecasting performance. Specifically, the loss function is used to show the gap between the forecast and the actual data. Training loss is

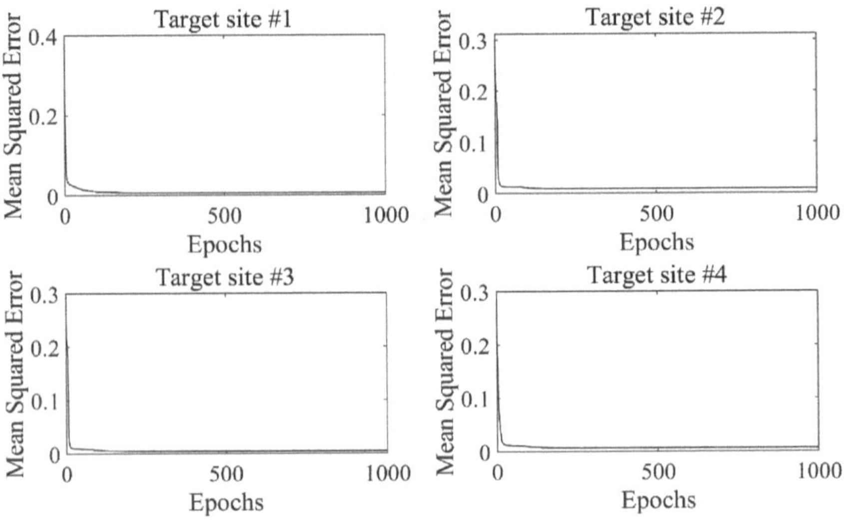

Figure 8.21 Mean squared error of SAE during the training process of four target sites.

used to evaluate the loss on the training set, and validation loss is utilized to evaluate the loss on the validation set. The most satisfactory result is that both the training error and the validation error decrease and converge. In other cases, underfitting and overfitting could happen. For instance, if the training error decreases but the validation error increases, the model gradually tends to overfit. If the training error and validation error do not converge in the end, the model gradually tends to underfit.

Fig. 8.22 shows the loss during the training process of LSTM. Fig. 8.23 shows the loss during the training process of BILSTM. Based on the results, the LSTM and BILSTM models both have satisfactory convergence performances. Their training and validation loss rapidly decreased and eventually converged. The final values of loss are close to zero. Besides, the LSTM and BILSTM models can converge on different datasets, which demonstrates that the models are robust to different data.

8.5.3.2 Results of deep learning spatial forecasting models

To evaluate the proposed deep learning spatial analysis method, two groups of models are established and compared. The first group of models includes MI-LSTM, MI-BILSTM, and MI-DeepESN. They utilize the wind speed data at the target site and adjacent sites selected by MI to achieve spatial forecasting. To analyze the results of the spatial feature selection method, the second group of models includes MI-SAE-LSTM, MI-SAE-BILSTM,

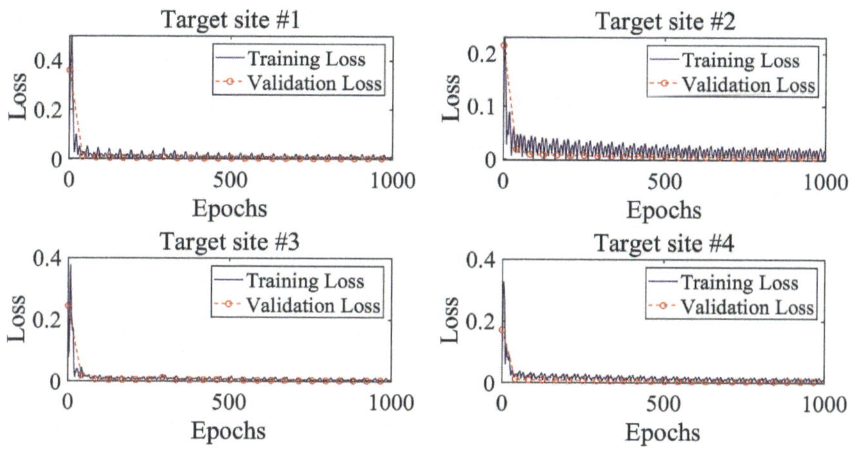

Figure 8.22 Training and validation loss during the training process of LSTM.

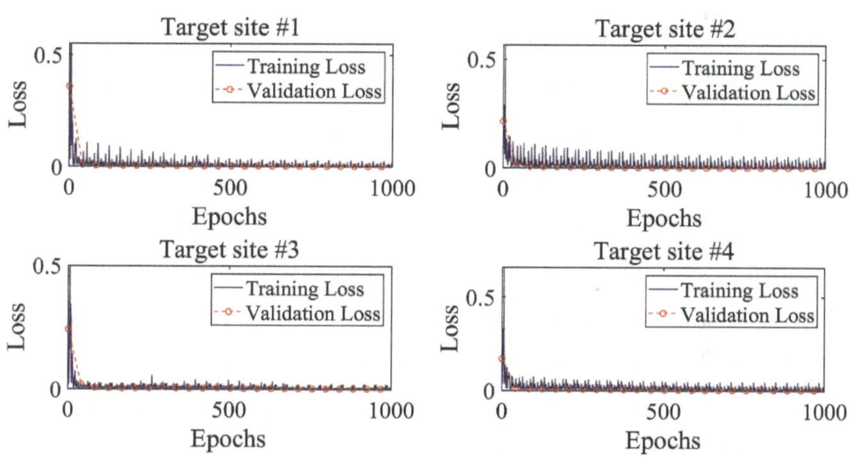

Figure 8.23 Training and validation loss during the training process of BILSTM.

and MI-SAE-DeepESN. They adopt SAE to further extract features of selected sites. The 1-step ahead forecasting results of the above models for targets #1–#4 are shown in Figs. 8.24–8.27. The top subgraph represents the overall forecasting results in the whole testing set. The bottom subgraph represents the local forecasting results at the 1451th–1550th observations in the testing set. According to the figures, conclusions can be summarized as follows:

(a) According to the overall forecasting results in the whole testing set, all deep learning forecasting models have graphically satisfactory

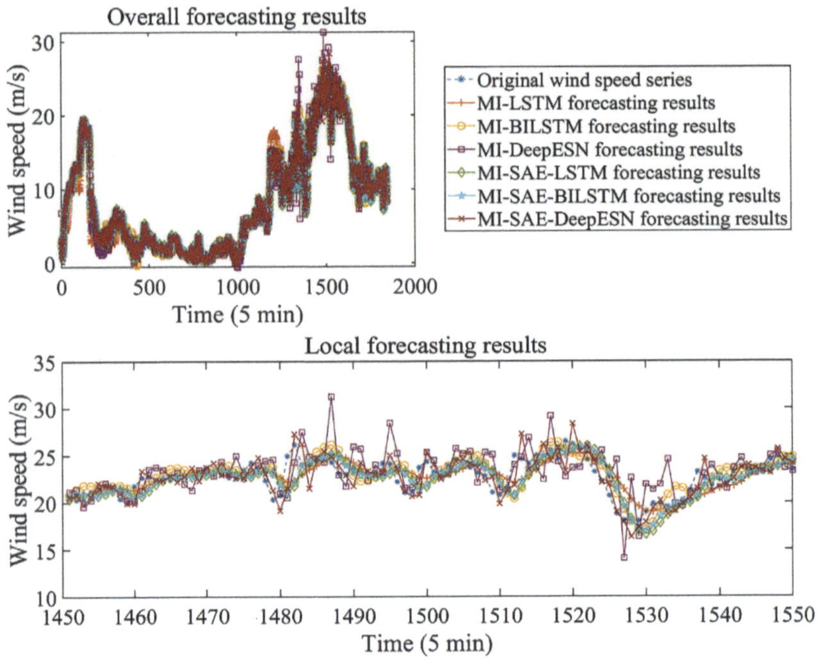

Figure 8.24 The 1-step ahead results of deep learning spatial forecasting models for target site #1.

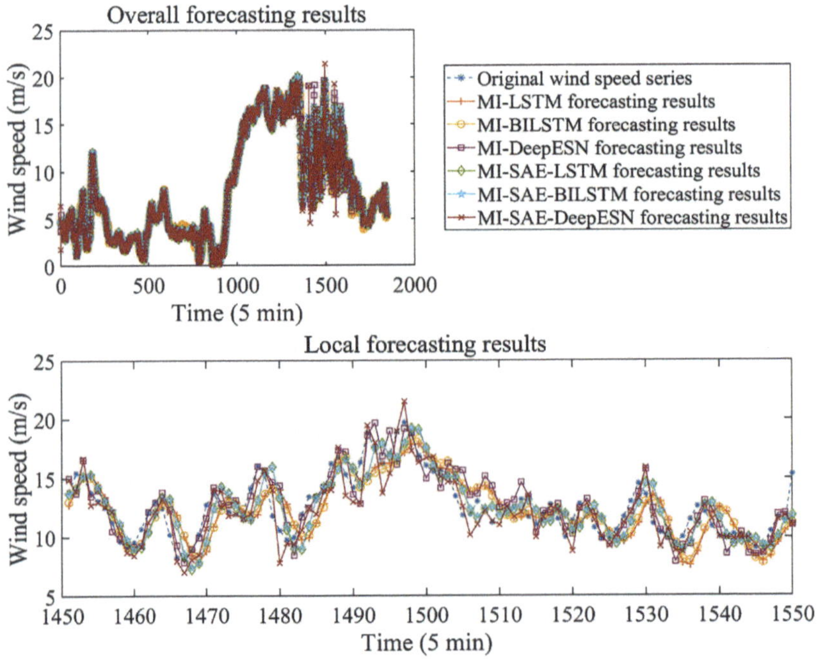

Figure 8.25 The 1-step ahead results of deep learning spatial forecasting models for target site #2.

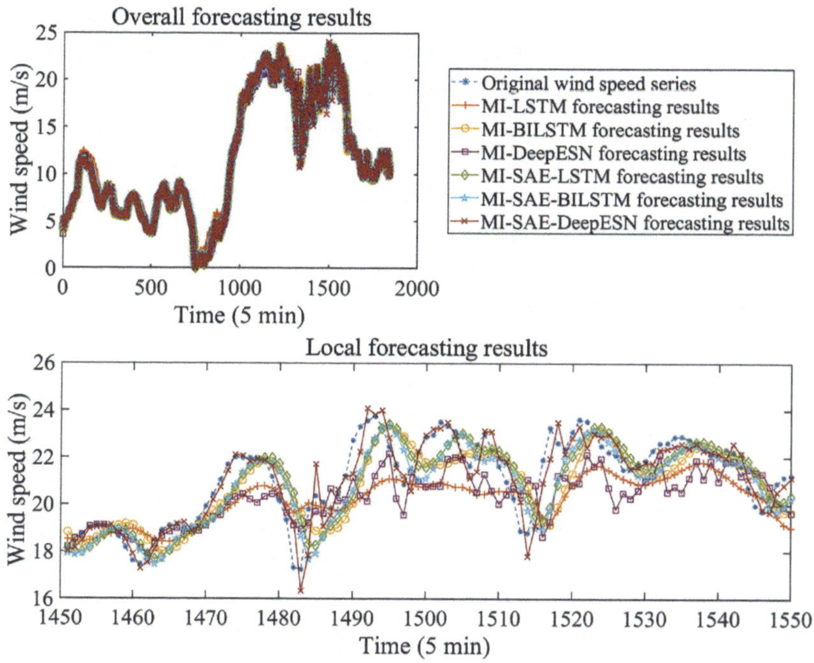

Figure 8.26 The 1-step ahead results of deep learning spatial forecasting models for target site #3.

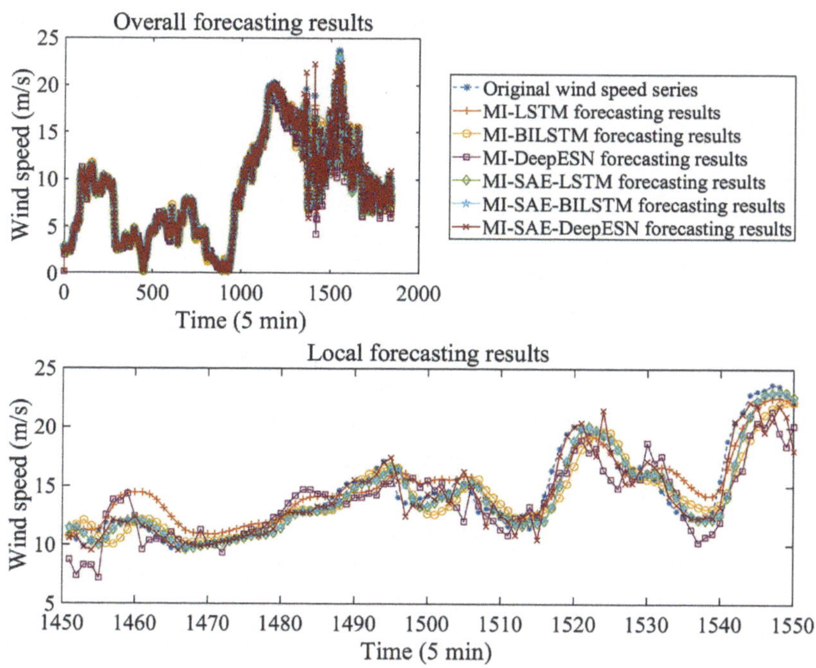

Figure 8.27 The 1-step ahead results of deep learning spatial forecasting models for target site #4.

forecasting performances. The 1-step forecasting results of six models are consistent with the actual wind speed series in the general trend.

(b) For the local forecasting results, the MI-LSTM and MI-DeepESN models sometimes do not have satisfactory performances, especially in target sites #1 and #3. The result is consistent with the forecasting results of intelligent models. The feature selection method is significant to reduce the dimensionality of spatial features.

(c) The MI-SAE-DeepESN model performs the best in tracking the abrupt changes of wind speed series. It can not only capture the trend of wind speed series but also generate forecasting values closer to the real observation values. For instance, the MI-SAE-DeepESN model performs significantly better than the other five models in the 1-step ahead results of deep learning spatial forecasting models for target site #3.

(d) The forecasting results of MI-SAE-LSTM and MI-SAE-BILSTM models show an obvious delay phenomenon, especially in target site #3. The delay phenomenon is commonly encountered in time series prediction because the forecasting of future wind speed is dependent on historical observations. However, the obvious delay phenomenon demonstrates poor forecasting performances from another point of view. The persistence model directly utilizes the observation at a previous time lag to forecast the future wind speed. If the deep learning models perform worse than the persistence model, it indicates that the deep learning models do not learn anything from historical data. In the study, the delay phenomenon of MI-SAE-LSTM and MI-SAE-BILSTM models is obvious in target site #3. This is because the wind speed in target site #3 is particularly variable. However, both LSTM and BILSTM are designed to learn the long-term dependence of data. Hence, their unsatisfactory performance is reasonable in the case of short-term dependence.

To quantitatively analyze the forecasting performance of deep learning spatial forecasting models, the evaluation indices of MI-SAE-LSTM, MI-SAE-BILSTM, and MI-SAE-DeepESN models are provided in Table 8.6. Besides, the evaluation indices of MI-LSTM, MI-BILSTM, and MI-DeepESN models are provided in Table 8.7 as a comparison. According to Tables 8.6 and 8.7, the conclusions are as follows:

(a) The deep learning spatial forecasting models with SAE algorithms significantly outperform the models without SAE algorithms for all target sites and forecasting steps. The 3-step MAPE, MAE, RMSE

Table 8.6 Evaluation indices of deep learning spatial forecasting models without SAE.

Target site	Model	Step	MAE(m/s)	MAPE(%)	RMSE(m/s)
#1	MI-LSTM	1-step	1.012	25.076	1.431
		2-step	1.069	25.950	1.506
		3-step	1.155	27.695	1.623
	MI-BILSTM	1-step	0.575	12.626	0.853
		2-step	0.719	15.562	1.076
		3-step	0.846	18.201	1.261
	MI-DeepESN	1-step	0.597	13.972	1.000
		2-step	0.720	16.179	1.195
		3-step	0.839	18.745	1.365
#2	MI-LSTM	1-step	0.461	8.118	0.741
		2-step	0.586	10.275	0.956
		3-step	0.703	12.619	1.132
	MI-BILSTM	1-step	0.450	7.872	0.727
		2-step	0.575	9.881	0.945
		3-step	0.692	12.060	1.122
	MI-DeepESN	1-step	0.267	3.739	0.558
		2-step	0.429	6.382	0.844
		3-step	0.582	9.077	1.072
#3	MI-LSTM	1-step	0.626	14.074	0.840
		2-step	0.671	14.823	0.905
		3-step	0.714	15.866	0.962
	MI-BILSTM	1-step	0.388	5.764	0.631
		2-step	0.467	6.883	0.755
		3-step	0.541	8.372	0.856
	MI-DeepESN	1-step	0.441	6.691	0.702
		2-step	0.512	7.973	0.802
		3-step	0.575	9.149	0.889
#4	MI-LSTM	1-step	0.634	9.793	0.963
		2-step	0.718	11.039	1.096
		3-step	0.790	12.206	1.207
	MI-BILSTM	1-step	0.506	6.938	0.816
		2-step	0.621	8.536	1.005
		3-step	0.727	10.204	1.161
	MI-DeepESN	1-step	0.612	8.961	0.941
		2-step	0.706	10.470	1.094
		3-step	0.792	11.909	1.232

of MI-LSTM model for target site #1 are 27.695%, 1.155 m/s, 1.623 m/s, respectively. The 3-step MAPE, MAE, RMSE of MI-BILSTM model for target site #1 are 18.201%, 0.846 m/s, 1.261 m/s, respectively. The 3-step MAPE, MAE, RMSE of MI-LSTM model

Table 8.7 Evaluation indices of deep learning spatial forecasting models with SAE.

Target site	Model	Step	MAE(m/s)	MAPE(%)	RMSE(m/s)
#1	MI-SAE-LSTM	1-step	0.476	9.902	0.765
		2-step	0.636	13.348	0.996
		3-step	0.771	16.396	1.183
	MI-SAE-BILSTM	1-step	0.413	7.782	0.688
		2-step	0.581	11.051	0.939
		3-step	0.735	14.600	1.151
	MI-SAE-DeepESN	1-step	0.207	3.996	0.420
		2-step	0.432	8.517	0.793
		3-step	0.637	12.735	1.087
#2	MI-SAE-LSTM	1-step	0.284	4.496	0.562
		2-step	0.456	7.543	0.849
		3-step	0.606	10.687	1.054
	MI-SAE-BILSTM	1-step	0.284	4.589	0.562
		2-step	0.457	7.704	0.849
		3-step	0.599	10.680	1.045
	MI-SAE-DeepESN	1-step	0.206	2.661	0.504
		2-step	0.412	5.597	0.928
		3-step	0.591	8.419	1.208
#3	MI-SAE-LSTM	1-step	0.305	3.738	0.536
		2-step	0.416	5.387	0.702
		3-step	0.486	6.512	0.804
	MI-SAE-BILSTM	1-step	0.293	3.900	0.507
		2-step	0.395	5.316	0.668
		3-step	0.489	6.739	0.803
	MI-SAE-DeepESN	1-step	0.128	1.423	0.262
		2-step	0.272	3.139	0.535
		3-step	0.394	4.763	0.737
#4	MI-SAE-LSTM	1-step	0.332	4.482	0.600
		2-step	0.485	6.865	0.844
		3-step	0.626	9.224	1.040
	MI-SAE-BILSTM	1-step	0.334	4.946	0.588
		2-step	0.491	7.153	0.845
		3-step	0.633	9.518	1.043
	MI-SAE-DeepESN	1-step	0.194	2.228	0.463
		2-step	0.420	4.838	0.929
		3-step	0.635	7.387	1.343

for target site #1 are 18.745%, 0.839 m/s, 1.365 m/s, respectively. However, the 3-step MAPE, MAE, RMSE of MI-SAE-LSTM model for target site #1 are 16.396%, 0.771 m/s, 1.183 m/s, respectively. The 3-step MAPE, MAE, RMSE of MI-SAE-DeepESN model for target site #1 are 12.735%, 0.637 m/s, 1.087 m/s, respectively. The error indices of models with SAE algorithms are much smaller than the models without SAE algorithms. The results are consistent with the local forecasting results.

(b) According to the horizontal comparison of three deep learning models, the MI-SAE-DeepESN model significantly outperforms the MI-SAE-LSTM and MI-SAE-BILSTM models. The forecasting results of MI-SAE-DeepESN have the lowest MAPE, MAE, and RMSE in most cases. Taking target site #4 as an example, the 3-step MAPEs of MI-SAE-LSTM and MI-SAE-BILSTM models are 9.224% and 9.518%, respectively. While the 3-step MAPE of MI-SAE-DeepESN is 7.387%. The results are consistent with the conclusions drawn from forecasting figures. The MI-SAE-DeepESN model can perfectly track the abrupt changes, while the MI-SAE-LSTM and MI-SAE-BILSTM models show an obvious delay phenomenon. The result demonstrates that DeepESN outperforms the LSTM and BILSTM in the case of model setup in this study.

(c) The evaluation indices are significantly affected by the characteristics of the dataset. For instance, target site #1 has the highest standard deviation. As a result, the MI-SAE-DeepESN shows higher forecasting errors than in other target sites. For instance, the MAPEs of 3-step forecasting results in target sites #1−4 are 12.735%, 8.419%, 4.763%, and 7.387%, respectively. The results make sense because when the abrupt change of wind speed is large and the data distribution is discrete, the future wind speed would be more difficult to predict.

8.6 Summary and outlook

This chapter analyzed the data-driven spatial wind forecasting methods along railways. The dataset includes wind speed series collected from 415 adjacent monitoring sites along railways. Four target sites are selected among them to test the performances of proposed models. The way to consider the spatial relationship of wind speed is to first use MI to analyze the relationship of wind speed series between the target monitoring site and the adjacent monitoring sites and then screen out the spatially related sites.

For intelligent and deep learning spatial wind forecasting methods, four binary optimization algorithms (BPSO, BDE, BGWO, BHHO) and a deep feature extractor (SAE) are applied to achieve further dimensionality reduction.

As for the wind forecasting methods, six models are applied to achieve multi-step forecasting, including statistical models (ARIMAX and GLR), intelligent models (ORELM), and deep learning models (LSTM, BILSTM, and DeepESN). Based on the simulation results, the conclusions are drawn as follows:

(a) MI is effective in selecting spatially related sites and constructing spatial models. According to the results, only a few but representative features remain. The selected monitoring sites are all located near the target stations.

(b) All spatial forecasting models have graphically satisfactory forecasting performances as for the whole testing set. The 1-step forecasting results of the above models are consistent with the actual wind speed series in the general trend.

(c) The performances of the spatial forecasting model vary greatly among different target sites. The local forecasting results for target #1 achieve higher errors than other target sites. This is because the wind speed series in target site #1 has a high range, standard deviation, and kurtosis, which brings extra difficulty to prediction.

(d) According to the results of binary optimization algorithms, the selected features concentrate on the target site and short-term time lags. The *t-3*, *t-2*, and *t-1* features are frequently selected. The results demonstrate that wind speed series tends to have a short-term dependency. Short-term lagged features are usually more significant than relatively long-term lagged features.

Besides, there is much work to be done in the field of spatial wind forecasting. Other influential factors of wind such as temperature, humidity, and wind direction can be considered. The relationship between model complexity and input dimensions can also be studied.

References

[1] R. Yu, Z. Liu, X. Li, et al., Scene learning: deep convolutional networks for wind power prediction by embedding turbines into grid space, Appl. Energy 238 (2019) 249–257.

[2] P. Jiang, Y. Wang, J. Wang, Short-term wind speed forecasting using a hybrid model, Energy 119 (2017) 561–577.

[3] A. Pourhabib, J.Z. Huang, Y. Ding, Short-term wind speed forecast using measurements from multiple turbines in a wind farm, Technometrics 58 (2016) 138–147.
[4] Q. Zhu, J. Chen, D. Shi, et al., Learning temporal and spatial correlations jointly: a unified framework for wind speed prediction, Ieee Trans. Sustain. Energy 11 (2019) 509–523.
[5] S. Velázquez, J.A. Carta, J. Matías, Influence of the input layer signals of ANNs on wind power estimation for a target site: a case study, Renew. Sustain. Energy Rev. 15 (2011) 1556–1566.
[6] Y. Noorollahi, M.A. Jokar, A. Kalhor, Using artificial neural networks for temporal and spatial wind speed forecasting in Iran, Energy Convers. Manag. 115 (2016) 17–25.
[7] M.A. Khanesar, M. Teshnehlab, M.A. Shoorehdeli, A novel binary particle swarm optimization, in: 2007 Mediterranean Conference on Control & Automation, 2007, pp. 1–6.
[8] G. Pampara, A.P. Engelbrecht, N. Franken, Binary differential evolution, in: 2006 IEEE International Conference on Evolutionary Computation, 2006, pp. 1873–1879.
[9] E. Emary, H.M. Zawbaa, A.E. Hassanien, Binary grey wolf optimization approaches for feature selection, Neurocomputing 172 (2016) 371–381.
[10] J. Too, A.R. Abdullah, N.M. Saad, A new quadratic binary Harris hawk optimization for feature selection, Electronics 8 (2019) 1130.
[11] K. Zhang, M. Luo, Outlier-robust extreme learning machine for regression problems, Neurocomputing 151 (2015) 1519–1527.
[12] H. Jaeger, H. Haas, Harnessing nonlinearity: predicting chaotic systems and saving energy in wireless communication, Science 304 (2004) 78–80.

Index

Note: 'Page numbers followed by "*f*" indicate figures and "*t*" indicate tables.

MIX
Papier | Fördert gute Waldnutzung
FSC® C083411

Kolibri 360 GmbH
Ferdinand-Jühlke-Straße 7
99095 Erfurt, Deutschland
produktsicherheit@kolibri360.de